BIODIVERSITY, FOOD AND NUTRITION

This book examines the challenges and impacts of poor diets and nutrition from current food systems and the potential contribution of biodiversity and ecosystem services in addressing these problems.

There is a strong need for a multi-level, cross-sectoral approach that connects food biodiversity conservation and sustainable use to address critical problems in our current food systems, including malnutrition. Building on research from the Biodiversity for Food and Nutrition Project (BFN), which aims to better link biodiversity, diets and nutrition, the book presents a multi-country, cross-sectoral analysis of initiatives that have promoted local food biodiversity in four countries: Brazil, Kenya, Turkey and Sri Lanka. This book offers a comprehensive summary of the BFN Project results in each of the four countries along with lessons learned and how this work could be upscaled or applied in other regions. It argues that the strategic promotion and use of food biodiversity is critical in uniting attempts to address conservation, nutrition and livelihood concerns. The book is structured around chapters and case studies encompassing the BFN Project with specific experiences related by partners who played key roles in the work being done in each country. By offering a comparative view capable of furthering dialogue between the respective countries, it is also meant to connect the individual cases for a "greater than the sum of its parts" effect. This means consideration of how localized activities can be adapted to more countries and regions. Therefore, the book addresses global issues with a foot planted firmly in the grounded case study locations.

This book will be of great interest to policymakers, practitioners and NGOs working on food and nutrition, as well as students and scholars of agriculture, food systems and sustainable development.

Danny Hunter is Senior Scientist at Bioversity International and the Global Project Coordinator for the BFN Project.

Teresa Borelli is a Programme Specialist at Bioversity International, providing research and communication support to the BFN Project.

Eliot Gee is a Research Fellow at Bioversity International, providing research and communication support to the BFN Project.

Issues in Agricultural Biodiversity

Series editors: Michael Halewood and Danny Hunter

This series of books is published by Earthscan in association with Bioversity International. The aim of the series is to review the current state of knowledge in topical issues associated with agricultural biodiversity, to identify gaps in our knowledge base, to synthesize lessons learned and to propose future research and development actions. The overall objective is to increase the sustainable use of biodiversity in improving people's well-being and food and nutrition security. The series' scope is all aspects of agricultural biodiversity, ranging from conservation biology of genetic resources through social sciences to policy and legal aspects. It also covers the fields of research, education, communication and coordination, information management and knowledge sharing.

Published titles:

Farmers and Plant Breeding
Current Approaches and Perspectives
Edited by Ola T. Westengen and Tone Winge

Agrobiodiversity, School Gardens and Healthy Diets
Promoting Biodiversity, Food and Sustainable Nutrition
Edited by Danny Hunter, Emilita Monville Ora, Bessie Burgos, Carmen Nyhria Roel, Blesilda M. Calub, Julian Gonsalves and Nina Lauridsen

Biodiversity, Food and Nutrition
A New Agenda for Sustainable Food Systems
Edited by Danny Hunter, Teresa Borelli and Eliot Gee

For more information about this series, please visit: www.routledge.com/Issues-in-Agricultural-Biodiversity/book-series/ECIAB

BIODIVERSITY, FOOD AND NUTRITION

A New Agenda for Sustainable Food Systems

Edited by Danny Hunter, Teresa Borelli and Eliot Gee

Routledge
Taylor & Francis Group
LONDON AND NEW YORK

earthscan
from Routledge

First published 2020
by Routledge
2 Park Square, Milton Park, Abingdon, Oxon OX14 4RN

and by Routledge
52 Vanderbilt Avenue, New York, NY 10017

Routledge is an imprint of the Taylor & Francis Group, an informa business

British Library Cataloguing-in-Publication Data
A catalogue record for this book is available from the British Library

Library of Congress Cataloging-in-Publication Data
Names: Hunter, Danny, editor. | Borelli, Teresa, editor. | Gee, Eliot, editor.
Title: Biodiversity, food and nutrition : a new agenda for sustainable food systems / edited by Danny Hunter, Teresa Borelli and Eliot Gee.
Description: Abingdon, Oxon ; New York, NY : Routledge, 2020. | Includes bibliographical references and index.
Identifiers: LCCN 2019051493 | ISBN 9780367141509 (hardback) | ISBN 9780367141516 (paperback) | ISBN 9780429030574 (ebook)
Subjects: LCSH: Agrobiodiversity. | Sustainable agriculture. | Food security–Environmental aspects. | Food supply–Environmental aspects. | Nutrition–Environmental aspects.
Classification: LCC S494.5.A43 B548 2020 | DDC 631.5/8–dc23
LC record available at https://lccn.loc.gov/2019051493

ISBN: 978-0-367-14150-9 (hbk)
ISBN: 978-0-367-14151-6 (pbk)
ISBN: 978-0-429-03057-4 (ebk)

Typeset in Bembo
by Swales & Willis, Exeter, Devon, UK

This book is dedicated to Chef Shantha Mayadunne who passed away in April 2019 in the Sri Lanka attacks. Ms. Mayadunne, who was Sri Lanka's first TV chef and was renowned for her cookbooks and cooking school, gracefully presided over the BFN Food Festival and cooking competition held in Peradeniya, Sri Lanka, in October 2017 as part of celebrations for World Food Day. With our immense gratitude and admiration for her work.

CONTENTS

FIGURES

TABLES

CONTRIBUTORS

Daniela Moura de Oliveira Beltrame holds a BSc in Nutrition from the Federal University of Viçosa, and an MSc and a PhD in Nutrition (Food Science) from the University of São Paulo, Brazil. As the National Coordinator of the BFN Project in Brazil, Daniela led the implementation of country activities and collaborated with universities and national research agencies to carry out work such as determining the nutrition composition of target species and developing an online database with 70 underutilized fruit species native to Brazil. As part of her work, she liaised with relevant sector ministries for the inclusion of food biodiversity in public policies and programmes, such as nutrition education, food procurement and school feeding.

Tracy Berno is an Associate Professor at AUT University in Auckland, New Zealand. Her research interests include the relationship between agriculture, tourism and cuisine, sustainable food systems and food politics. Tracy has lived and worked in the South Pacific region for over 30 years. She has researched and published widely on agriculture, culture, cuisine and tourism in the South Pacific and has co-authored two international award-winning books in this area, including one (*Me'a Kai: The Food and Flavours of the South Pacific*) which won best cookbook in the world award in 2010.

Teresa Borelli is a Scientist providing research support to the Healthy Diets from Sustainable Food Systems Initiative of Bioversity International. She has over 18 years' experience working within the CGIAR system on research and development of sustainable agri-food systems in low- and middle-income countries. Over the years, she has acquired a cross-disciplinary set of skills spanning from soil biology, agroforestry, agrobiodiversity and crop wild relatives' conservation, food security and dietary diversity as well as policy development and awareness raising. She holds a BSc in Zoology from Bangor University, UK, and an MSc in Advanced Ecology from Durham University, UK. Earlier contributions to this series include the co-editing of *Diversifying foods and diets: using agricultural biodiversity to improve nutrition and health.*

Barbara Burlingame, PhD, is a Scientist and Professor of Nutrition and Food Systems in New Zealand. She is Co-chair of the Sustainable Diets Task Force of the International Union of Nutritional Sciences, Specialty Chief Editor of Frontiers in Nutrition, member of the Committee on World Food Security's High Level Panel of Experts, and Independent Consultant to several regional and global programmes. In the recent past, she was Chief of Nutrition at the Food and Agriculture Organization of the United Nations (FAO) where she worked for 16 years. She is the author of hundreds of scientific papers and technical publications. Her areas of work include food composition, dietary assessment, biodiversity for food and nutrition, and sustainable diets.

Chantal Wei-Ying Clément is Deputy Director of the International Panel of Experts on Sustainable Food Systems. She holds a PhD in Political Science from Carleton University, where she researched collaborative governance schemes to support sustainable food system development at the municipal level in Canada and the EU. She obtained her MA in German and European Studies from the School of Foreign Service at Georgetown University. Dr. Clément has over 10 years' experience in food system research and practice including governance methods for food system transition, local food systems and community food security.

Bruce Cogill, PhD, is a Research Leader, Global Coordinator, Project Director and a Food, Health and Nutrition Security Specialist and Advisor. He has over 30 years of emergency and development related experience with UN agencies, governments, universities and foundations, mostly in Sub-Saharan Africa, South Asia and the South Pacific where he has lived and worked. He directed a global team of transdisciplinary researchers at Bioversity International, studying agricultural biodiversity, dietary diversity, sustainable food systems/diets, health and nutrition. He is a consultant and advisor to the Department of Foreign Affairs, Government of Australia, a technical advisor for the Power of Nutrition, Wellcome Trust Foundation, universities and technical groups. He is Associate Editor for nutrition for Global Health, Science and Practice Journal. He holds a PhD and Masters in nutrition and agricultural economics from Cornell University and a Bachelor's degree in food science and technology from the University of Queensland.

Lidio Coradin is the National Project Director of the BFN Project in Brazil. He was Research Director of the Genetic Resources and Biotechnology Research Centre of the Brazilian Agriculture Research Corporation (Embrapa), and was in charge of the Species Conservation Division at the Ministry of the Environment. Lidio has implemented actions related to the conservation and sustainable use of *in situ, ex situ* and on-farm genetic resources as well as conservation and mapping of land races and crop wild relatives, control and eradication of invasive alien species, and conservation of threatened species. Lidio is responsible for the establishment of the Plants for the Future Initiative in Brazil and is the main editor of the *Plants for the Future* books. In the past, he has represented Brazil during the meetings of the Convention on Biological Diversity and FAO's International Treaty on Plant Genetic Resources for Food and Agriculture.

Florence Egal is a Food Security, Nutrition and Livelihoods Expert. She spent most of her career in the UN Food and Agriculture Organization and was actively engaged in the promotion of underutilized foods and biodiversity, indigenous food systems and the Mountain Partnership initiative, as well as urban-related issues in both emergency and development contexts and at both local and global levels. She worked in close collaboration with other UN agencies and the Consultative Group on International Agricultural Research (CGIAR), and in particular Bioversity International. Since her retirement in 2013, she has pursued the promotion of sustainable diets and food systems through a variety of networks and initiatives.

Thomas Forster has worked for over 30 years to strengthen urban and rural linkages in food systems. He is a Senior Fellow and UN Representative for EcoAgriculture Partners that focuses on integrated landscape management as a pathway for integrating the Sustainable Development Goals (SDGs). He was part of a team facilitating the creation of an international knowledge and policy platform on "city region food systems" hosted by FAO. He led the technical team that drafted the Milan Urban Food Policy Pact and coordinated the first three years of the Milan Pact Awards for food policy and practice. In 2017, he joined the FAO Liaison Office in New York to review SDG2. Since 2018, he has been Lead Consultant for the UN Habitat project Guiding Principles for Urban Rural Linkages. Thomas also teaches food systems policy and governance in the New School for Public Engagement in New York where he lives with his family.

Emile Frison, PhD, is a member of the International Panel of Experts on Sustainable Food Systems. He spent his entire career in international agricultural research for development. In 2003, he became Director General of Bioversity International and developed a strategy entitled "Diversity for Well-being" focusing on the contribution of agricultural biodiversity to the nutritional quality of diets and the sustainability, resilience and productivity of smallholder agriculture. Dr. Frison is lead author of the IPES-Food report "From Uniformity to Diversity: a paradigm shift from industrial agriculture to diversified agroecological systems" and is Chair of the Board of Directors of EcoAgriculture Partners.

Eliot Gee is a Research Fellow in Bioversity International's Healthy Diets from Sustainable Food Systems Initiative, and a member of the BFN Project since 2018. As a Food Anthropologist, he is attuned to the gastronomic and cultural implications of food research, policy and communication. Drawing on experience ranging from Chinese culinary education to ethnographic photography, he employs diverse approaches to connect research and people for more sustainable food systems. Eliot holds degrees in the Anthropology of Food from SOAS, University of London, and Anthropology and Visual Arts from Princeton University, and has been involved in the cross-disciplinary platforms IFSTAL, Eating City and the Future Food Institute.

Birgül Güner holds a BSc and MSc from the Faculty of Agriculture, Department of Soil, from the universities of Selçuk and Ankara respectively. Since 2001, she has worked as an Agricultural Engineer in the General Directorate of Agricultural Research and Policies under the Turkish Ministry of Agriculture and Forestry. She participated in the planning process of the BFN Project back in 2009 and was Deputy Coordinator for the project since its inception. She is currently Head of the Biosafety Unit in the Ministry of Agriculture and Forestry.

Nurcan Ayşar Güzelsoy is a Food Engineer working in the Food Authenticity Department of the Central Research Institute of Food and Feed Control in Turkey. She holds an MSc in Analytical Chemistry and is currently a PhD student in Food Engineering, working on food allergens. She carries out food composition analysis as well as research on reducing food losses. She supports the management of the Turkish Food Composition Database "TURKOMP". In the BFN Project, she was the Food and Nutrition Analyses Coordinator responsible for organizing the food composition analysis of prioritized species and data compilation for national and international databases.

Libby Hattersley is an Independent Consultant with over 15 years' experience working on nutrition and food systems issues. She consults to UN agencies and other international organizations, civil society organizations and businesses. She previously worked in academia for nine years, in Canada and Australia. She completed her PhD and post-doctoral studies at the Australian National University, where her research focused on food system approaches to preventing obesity and diet-related chronic diseases. She also holds a Master of International Public Health from the University of Sydney and a Bachelor of Science in Human Nutrition from the University of Otago.

Clare Hindley is a Professor in the Language and Communication Department at IUBH International University of Applied Sciences. She lectures primarily in research methods and communication modules and is a University Research Coordinator. Her international career in business and education has been diverse in location and focus, but language, culture and communication have remained the guiding lights with an interest in cross-disciplinary studies. Her recent research publications focus on areas of hospitality, tourism, management education, culture and sociology.

Danny Hunter is Senior Scientist in the Healthy Diets from Sustainable Food Systems initiative at Bioversity International. He has more than 25 years' experience working in over 30 countries, covering most regions of the world, specialising in the conservation and sustainable use of crop and tree genetic resources and their role in linking sustainable agriculture, environment, health and nutrition. He is the Agrobiodiversity Research Theme Leader for the Plant and AgroBiosciences Centre (PABC) at the National University of Ireland Galway, Ireland and Adjunct Associate Professor in the School of Agriculture and Wine Sciences, Charles Sturt University, Australia. He is Series Editor of *Issues in Agricultural Biodiversity*, a series of books published by Earthscan from Routledge in association with Bioversity International and Lead Editor of the *Routledge Handbook of Agricultural Biodiversity*.

Nick Jacobs is the Director of the International Panel of Experts on Sustainable Food Systems (IPES-Food), having previously worked in the support team to the UN Special Rapporteur on the right to food, and as an Agri-food Journalist. Nick holds an MA in Modern Languages from the University of Cambridge and an MSc in Globalization and Development from the University of Antwerp. Nick has led editorial and research work on the IPES-Food reports *Towards a Common Food Policy for the EU* (2019), *Unravelling the Food-Health Nexus* (2017) and *From Uniformity to Diversity* (2016). He was a member of the Food 2030 Independent Expert Group convened by the European Commission.

Sevinç Karabak is a Researcher in the Field Crops Central Research Institute of the Turkish Ministry of Agriculture and Forestry. Holding a degree in Agricultural Economics from Cumhuriyet University, she joined the BFN Project in 2013, offering her expertise in agricultural economics, extension, marketing, management, policy and rural development. She leads several national and international projects, while working as a consultant in agricultural economics. For 10 years, she was also the Head of the Department of Agricultural Economics Research in the General Directorate of Agricultural Research and Policies of the Ministry of Agriculture and Forestry.

Gina Kennedy, PhD, is a Senior Scientist with Bioversity International and team leader for the Healthy Diets from Sustainable Food Systems Initiative. She is the Chair of the IUNS Task Force on Indigenous and Traditional Food Systems and active in the CGIAR research flagship Agriculture for Nutrition and Health: Food Systems for Healthier Diets. She has 25 years of experience working on public health and nutrition issues in low- and middle-income countries. Her work has focused on measurement of dietary diversity and diet quality, and assessment of the contribution of agricultural biodiversity on nutrient intake and healthy, diverse diets.

Gabriel C. M. Laeis has a keen interest in gastronomy, food and sustainability. He worked for a number of restaurants, hotel chains and hospitality management consultancies and holds a BA in Hotel Management and an MSc in Organic Agriculture and Food Systems. For his PhD in Development Studies at Massey University he looked at the role of cuisine in tourism-agriculture linkages in Fiji. Gabriel has co-organized the first conference on tourism and the SDGs in Auckland, New Zealand, and published in leading hospitality journals and textbooks. He currently lectures at IUBH International University of Applied Sciences.

Hyunju Lee is a Public Health Nutritionist and an Environmental Epidemiologist focusing on linkages between malnutrition, food systems and climate change. She has a great interest in the public health, environmental and nutrition imperative of moving towards sustainable diets and in contributing to building a healthy environment, which helps everyone gain easy access to healthy and sustainable food and lifestyles. She obtained her MSc in Nutrition for Global Health at the London School of Hygiene and Tropical Medicine. She is a food system professional trained through IFSTAL, a registered associate nutritionist and a member of the UK & Ireland Occupational and Environmental Epidemiology Society. She has consulted for healthy and sustainable food systems to international organizations and governments, including the World Health Organization and the Korean Ministry of Food and Drug Safety.

Willy Legrand is a Professor of Hospitality Management at the IUBH International University of Applied Sciences located in Bad Honnef – Bonn, Germany. He is the lead author of *Sustainability in the Hospitality Industry: Principles of Sustainable Operations* and the co-editor of the *Routledge Handbook of Sustainable Food and Gastronomy*. Prof. Legrand is Guest Chief Editor of the *Hotel Yearbook Sustainable Hospitality 2018* and *2020 Special Editions*. Both editions are free-of-charge, and feature industry leaders, academic researchers and consultants sharing their views on a variety of key topics on sustainability in hospitality.

Aurillia Manjella Ndiwa is currently a PhD student in Agricultural and Applied Economics at the University of Nairobi, Kenya. Her research focuses on value chain analysis of African leafy vegetables. She holds a BSc and MSc in Agricultural Economics and Resource Management from Moi University, Kenya. In 2015, she joined Bioversity International as a Research Consultant funded by the Australian Centre for International Agricultural Research (ACIAR). For four years, she supported field research and development activities in Busia, Western

Kenya, aimed at linking farmers to institutional markets by testing a direct procurement approach, which encouraged the incorporation of underutilized, nutrient-rich crops in school meals.

Camila Neves Soares Oliveira is an Environmental Analyst in the Department of Species Conservation and Management of the Brazilian Ministry of Environment. She acted as the National Project Manager of the BFN Project in Brazil. She is a Forest Engineer graduate at the University of Brasília, Brazil, and participated in the Managing Global Government Programme in 2013, an initiative of the German Ministry for Economic Cooperation and Development (BMZ), implemented by the German Development Institute (DIE).

Kürşad Özbek holds a PhD in Soil Science from the Faculty of Agriculture of Ankara University in Turkey. He is the founder and currently the Head of the Department of Biodiversity and Plant Genetic Resources while managing the Turkish Seed Genebank. He has worked on plant biodiversity for more than 15 years, managing more than 40 national and international projects aimed at the collection, conservation and characterization of plant genetic resources, particularly of land races, endemic species and edible wild plants.

Gamini Samarasinghe was the BFN National Project Coordinator for Sri Lanka from 2014 to 2019. He holds a BSc in Agriculture and a PhD in Biotechnology from the University of Peradeniya, Sri Lanka. Currently serving as the Director of the Horticultural Research and Development Institute of the Department of Agriculture, he was the Head of the Plant Genetic Resources Center. He has 29 years' experience in the conservation and management of plant genetic resources, molecular and morphological characterization and molecular marker-assisted breeding. Also Visiting Lecturer at the Post Graduate Institute of Agriculture, University of Peradeniya, he supervises many students. He has contributed to 10 peer-reviewed publications and 22 short communications and acted as country representative on several international commissions on plant genetic resources such as the ITPGRFA and the Intergovernmental Committee on Intellectual Property and Genetic Resources, Traditional Knowledge and Folklore.

Luana F. J. Swensson is a Legal Specialist and International Consultant in the Nutrition and Food Systems Division of the Food and Agriculture Organization of the United Nations in Rome. With eight years' experience in the development field, her work focuses on multidisciplinary policy-oriented research and technical support for the design and implementation of country projects on the areas of sustainable public food procurement and home-grown school feeding programmes. Before joining FAO in 2013, she was a Post-doctoral Researcher at the European University Institute in Florence and managed an international research project funded by the Inter-American Development Bank on market access and value chains. Luana is a Lawyer and holds a PhD in Economics and Business Law (Comparative Law) from the Sapienza University of Rome.

Ayfer Tan holds a PhD in Genetics from the Faculty of Agriculture of the Aegean University in Turkey. Until her retirement in 2014, she was Head of the Department of Biodiversity and Plant Genetic Resources, and simultaneously managed the National Genebank and the National Biodiversity and Plant Genetic Resources Programme. In her career, she has led various *ex situ*, *in situ* (GEF Project), on-farm (Global Project, Coordinated by Bioversity International) and ecosystem conservation projects (Life Program of EU). As part of the BFN Project, she was the Regional Coordinator for the Aegean Region. She also continues her work as a Volunteer, promoting biodiversity for food and nutrition with local initiatives and NGOs.

Florence Tartanac is Senior Officer in the Nutrition and Food Systems Division and Group Leader of the Market Linkages and Value Chain Group in FAO in Rome. Her areas of expertise include: sustainable value chain development and inclusive business models; voluntary standards and geographical indications; institutional procurements; small and medium food enterprises development. Prior to joining FAO in 2001, she worked for 10 years in Guatemala for the French Cooperation, for the Institute of Nutrition for Central America and Panama (INCAP) and the United Nations Industrial Development Organization (UNIDO). Florence is a Food Engineer and holds a PhD in Economical Geography from Paris University.

Saadet Tuğrul Ay holds a degree from the Faculty of Agriculture of the Aegean University in Turkey. She then started working as a Research Assistant within the Department of Field Crops at Akdeniz University, while pursuing an MSc on dye plants. In 2005, she completed her PhD on Medicinal and Aromatic Plants from Akdeniz University. She currently works at the Batı Akdeniz Agricultural Research Institute where she leads and works as a consultant for various national, EU and international projects. She has co-authored more than 70 publications. Within the BFN Project, she was Regional Coordinator for the Mediterranean Region.

Victor Wasike was the National Project Coordinator of the BFN Project in Kenya from 2012 to 2019. In collaboration with the County Government of Busia, he developed the first Biodiversity Conservation Policy for any of Kenya's 47 counties. Earlier, Victor was the Project Agronomist for the GTZ Soybean Kenya project that undertook pioneering soybean promotion work that led to the release of the current soybean varieties under cultivation in Kenya. He holds a PhD in Soil Microbiology from Egerton University, Kenya, and an MSc in Soil Science from the University of Nebraska, USA. He is currently Centre Director for the Genetic Resources Research Centre in Muguga, Kenya.

ABOUT BIOVERSITY INTERNATIONAL

Bioversity International is a global research-for-development organization. We have a vision – that agricultural biodiversity nourishes people and sustains the planet.

We deliver scientific evidence, management practices and policy options to use and safeguard agricultural and tree biodiversity to attain sustainable global food and nutrition security. We work with partners in low-income countries in different regions where agricultural and tree biodiversity can contribute to improved nutrition, resilience, productivity and climate change adaptation.

Bioversity International is a CGIAR Research Centre. CGIAR is a global research partnership for a food-secure future.

www.bioversityinternational.org

FOREWORD

Braulio Ferreira de Souza Dias

On 27 November 2017 during the "International Symposium on Biodiversity for Food and Nutrition" organized in Brasília by the Biodiversity for Food and Nutrition (BFN) Project[1], supported by the Global Environment Facility (GEF) and coordinated by Bioversity International and partners, I had the opportunity to deliver a presentation titled "Biodiversity for Food and Nutrition, a historical perspective".

I thought it would be appropriate to share the information I presented then in the foreword to this new book, on the lessons learned during the implementation of the Cross-cutting initiative on biodiversity for food and nutrition of the Convention on Biological Diversity (CBD)[2] led by Bioversity International and the Food and Agriculture Organization of the United Nations (FAO), and resulting in the implementation of the GEF-supported BFN Project in Brazil, Turkey, Kenya and Sri Lanka. This new book is a welcome follow-up to Bioversity International's 2013 book *Diversifying Food and Diets: Using agricultural biodiversity to improve nutrition and health* (Fanzo et al. 2013), for which I also provided the foreword.

In 1996, during the second meeting of the Subsidiary Body on Scientific, Technical and Technological Advice (SBSTTA 2) of the CBD, Brazil submitted a proposal for a work programme on agrobiodiversity with two aims: to reduce the impacts of agriculture on biodiversity and to enhance biodiversity's support to agriculture, while also supporting the three initiatives on pollinators, soil biodiversity and biological control[3]. Based on the Brazilian proposal and the SBSTTA 2 recommendations, at the Third Conference of the Parties to the CBD (COP 3) in Buenos Aires, the CBD adopted Decision III/11 establishing the Work Program on Agricultural Biological Diversity (Agrobiodiversity)[4], further detailed in Decisions V/5 in 2000 and VI/5 in 2002.

In 2004, Emile Frison, then head of the International Plant Genetic Resources Institute (IPGRI, now Bioversity International), gave a keynote plenary speech at CBD COP 7 in Kuala Lumpur on the links between biodiversity and nutrition. In 2005, Brazil hosted a consultation meeting with the CBD Secretariat, FAO, IPGRI and the UN Standing Committee on Nutrition on the links between biodiversity, food and nutrition resulting in proposals that led to the adoption in 2006 in Curitiba, Brazil, of the CBD Decision VIII/23 that led to the establishment of the "Cross-cutting initiative on biodiversity for food and nutrition". The overall aim of the Initiative on Biodiversity for Food and Nutrition is to promote and improve the sustainable use of biodiversity in programmes contributing to food security and human nutrition, as a contribution, initially, to the achievement of the then Millennium Development Goal 1, Goal 7 and related goals and targets, and thereby, to raise awareness of the importance of biodiversity, its conservation and sustainable use.

The Cross-cutting initiative on biodiversity for food and nutrition has four elements:

Element 1. Developing and documenting knowledge – To substantiate the links between biodiversity, food and nutrition, in particular clarifying the relationship between biodiversity, dietary diversity and food preferences, and the relevant links between human health and ecosystem health.

Element 2. Integration of biodiversity, food and nutrition issues into research and policy instruments – To mainstream the conservation and sustainable use of biodiversity into agendas, programmes and policies related to nutrition, health, agriculture and hunger and poverty reduction.

Element 3. Conserving and promoting wider use of biodiversity for food and nutrition – To counter the loss of diversity in human diets, and in ecosystems, by conserving and promoting the wider use of biodiversity for food and nutrition.

Element 4. Public awareness – To raise awareness of the links between biodiversity, food and nutrition, and the importance of biodiversity conservation to meeting health and development objectives, including the elimination of hunger.

The Brazilian Government worked with UN Environment Programme (UNEP), FAO and Bioversity International to propose a concept for a GEF global project on biodiversity, food and nutrition (that eventually became the BFN Project). The concept was approved by the GEF Council in November 2011, including four countries – Brazil, Kenya, Sri Lanka and Turkey – and was launched in April 2012 in Rio de Janeiro during the World Nutrition Congress. In 2012, Brazil hosted the UN Conference on Sustainable Development (Rio+20), which resulted in the outcome "The Future We Want", and which formed the basis for the UN General Assembly adoption in 2015 of the "2030 Agenda on Sustainable Development" and its 17 Sustainable Development Goals (SDGs), including SDG 2 to eradicate hunger, ensure food security and promote sustainable agriculture. In 2013, Brazil also hosted a meeting in Rio de Janeiro jointly organized by Bioversity International, the

Brazilian Foundation for Sustainable Development (FBDS) and the Brazilian Ministry of the Environment which led to the establishment of the "Bridging Agriculture and Conservation Initiative (BACI)"[5].

As the Executive Secretary of the CBD, I worked from 2012 to 2016 to strengthen cooperation between the CBD and the World Health Organization (WHO), which led to the launching in 2015 of the joint assessment on the "Interlinkages between biodiversity and human health" (WHO and SCBD, 2015) and CBD COP Decisions XII/21 in 2014 in Pyeongchang and XIII/6 in 2016 in Cancun (also complemented with Decision XIV/4 in 2018). Also in 2014 and 2015, I had the privilege to join "The Rockefeller Foundation-Lancet Commission on planetary health", which produced the Planetary Health Report, highlighting the relationships between human health and ecosystem health (Whitmee et al., 2015).

As the Executive Secretary of the CBD, between 2012 and 2016, I also worked to strengthen collaboration between the CBD and FAO, which paved the way for the CBD COP Decision XIII/3 on mainstreaming biodiversity in agriculture, forestry, fisheries and tourism[6]. Since 2014, FAO has been producing key documents on the promotion of sustainable agriculture and the conservation and use of biodiversity (FAO, 2014, 2015a, 2015b, 2016, 2018a, 2018b), culminating, in 2019, in the publication of the first ever and comprehensive "State of the World's Biodiversity for Food and Agriculture" (FAO, 2019a) and in the Proceedings of the 2nd FAO International Symposium on Agroecology "Scaling Up Agroecology to Achieve the Sustainable Development Goals", held in Rome in April 2018 (FAO, 2019b).

Other recent reports of much relevance to the global biodiversity, nutrition and health agenda are: the "Healthy Environment, Healthy People" report (UNEP, 2016); the TEEB AgriFood 'Scientific and Economic Foundations' report (TEEB, 2018); the report "From uniformity to diversity" (IPES-Food, 2016); the "2nd Note on Critical and Emerging Issues for Food Security and Nutrition" (HLPE, 2017a) and the "Report on Nutrition and Food Systems" (HLPE, 2017b), both by the High Level Panel of Experts on Food Security and Nutrition of the Committee on World Food Security.

In 2017, I joined the board of Bioversity International and had the opportunity to sign the foreword to Bioversity's key 2017 publication "Mainstreaming Agrobiodiversity in Sustainable Food Systems: Scientific Foundations for an Agrobiodiversity Index" and to support the development in 2019 of a joint Bioversity and International Center for Tropical Agriculture (CIAT) Alliance proposal on "Including Food Systems, Biodiversity, Nutrition and Health in the Post-2020 Global Biodiversity Framework"[7].

As the reader will appreciate in this new book, there are rich and diverse lessons learned from the implementation of the GEF supported Biodiversity for Food and Nutrition Project in Brazil, Turkey, Kenya and Sri Lanka, and more broadly from other countries. What is next for the BFN Project and the Cross-cutting initiative on biodiversity for food and nutrition? My answers to this question back in November 2017 and now are the same: (1) It is critical not to lose momentum; (2) Share experiences among countries; (3) Enhance inter-sectoral coordination and cooperation; (4) Establish public policies to scale-up; (5) Integrate with national implementation of the SDGs.

Further, we should build on the major motivators for policy change, such as: (1) Holistic, inclusive, cross-sectoral approaches such as One Health; (2) Opportunities to address the common drivers of biodiversity loss and poor human health; (3) Opportunities to link policies for conservation as a delivery mechanism for human health; (4) Evaluate and consider implications of ecosystem degradation for infectious diseases and non-communicable diseases; (5) Opportunities to integrate health in biodiversity strategies, plans and actions and vice versa.

I hope the lessons learned from the BFN Project contained in this book and the arguments and proposals contained in the Bioversity and CIAT Alliance policy paper on "Including Food Systems, Biodiversity, Nutrition and Health in the Post-2020 Global Biodiversity Framework"[8] will motivate governments, farmers and the business sector to further promote the approaches and practices developed and tested to further enhance food and nutrition security based on the implementation of a wider diversity of food systems, including a broader use of underutilized food species and varieties.

Braulio Ferreira de Souza Dias, Associate Professor of Ecology at the University of Brasilia, Vice-chair of the board of trustees of Bioversity International, member of the advisory committee of the Future Earth global research programme on sustainability, former National Secretary for Biodiversity and Forests in Brazil, former Executive Secretary of the UN Convention on Biological Diversity.

Notes

1 Biodiversity for Food and Nutrition GEF Project www.b4fn.org/
2 www.cbd.int/decision/cop/?id=11037
3 www.cbd.int/doc/meetings/sbstta/sbstta-02/information/sbstta-02-inf-18-en.pdf
4 www.cbd.int/agro/pow.shtml
5 Bridging Agriculture and Conservation Initiative (BACI) www.bioversityinternational.org/bridging-agriculture-conservation/
6 www.cbd.int/doc/decisions/cop-13/cop-13-dec-03-en.pdf
7 www.cbd.int/api/v2013/documents/6CD5FB27-3475-1E09-291D-41C0F41B834C/attachments/ABI-2.pdf
8 www.bioversityinternational.org/fileadmin/user_upload/campaigns/CBD/Food_Systems_Paper_2020.pdf

References

Bioversity International. (2017) *Mainstreaming Agrobiodiversity in Sustainable Food Systems: Scientific Foundations for an Agrobiodiversity Index*. Bioversity International, Rome. xv +157p. www.bioversityinternational.org/fileadmin/user_upload/campaigns/CBD/Mainstreaming_Agrobiodiversity_Sustainable_Food_Systems_Summary.pdf

FAO. (2014) *Building a Common Vision for Sustainable Food and Agriculture – Principles and Approaches*. Rome. 56p. www.fao.org/3/a-i3940e.pdf (Accessed 7 October 2019).

FAO. (2015a) *Incentives for Ecosystem Services in Agriculture (IES)*. Rome. www.fao.org/3/a-i4702e.pdf (Accessed 7 October 2019).

FAO. (2015b) *Guidelines for Developing a National Strategy for Plant Genetic Resources for Food and Agriculture*. Rome. www.fao.org/3/a-i4917e.pdf (Accessed 7 October 2019).

FAO. (2016) *Sustainable Agriculture for Biodiversity/Biodiversity for Sustainable Agriculture*. Rome. 48p. www.fao.org/3/a-i6602e.pdf (Accessed 7 October 2019).

FAO. (2018a) *Transforming Food and Agriculture to Achieve the SDGs: 20 Interconnected Actions to Guide Decision-Makers*. FAO, Rome. www.fao.org/3/I9900EN/i9900en.pdf (Accessed 7 October 2019).

FAO. (2018b) *Biodiversity for Sustainable Agriculture: FAO's Work on Biodiversity for Food and Agriculture*. FAO, Rome. www.fao.org/3/ca2227en/CA2227EN.pdf (Accessed 7 October 2019).

FAO. (2019a) Bélanger, J. & Pilling, D. (eds) *The State of the World's Biodiversity for Food and Agriculture*. FAO Commission on Genetic Resources for Food and Agriculture Assessments, Rome. 572p. www.fao.org/3/CA3129EN/CA3129EN.pdf (Accessed 7 October 2019).

FAO. (2019b) Scaling Up Agroecology to Achieve the Sustainable Development Goals. Proceedings of the 2nd FAO International Symposium on Agroecology, 3–5 April 2018, Rome. FAO, 2019, 416p. License: CC BY-NC-SA 3.0 IGO. www.fao.org/3/ca3666en/ca3666en.pdf (Accessed 7 October 2019).

HLPE. (2017a) *2nd Note on Critical and Emerging Issues for Food Security and Nutrition. A note by the High Level Panel of Experts on Food Security and Nutrition of the Committee on World Food Security*. Rome. www.fao.org/fileadmin/user_upload/hlpe/hlpe_documents/Critical-Emerging-Issues-2016/HLPE_Note-to-CFS_Critical-and-Emerging-Issues-2nd-Edition__27-April-2017_.pdf (Accessed 7 October 2019).

HLPE. (2017b) Nutrition and food systems. A report by the High Level Panel of Experts of Food Security and Nutrition of the Committee on World Food Security. Report 12. Rome. www.fao.org/3/a-i7846e.pdf (Accessed 7 October 2019).

IPES-Food. (2016) *From Uniformity to Diversity: A Paradigm Shift from Industrial Agriculture to Diversified Agroecological Systems*. IPES-Food – International Panel of Experts on Sustainable Food Systems, Brussels. www.ipes-food.org/_img/upload/files/UniformityToDiversity_FULL.pdf (Accessed7 October 2019).

TEEB. (2018) *TEEB for Agriculture & Food: Scientific and Economic Foundations*. UN Environment, Geneva. http://teebweb.org/agrifood/wp-content/uploads/2018/11/Foundations_Report_Final_October.pdf (Accessed 7 October 2019).

UNEP. (2016) *Healthy Environment, Healthy People*. UNEP. Nairobi. 56 p. https://wedocs.unep.org/bitstream/handle/20.500.11822/17602/K1602727%20INF%205%20Eng.pdf?sequence=1&isAllowed=y (Accessed 7 October 2019).

Whitmee, S., Haines, A., Beyrer, C., Boltz, F., Capon, G.A., de Souza Dias, B.F. et al. (2015) Safeguarding Human Health in the Anthropocene epoch: report of The Rockefeller Foundation–Lancet Commission on Planetary Health. *The Rockefeller Foundation–Lancet Commission on Planetary Health*. 386(10007), 1973–2028. DOI: 10.1016/S0140-6736(15)60901-1. www.thelancet.com/journals/lancet/article/PIIS0140-6736(15)60901-1/fulltext

WHO and SCBD. (2015) *Connecting Global Priorities: Biodiversity and Human Health: A State of Knowledge Review*. WHO Press, Geneva, Switzerland. www.cbd.int/health/SOK-biodiversity-en.pdf

ACKNOWLEDGEMENTS

Editors

The compilers and editors of the book are Danny Hunter, Senior Scientist and Global Project Coordinator of the BFN Project, Teresa Borelli, Programme Specialist, BFN Project and Eliot Gee, Research Fellow, BFN Project. The editors all work within the Healthy Diets from Sustainable Food Systems Initiative at Bioversity International headquarters in Rome, Italy. Further editing and layout support was provided by Olga Spellman, Louis Aburto and Nina Olsen Lauridsen.

Case studies and other contributors

The work described in Chapter 5 of this book is the result of the combined efforts of many people who were closely involved with the UNEP/FAO/GEF supported project "Mainstreaming Biodiversity Conservation and Sustainable Use for Improved Human Nutrition and Well-being" (BFN Project) since its beginning in 2012. It has involved significant contributions from individuals representing national and international organizations who collaborated closely with the project. Their dedication and commitment to the planning and implementation of the project and to the realization of this chapter is gratefully recognized and appreciated.

The list of national partners who directly contributed to project activities is incredibly extensive and is presented in a detailed table in the Annex. Naming all the people who contributed to the realization of the BFN Project is not an easy task, which carries with it the risk that someone may be overlooked. Apologies are expressed in advance to anyone who provided assistance but whose name has been omitted from the list. Falling outside the table, but not outside our thoughts and gratitude, are those friends and colleagues who took part in international steering committee meetings or contributed their expertise and were equally important for shaping the project. Some were with us for a brief time in different capacities, but each brought their unique and valuable knowledge and perspectives to the BFN strategy and ensured the project's long-running success. To them we are greatly indebted and now consider them as part of the wider BFN Family. By country and in alphabetical order, the authors would like to acknowledge and express their most sincere gratitude to:

Brazil

Jaime Aguiar (National Institute of Amazonian Research); Deborah Bastos (School of Public Health, São Paulo State University); Braulio Dias (Department of Ecology, University of Brasília); Nathalia Dreyer (FUNBIO); Sydney Fonseca (SENAC); Juliana Furlaneto; Andrea Guerra (Mackenzie University); Fábio Leite (FUNBIO); Marília Marini (Ministry of Environment); Joanna Martins (Paulo Martins Institute); Semiramis Martins (Federal University of São Paulo); Francisco Nascimento (Federal University of Pará); Ilana Oliveira (FUNBIO); Marco Pavarino (Ministry of Agriculture); Veridiana Rosso (Federal University of São Paulo); Tiago Rusin; Raquel Santiago (Federal University of Goiás); Alessandra Santos (UniCEUB); Hetel Santos; Carlos Scaramuzza; Alberto Silva; Vanuska Silva (Federal University of Rio Grande do Sul); Adriana Siqueira (Federal University of Ceará); partner members of the Federal Institute of Goiás; State University of Ceará; the Collaborating Center for the School Feeding Programme (CECANE) of the Federal University of Santa Catarina and of the Federal University of Paraná (UFPR); Centre for Excellence in Tourism of the University of Brasília; EMBRAPA Tropical Fruits; EMBRAPA Temperate Climate; EMBRAPA Vegetables; EMBRAPA Products and Market; all members of regional teams from partner universities and research institutes; and partner members of the following federal initiatives: the Food Procurement Programme (PAA); National School Feeding Programme (PNAE); Food and Nutrition National Policy (PNAN); National Plan for Promotion of Socio-biodiversity Product Chains (PNPSB); National Plan of Organic Production and Agroecology (PLANAPO); Minimum Price Guarantee Policy on Biodiversity Products (PGPM-Bio); as well as the Information System on Brazilian Biodiversity (SiBBr).

Kenya

Mary Abukutsa-Onyango (Jomo Kenyatta University of Agriculture and Technology – JKUAT); Ruth Adeka (National Museums of Kenya, NMK); Janerose Ambuchi (Ministry of Health and Sanitation – MoH); Thadeus Awuor

(MoE – Busia County); William Buluma and the farmers from the Sustainable Income Generating Investment community-based organization of Busia County; Rebecca Butalanyi (County Director of Education, Busia County); Nelson Etyang (Early Childhood Development Education); James Gachungi (County Director of Education, Busia County); Elsie Kangai (Centre for African Bio-entrepreneurship – CABE); Samson Khachina (Ministry of Agriculture Livestock, Fisheries and Irrigation – MoA, Busia County); Florence Kigunzu (MoA, Busia County); Judith Kimiywe (Kenyatta University); Violet O. Kirigua (Kenya Agricultural and Livestock Research Organization – KALRO); Jane Kiugu (KALRO); Peter Kubebea (Busia County); Vincent Kwena (MoH, Busia County); Melisa Lutomia (MoH, Busia County); Felister Makini (KALRO); Anselimo Makokha (JKUAT); Maureen Kemunto Miruka (KALRO); Gladys Mugambi (MoH); Moses Mulomi (Deputy Governor-Busia County); Festus Murithi (KALRO); Leonard Murwai (Busia county); Eunice N. Mutemi; John Mwai (MoH); Paul Mwongera (Ministry of Education – MoE); Scholastic Nabade (MoH); Jane Nasaka (Agricultural Sector Development Support Programme); John M. Ndungu (KALRO, Thika); Desterio O. Nyamongo (Genetic Resources Research Institute); Bernard Nyongesa (Principal St. Mary Mundika Secondary School); Mark Obonyo (former Principal St. Mary Mundika Secondary School); Hannington Odame (CABE); Frederick Odeyo (MoE, Busia County); Japheth Odhiambo (MoE); Wilson Oduori (MoA); David Ombalo (Ministry of Agriculture, Livestock and Forestry, MoALF); Daniel Ombede (Centre for African Bio-Entrepreneurship); Moses Osia (MoA, Busia County); Miriam Otipa (KALRO, Kabete); Mary Rapando (Principal of St Stephen's Lwanya Girls High School); Ibrahim Rugut (Teachers Service Commission, Busia County); Maurice Siminyu (MoH, Busia County); Lorna Simiyu (Busia County); Lusike Wasilwa (KALRO, Nairobi); Lydiah Waswa (Egerton University).

Sri Lanka

R.H.M.P. Abeykoon (Biodiversity Secretariat, Ministry of Mahaweli Development and Environment – MMDE); Thilini Abeywickrama (BFN National Project Management Unit – NPMU); A.A.R.S.K. Bakmeedemiya (MMDE); B.M.U.D. Basnayake (MMDE); the Community Development Centre (CDC) of Aranayaka; Jayalal Chandrasiri (Policy Consultant); Ranjan Dharmawardena (Presidential Secretariat); D. Godamulla (Director – CDC Aranayaka); L.H.P. Gunaratne (University of Peradeniya); K. Karawita (Provincial Director, Department of Agriculture – DoA); Rohan Karawita (Director – National Food Promotion Board); Kethakie Kulapala (Department of Ayurveda); Amela Kulasekera (BFN NPMU); Dharshana Kumara (MMDE); Terrence Madhujith (University of Peradeniya); H.M. Jayantha Ilankoon Menike (Extension & Training Centre, DoA); P. Pallemulla (Provincial Director, DoA); S. Periyasamy (Agriculture Information and Communication

Center, DoA); D.K.N.G. Pushpakumara (University of Peradeniya); Nirma Rajapaksha (DoA); Disna Rathnasinghe (Extension & Training Centre, DoA); Piumi Rathnayake (BFN Project Assistant); R.S.S. Ratnayake (MMDE); S.B. Semasinghe (Provincial Director, DoA); Ajith Silva (MMDE); Renuka Silva (Wayamba University); R.D. Siripala (Agriculture Information and Communication Center, DoA); Sudeepa Sugathadasa (Department of Ayurveda); Charitha Tennakoon (Development Officer); W.A.G. Weerasinghe (DoA); Anura Wijesekara (DoA); A. Wijewardena (MMDE); W.M.A.D.B. Wickramasinghe (DoA). Special thanks are expressed to Azra Sartaj and Nethmini Samaradiwakara for their untiring technical and administrative support to the project and to the National Project Coordinator of BFN Sri Lanka.

Turkey

Neşe Adanacıoğlu (Aegean Regional Team); Prof. Ahmet Aksoy (Akdeniz University); Alaçatı Arts Association; Yusuf Arslan (General Directorate of Agricultural Research and Policies – TAGEM); Hilal Yüce Arslan (National Project Management Unit, NPMU, TAGEM); Fırat Ayas (Mediterranean Regional Team); Serdar Aydemır (NPMU, TAGEM); Lerzan Aykas (Aegean Regional Team); Merve Bolat (Socio-economy team, Field Crops Central Research Institute); Masum Burak (TAGEM); Nermin Çelikay (Ministry of Health); Prof. Vedat Ceyhan (Ondokuz Mayıs University); Çeşme Municipality; Ahu Çınar and Orçun Çınar (Mediterranean Regional Team); Durmuş Deniz (Black Sea Regional Team); Süha Dinçer (NPMU, TAGEM); Orhan Eren (Food Composition Team); Prof. Kuddusi Ertuğrul (Selçuk University); Hasan Gezginç (NPMU, TAGEM); Safinaz Yaşak Elmasulu (Mediterranean Regional Team); Vehbi Eser (NPMU, TAGEM); Bengü Koyuncu (NPMU, TAGEM); Serdar Burak Öksüz (NPMU, TAGEM); Isa Özkan (NPMU, TAGEM); Ayfer Şahın (TAGEM); Ali Osman Sarı; Erüst Tarım; Rahmi Taşci (Socio-economy team, Field Crops Central Research Institute); Tefik Taylan (Aegean Regional Team); Ergül Terzioğlu (Ministry of Agriculture and Forestry); Emre Tokat (Food Composition Team); Özgül Uçurum (Food Composition Team); Arzu Ünal (TAGEM); Osman Uysal (Alata Horticultural Research Institute); Fatih Alpay Vuran (Mediterranean Regional Team); and Kadriye Yüksel (Mediterranean Regional Team).

International Partners

The contributions of the following individuals from the project's international partner organizations are gratefully acknowledged. First, our appreciation and gratitude go to our lead technical officers and experts within the two co-implementing agencies, United Nations Environment Programme (UNEP) and FAO, who offered guidance and support from project start to finish. These are Janice Albert (FAO); Barbara Burlingame (FAO); Ruth Charrondière (FAO); Marieta Sakalian (UNEP/DGEF);

Kuena Morebotsane (FAO/GEF); and Florence Tartanac (FAO). We would also like to extend our appreciation to Ezzeddine Boutrif (FAO); Linda Collette (FAO); Sandro Dernini (FAO); Angela Kimani (FAO); Catherine Leclerq (FAO); Stefano Mondovi (FAO); Balakrishna Pisupati (UNEP); and Luana F. J. Swensson (FAO) who were involved in several stages of the project.

For their participation and expertise we would like to thank Rosemary Babu (World Food Programme – WFP); Akele Bereket (Ministry of Education, Ethiopia); Sinead Boylan (University of Sydney, Australia); Carmen Burbano (WFP); Lesley Drake (Partnership for Child Development); Thomas Dubois (WorldVeg); Michael Hermann (Crops for the Future); Katja Kehlenbeck (World Agroforestry Centre – ICRAF); Richard Markham (Australian Centre for International Agricultural Research – ACIAR); Ben Mbaya (Feed the Children); Stepha McMullin (ICRAF); Pauline Mundia (Biovision Africa Trust); Leah Ndungu (ACIAR); Eric Ngethe (ICRAF); Justus Ochieng (WorldVeg); Liz Ogutu (ACIAR); Kennedy Osano (ACIAR); Annie Sanderson (ACIAR); Tsvetelina Stoilova (WorldVeg); Mellissa Wood (ACIAR); and Ray Yu Yang (WorldVeg).

Last but not least, we would like to thank all Bioversity staff who have offered administrative, communication and scientific support to the Global Project Management Unit throughout these seven years. In alphabetical order, we express our most sincere gratitude to our colleagues past and present who represent an important piece of the BFN puzzle: Vittorio Agueci; Susan Ajambo; Lilian Aluso; Enrico Baccioni; Sarah Backstrand; Valentina Barbiero; Mauricio Bellon; Nadia Bergamini; Julia Boedecker; Nora Capozio; Roberta Castellani; Zhuo Chai; Samantha Collins; Bruce Cogill; Celia da Silva; Allison de Luise; Beatrice Ekesa; Pablo Eyzaguirre; Jessica Fanzo; Emile Frison; Pablo Gallo; Mestawet Gebru; Hannah Gentle; Alessandra Grasso; Michael Halewood; Toby Hodgkin, Matthias Jäger; Gina Kennedy; Hugo Lamers; Shawn Landersz; Michelle Lopez; Diana Luna; Mary Margaret McRae; Maria Merzouk; Maureen Miruka; Simone Mori; Yasusuki Morimoto; Dunja Mijatovic; Deborah Nabuuma; Obiero Nicanor; Giulia Rota Nodari; Francis Oduor; Stefano Padulosi; Daniela Penafiel; Riccardo Pessina; Luca Pierotti; Roseline Remans; Lars Scheerer; Olga Spellman; Celine Termote; Ann Tutwiler; and Laurien Van Heupen. Our special thanks go to Laura Vuerich and Maria Gehring from the budget office and to Trang Nguyen from the Healthy Diets from Sustainable Food Systems Initiative for keeping us ticking and for their patience, understanding and unfailing assistance.

Much has happened in the past seven years and acknowledgements are due to many more. Among these Teresa would like to single out her immediate family: first and foremost her husband, Fabio Luise, for putting up with the long working hours and for his abiding love and encouragement. We met and fell in love again just as the BFN Project was starting, married in between, and have just moved into our new home. Her gratitude also goes to her beloved parents, Giacinto and Rosalind, who guide her until today and to her brothers, Stefano and Simone, for their closeness. Lastly, she would like to thank her mentor, Danny, for the many laughs despite the distance, the music and books, and for his unwavering support through good and bad times, and to Nina and Eliot for their patience and support in the last leg of BFN.

Danny would like to thank family and friends who have been supportive and generous not only during the preparation of this book, but during the lifetime of the BFN Project. He would especially like to thank his two children, Imogen and Callum, for enduring the many challenges and sacrifices they have made during the course of this work. He would also like to thank Teresa and Eliot for their friendship, creativity and constant support. Finally, he owes a large debt of gratitude to colleague and friend, Pablo Eyzaguirre, a driving force behind the initiation of the BFN Project and the biodiversity and nutrition agenda within Bioversity International.

Eliot is grateful to have been welcomed so warmly by Danny, Teresa, Nina and the rest of the BFN Family during the project's final phase. He thanks his parents for their inspiration, the Orlandos for their generosity, friends both near and far for sharing perspective and laughter, and of course Sofia for making it all make sense.

Our collective, sincere gratitude goes to the core BFN Family – Victor, Lusike, Aurillia, Desterio, Isa, Hasan, Birgul, Ayfer, Saadet, Sevinç, Burak, Serdar, Hilal, Anura, Gamini, Azra, Nethmini, Camila, Daniela, Deborah and Lidio (who exists) of course – for their friendship and support through moments of loss and joy, for the dances and biodiverse dinners, and to our young Bioversity friends and colleagues Ana, Lars, Ines, Lukas, Min and Roberta for bringing happiness to our Rome workplace.

Publishing and production

We would like to extend our thanks to Amy Johnston and Hannah Ferguson and other staff at Earthscan for their guidance. We also sincerely thank Louis Aburto, Olga Spellman and Nina Olsen Lauridsen at Bioversity International, who have provided untiring support to the production of this publication.

Financial support

The development and publication of this book was made possible first and foremost through support from the Global Environment Facility, United Nations Environment Programme (UNEP) and the Food and Agriculture Organization of the United Nations within the framework of the global project 'Mainstreaming Biodiversity Conservation and Sustainable Use for Human Nutrition and Well-being'. Co-financing support was made available from national and international partners, while additional funding was provided by the CGIAR Research Program on Agriculture for Nutrition and Health (A4NH) and by the

Australian Centre for International Agriculture Research for the work in Kenya described in Chapter 5. Finally, we would like to express our sincere thanks to the various host organizations that supported the various authors and contributors to this book.

Copyright permissions

To the best of our knowledge, materials presented in this manual have been accurately referenced and sourced. We have made every effort to ensure that the original source of copyright material has been provided within the text. If any errors or omissions are noted, we would be pleased to rectify them.

ABBREVIATIONS

ACIAR	Australian Centre for International Agricultural Research
ALVs	African Leafy Vegetables
BACC	Mainstreaming Agrobiodiversity Conservation and use in Sri Lankan Agro-ecosystems for Livelihoods and Adaptation to Climate Change, Sri Lanka
BCC	Behavioural Change Communication
BFN	Biodiversity for Food and Nutrition
BMI	Body Mass Index
CBD	Convention on Biological Diversity
CBOs	Community-based organizations
CECANEs	Collaboration Centers on Food and Nutrition (Brazil)
CFS	Committee on World Food Security
CGIAR	Consultative Group on International Agricultural Research
CGRFA	Commission on Genetic Resources for Food and Agriculture
CHEEF	Culture, Health, Environment, Economic and Food Security
CIHEAM	Centre International de Hautes Études Agronomiques Méditerranéennes
CITES	Convention on International Trade in Endangered Species
CIRAD	Centre de coopération internationale en recherche agronomique pour le développement
CONSEA	The National Food and Nutrition Security Council (Brazil)
COP	Conference of the Parties
CSA	Community Supported Agriculture
DASH	Dietary Approaches to Stop Hypertension
EDCs	Endocrine Disruptive Chemicals
EU	European Union

FAO	Food and Agriculture Organization of the United Nations
FBDGs	Food-based Dietary Guidelines
FBS	Farmer Business School
FCTs	Food Composition Tables
FNDE	National Fund for the Development of Education (Brazil)
GDP	Gross Domestic Product
GEF	Global Environment Facility
GHG	Greenhouse Gas
GPA	Global Plan of Action
GPMU	Global Project Management Unit (BFN Project)
GSPC	Global Strategy for Plant Conservation
GYBN	The Global Youth Biodiversity Network
HGSF	Home-grown School Feeding
IAASTD	International Assessment of Agricultural Knowledge, Science and Technology for Development
IBFN	Initiative on Biodiversity for Food and Nutrition
IBGE	Instituto Brasileiro de Geografia e Estatistica - Brazilian Institute of Geography and Statistics
ICN2	The 2nd International Conference on Nutrition
IFPP	Institutional Food Procurement Programmes
IFSTAL	Interdisciplinary Food System Teaching and Learning
INFOODS	International Network of Food Data Systems
INRA	Institut National de la Recherche Agronomique
INRIA	Institut National de Recherche en Informatique et en Automatique
IPBES	Intergovernmental Science-Policy Platform on Biodiversity and Ecosystem Services
IPCC	International Panel on Climate Change
IRD	Institut de Recherche pour le Développement
ISC	International Steering Committee
ITPGRFA	International Treaty on Plant Genetic Resources for Food and Agriculture
KALRO	Kenya Agricultural and Livestock Research Organization
KENRIK	Kenyan Resource Centre for Indigenous Knowledge
KEPHIS	Kenya Plant Health Inspectorate Service
KIPI	Kenya Industrial Property Institute
LOHAS	Lifestyle of Health and Sustainability
MEA	Millennium Ecosystem Assessment
MDGs	Millennium Development Goals
MMA	Ministerio do Meio Ambiente - Ministry of the Environment (Brazil)
MMDE	Ministry of Mahaweli Development and Environment (Sri Lanka)
MoH	Ministry of Health

MoMD&E	Ministry of Mahaweli Development and Environment of Sri Lanka
MUFPP	Milan Urban Food Policy Pact
NBSAP	National Biodiversity Strategy and Action Plan
NEMA	National Environment Management Authority
NGO	Non-governmental Organization
NICC	Nutrition Inter-Agency Coordinating Committee (Kenya)
NMK	National Museums of Kenya
NPC	National Project Coordinator (BFN Project)
NPMU	National Project Management Unit (BFN Project)
NSAPs	Nutrition-sensitive Agricultural Intervention Programmes
NUA	New Urban Agenda
NRA	National Restaurant Association
NRG4SD	Network of Regional Governments for Sustainable Development
NSC	National Steering Committee
OAP	Organic Action Plan
PA	Project Assistant
PAA	Programa de Aquisição de Alimentos - The Food Acquisition Programme (Brazil)
PAR	Platform for Agrobiodiversity Research
PFF	Plants for the Future Initiative (Brazil)
PFPP	Public Food Procurement Programmes
PGPM-Bio	Política de Garantia de Preços Mínimos para os Produtos da Sociobiodiversidade – Minimum Price Guarantee Policy for Biodiversity Products (Brazil)
PNAE	Programa Nacional de Alimentação Escolar - National School Feeding Programme (Brazil)
PNAN	Política Nacional de Alimentação e Nutrição - The National Food and Nutrition Policy (Brazil)
PNBSB	Plano Nacional para a Promoção dos Produtos da Sociobiodiversidade - National Plan for the Promotion of Socio-biodiversity Products (Brazil)
POW	Programme of Work
RCT	Randomised controlled trial
SDGs	Sustainable Development Goals
SiBBr	Sistemas de Informações sobre a Biodiversidade Brasileira - Brazilian Biodiversity Information Systems
SIDS	Small Island Developing States
SINGI	Sustainable Income Generating Investment group (Kenya)
SMEs	Small and Medium-sized Enterprises
SPC	The Secretariat of the Pacific Community
SPTO	South Pacific Tourism Organization
SRI	System of Rice Intensification
SUN	Scaling Up Nutrition

SWOT	Strengths, weaknesses, opportunities, and threats
SFYN	Slow Food Youth Network
TAGEM	Turkish Ministry of Agriculture and Forestry, General Directorate of Agricultural Research and Policies (Turkey)
TLFIX	The Timor-Leste Food Innovators Exchange
UCLG	United Cities and Local Governments
UN	United Nations
UNCITRAL	United Nations Commission for International Trade Laws
UNEP	United Nations Environment Programme
UNICEF	United Nations International Children's Emergency Fund
UNSCN	United Nations System Standing Committee on Nutrition
UPOV	International Union for the Protection of New Varieties of Plants
URL-GP	Urban-Rural Linkages Guiding Principles
URLs	Urban Rural Linkages
WHO	World Health Organization
WTO-TRIPs	World Trade Organization Trade-Related Aspects of Intellectual Property Rights
WWF	World Wide Fund for Nature
YPARD	Young Professionals for Agricultural Development

PART I

Why biodiversity for food and nutrition?

In this section:

- A comprehensive overview of the challenges and impacts of poor diets and nutrition from current food systems
- The potential contributions of biodiversity and ecosystem services in addressing these problems
- The current empirical evidence for the links between biodiversity, diets and nutrition
- The potential of biodiversity and agroecology to transform food systems, and key opportunities.

1

NOURISHING PEOPLE, NURTURING THE ENVIRONMENT

Biodiversity for food systems transformation and healthier diets

Danny Hunter, Eliot Gee and Teresa Borelli

> Globally, local varieties and breeds of domesticated plants and animals are disappearing. This loss of diversity, including genetic diversity, poses a serious risk to global food security by undermining the resilience of many agricultural systems to threats such as pests, pathogens and climate change (…) Fewer and fewer varieties and breeds of plants and animals are being cultivated, raised, traded and maintained around the world, despite many local efforts, which include those by indigenous peoples and local communities (…) The number of local varieties and breeds of domesticated plants and animals and their wild relatives has been reduced sharply as a result of land use change, knowledge loss, market preferences and large-scale trade (…) Many hotspots of agrobiodiversity and crop wild relatives are also under threat or not formally protected (…) The lands of indigenous peoples and local communities, including farmers, pastoralists and herders, are often important areas for *in situ* conservation of the remaining varieties and breeds (…) Ensuring the adaptive capacity of food production incorporates measures that conserve the diversity of genes, varieties, cultivars, breeds, landraces and species which also contribute to diversified, healthy and culturally-relevant nutrition.
>
> (IPBES Global Assessment Report on Biodiversity, Diaz et al., 2019)

Introduction

We face a stark picture of global biodiversity loss, as presented by the Intergovernmental Science-Policy Platform on Biodiversity and Ecosystem Services (IPBES) via its recent Global Assessment Report (Diaz et al., 2019). The outlook for the future is even starker unless we act now. The quotes above highlight the alarming state of biodiversity for food and agriculture: today's dominant agricultural model – focused on a uniformity of crops, livestock and limited genetic diversity – is the primary cause of land conversion and biodiversity loss. We currently rely on 12 plant and animal species for approximately 75% of our food needs, at the expense of many traditional crops and livestock, local varieties and breeds, and associated indigenous knowledge. Fewer plant varieties are being cultivated and fewer breeds raised; simultaneously, the global food system, large-scale trade and the market preferences that deliver our food are producing less diverse, less healthy diets. The globalization of diets and a general shift away from traditional food systems towards westernized diets, characterized by a prevalence of highly processed foods, is escalating nutrition problems and public health costs, not to mention environmental and climate change impacts (see Chapters 2 and 4). These diet transitions are themselves indirect drivers of food biodiversity loss, as they are decreasing global demand for traditional crops and livestock as well as local varieties and breeds.

The State of the World's Biodiversity for Food and Agriculture report (FAO, 2019), the first global assessment of biodiversity for food and agriculture worldwide, presents an equally alarming picture. The report warns that, despite the growing evidence of biodiversity's key role in food security and nutrition, the diversity of production systems worldwide is in decline. Of the thousands of plant species cultivated for food, fewer than 200 contribute substantially to global food output and only nine account for 66% of total crop production.

Food biodiversity – the global wealth of plants, animals and other organisms used for food, both cultivated and from the wild – is a critical resource in the fight against global malnutrition (Bioversity International, 2017). Typically rich in nutrients and adapted to local ecosystems, this biodiversity has been utilized by humanity throughout history, but is largely disregarded by our current agriculture and food systems. Instead, we are producing too much of the wrong kinds of food, to the extent that, at the global level, unhealthy diets are responsible for more deaths than any other risk (GBD 2017 Diet Collaborators, 2019), with all countries now suffering from undernutrition, micronutrient deficiencies or obesity (see Chapter 2). Healthy and sustainable food systems that re-integrate food biodiversity for improved diets and nutrition can provide appropriate, sustainable, local solutions to these challenges.

We have clearly reached a point where business as usual carries too high a cost and is no longer an option. What can the global community do in the face of such an onslaught? The Sustainable Development Goals (SDGs) and the UN Convention on Biological Diversity (CBD) post-2020 biodiversity framework both provide us with a renewed opportunity to commit and coalesce action towards the mainstreaming of biodiversity for sustainable agriculture, food systems and sustainable and healthy diets. However, this will only happen if there is real political understanding, will and leadership. While there has been some progress, there is still a long way to go. This is central if we are to achieve the SDGs by 2030.

We need to creatively use the wealth of food biodiversity in agriculture, food systems and supply chains. We need to explore innovative approaches that incentivize more demand and desirability driven by consumers (the focus of Parts II and III of this book). Only when we are actively mainstreaming food biodiversity for sustainable farming and food systems that deliver diverse, nutritious foods can we hope to effectively safeguard this natural wealth for future generations and reverse the alarming rate of biodiversity loss.

Indigenous peoples and local communities – custodians of food biodiversity

The foods we consume date back thousands of years to the domestication and origins of our crops and livestock. Over the course of this history, indigenous peoples and local communities have been stewards of diverse ecosystems and cultures (Ruiz and Vernooy, 2012; Sthapit et al., 2017), and remain important custodians of the world's food and genetic resources (Garnett et al., 2018; Diaz et al., 2019). Indigenous food systems often provide healthy and resilient diets with minimal impact on the environment, and for many generations have ensured food security and nutrition (WHO and CBD, 2015).

There are many examples of the complexity and remarkable diversity of food availability, utilization and associated knowledge derived from long-evolved cultures and patterns of living within local ecosystems. For example, a study led by the Centre for Indigenous Peoples' Nutrition and Environment at McGill University confirmed the diversity and complexity of indigenous peoples' food systems and diets across 12 communities in different global regions (Kuhnlein et al., 2009; Kuhnlein, 2017). Yet, indigenous food systems and land have been, and continue to be, seriously challenged by colonization, development and globalization (WHO and CBD, 2015). As the IPBES report (Diaz et al., 2019) highlights, the threatened lands of indigenous peoples and local communities are critical areas for maintaining varieties of crops, breeds of animals, crop wild relatives and the other elements of food biodiversity. These lands and communities must be the increased focus of support and resources if we are to tackle climate change and safeguard these critical building blocks for diversified agriculture, food systems and healthy diets. To do this effectively will require the development of a global network of partners, including indigenous peoples and local communities, to plan and implement conservation actions in the world's centres of biodiversity.

Indigenous food systems rich in biodiversity can be leveraged to improve diets and help reverse some of the negative food-related health outcomes highlighted above, through interventions that aim to identify nutritionally-rich traditional foods, and that promote, mobilize and deliver these foods to target populations, also ensuring that benefits are shared with the indigenous communities. Not only do these food-based approaches potentially improve nutrition and health in a sustainable manner, they also revive traditional knowledge, biocultural heritage and contribute to the conservation of biodiversity (WHO, CBD, 2015) (see Figures 1.1 and 1.2).

FIGURE 1.1 Diverse food ingredients during a traditional cooking demonstration, Sri Lanka, 2012
Source: S. Landersz

FIGURE 1.2 A traditional food festival draws representatives from groups across Busia County, Kenya, 2015
Source: John Ndungu

Biodiversity, diets and climate change

Without action and sweeping change on a global scale, food production and food security are at grave risk from rising global temperatures with the very real threat of severe food shortages, particularly in regions where agricultural and food production are already constrained. Climate change, driving changes in rainfall patterns and increases in extreme weather events, will likely reduce crop yields and animal productivity, contributing to lower incomes, higher food prices, decreased access to an adequate supply of nutrient-dense foods, insecurity for farmers and exacerbate malnutrition for certain households (Forum for the Future, 2014; Bioversity International, 2017; HLPE, 2017). The risk is greatest for the world's most vulnerable and marginal people. Rising temperatures in certain geographical regions, especially tropical areas, may lead to certain crops not being cultivated in the future (FAO, 2017). It is not just a matter of crop yields or quantity of food: studies have also shown that increased levels of carbon dioxide can affect the nutritional content of certain grains and legumes, in regards to levels of protein, iron, zinc and other micronutrients (Myers et al., 2014; FAO, 2017).

> Climate change creates additional stresses on land, exacerbating existing risks to livelihoods, biodiversity, human and ecosystem health, infrastructure, and food systems (high confidence). Increasing impacts on land are projected under all future GHG emission scenarios (high confidence). Some regions will face higher risks, while some regions will face risks previously not anticipated (high confidence). Cascading risks with impacts on multiple systems and sectors also vary across regions (high confidence).
>
> *IPCC Special Report on Climate Change and Land, 2019*

We know urgent climate action must be taken, and we even know what must be done. Many reports, most recently the Intergovernmental Panel on Climate Change (IPCC) Special Report on Climate Change and Land (2019), clearly articulates the need to change the types of food we grow and raise, the way in which we grow it and what we eat. One of the clear messages from the recent IPCC (2019) report is that a better future is within our reach: that is, if we transform food production and delivery to be more diversified, that improves soil health and is biodiversity- and environmentally-friendly. However, this will not happen without concerted political will. Some of the key urgent actions highlighted by the IPCC Special Report on Climate Change and Land (2019) include:

- Diversifying what we put on our plates to send a clear market signal that translates to more diversity of crops on the land:

  > Balanced diets, featuring plant-based foods, such as those based on coarse grains, legumes, fruits and vegetables, nuts and seeds, and animal-sourced food produced in resilient, sustainable and low greenhouse gas emission systems, present major opportunities for adaptation and mitigation while generating significant co-benefits in terms of human health (high confidence).

- Diversifying what we grow and produce through our food system, including a shift to more biodiversity-friendly approaches that enrich soils and soil biodiversity as well as agricultural landscapes:

> Diversification in the food system (e.g., implementation of integrated production systems, broad-based genetic resources and diets) can reduce risks from climate change (medium confidence).

- Diversifying the types and sources of foods in public procurement programmes to help realize multiple benefits across all spheres of environment, economy and society:

> Public health policies to improve nutrition, such as increasing the diversity of food sources in public procurement, health insurance, financial incentives, and awareness-raising campaigns, can potentially influence food demand, reduce healthcare costs, contribute to lower GHG emissions and enhance adaptive capacity (high confidence). Influencing demand for food, through promoting diets based on public health guidelines, can enable more sustainable land management and contribute to achieving multiple SDGs (high confidence).

Most importantly, as highlighted in the previous section and by the IPBES (Diaz et al., 2019) report, the IPCC Special Report on Climate Change and Land acknowledges the importance of indigenous peoples and their traditional knowledge as custodians of biodiversity and as stewards of territories and lands:

> Agricultural practices that include indigenous and local knowledge can contribute to overcoming the combined challenges of climate change, food security, biodiversity conservation, and combating desertification and land degradation.
>
> *IPCC, 2019*

The nutritional value of food biodiversity

What we know about the nutritional benefits of food biodiversity is but the tip of the iceberg. The limited knowledge that does exist indicates that biodiversity is often nutritionally superior to many foods that currently dominate our agricultural production, food systems and diets. Simply establishing the nutritional value of local neglected and underutilized food biodiversity can be a powerful incentive towards diversifying diets, production and food systems. At the same time, this information could shine a light on the role that indigenous peoples and local communities can play in reversing biodiversity loss and increasing resilience and adaptive capacity, not to mention the many benefits to global public health and nutrition.

The wealth of food biodiversity provides a natural richness of nutrients (macronutrients such as carbohydrates, proteins and fats, and micronutrients such as vitamins and minerals) and bioactive non-nutrients for healthy human diets and good nutrition (Toledo and Burlingame, 2006; Blasbalg et al., 2011; Fanzo et al., 2013; WHO, CBD, 2015; Bioversity International, 2017). They represent a natural wealth that remains underutilized by most, if not all, countries. Instead, perverse policies and incentives favour less nutritious crops and foods, locking countries into a pathway of dependency that supports large-scale industrial agriculture reliant on external inputs, and large-scale globalized trade that undermines the environment and human health and nutrition (see Chapter 4).

There are significant intra- and inter-specific differences in the nutrient content of most plant-source foods (Hunter et al., 2019) (see Figure 1.3). Often, these nutritional differences are statistically significant, with thousand-fold differences reported. Research in small Pacific islands, for example, has demonstrated the nutritional superiority of traditional varieties of locally important species (bananas, pandanus, breadfruit, taro and yams) compared to the more commonly consumed foods that dominate markets, many of which are imported at considerable cost (Englberger and Johnson, 2013). For instance, while the commonly consumed Cavendish banana contains around 25μg per 100g of pro-vitamin A carotenoid, it is completely outshone by certain local banana varieties that supply more than 8000μg per 100g (Englberger et al., 2003a, 2003b). This massive variation could be the determining factor in whether a young child experiences micronutrient deficiency or grows up healthy and well-nourished.

Considerable variety-specific differences in nutritional content have also been recorded for many other crops (Huang et al., 1999; Rubiang-Yalambing et al., 2014). At the species level, Kobori and Rodriguez-Amaya (2008) have demonstrated that carotenoid levels are higher in wild native Brazilian leafy green vegetables than commercial leafy vegetables. Furthermore, Fentahun and Hager (2009) have documented vitamin C levels in baobab fruits (*Adansonia digitata*) six times higher than in oranges, while McGarry and Shackleton

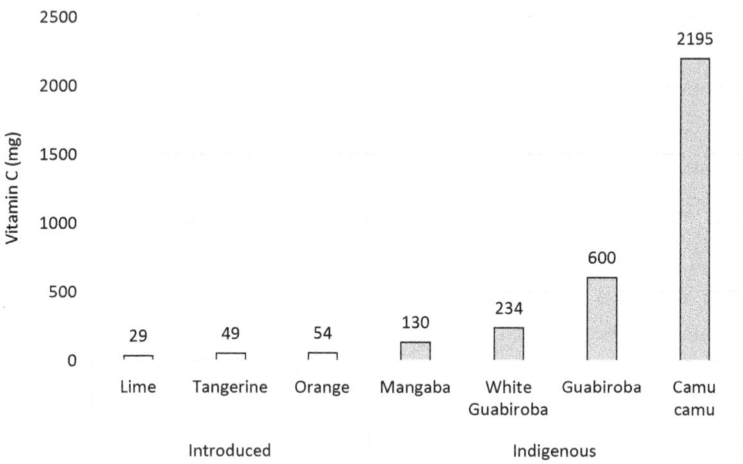

FIGURE 1.3 The vitamin C content of indigenous Brazilian species, compared to common introduced species

Source: The Biodiversity for Food and Nutrition Project, Hunter et al., 2019

(2009) demonstrated that amaranth, a widely used green leafy vegetable, has 200 times more vitamin A and ten times more iron than the same-sized portion of cabbage.

Nutritional composition differences of this nature can represent the difference between nutrient deficiencies and nutrient adequacy in populations and individuals. Furthermore, they offer more local, sustainable and culturally acceptable solutions to the multiple problems of malnutrition in many parts of the world (see Chapter 2).

Food biodiversity pathways to healthy, diverse diets

Biodiversity has been linked to food and nutrition at multiple levels; most directly, biodiversity supports nutrition by improving availability and consumption of a wide variety of nutrient-rich foods year-round. At the ecosystem level, biodiversity supports a range of ecosystem services that are essential to food production, including atmospheric regulation, pest regulation, pollination, water quality and soil fertility. At the farm, community and landscape levels, biodiversity can enhance resilience, reduce susceptibility to climate change, economic and other non-environmental shocks, support market diversification, and improve livelihoods, household income and food security, all of which are key determinants of nutrition. Despite this growing attention and conceptual basis, our understanding of how the relationship between biodiversity, diets and nutrition operates in practice remains limited, as highlighted by Hattersley and colleagues in Chapter 3.

How populations obtain their food is complex and context-specific (Bioversity International, 2017; Raneri and Kennedy, 2017; Jones et al., 2019). There are two principal pathways in which food biodiversity contributes to healthy, diverse diets: consumption via own production or foraged from the wild; and, the purchase of cultivated or wild foods through markets. Most often, households consume a varying mix of differently sourced foods from both of these pathways.

As highlighted in Chapter 3, the links between biodiversity and nutrition are long, complex, dynamic and can be influenced by multiple, potential confounding factors, creating considerable analytical and empirical challenges for researchers trying to establish that food biodiversity supports healthy, diverse diets. Studies to date provide only limited evidence of the role of biodiversity in diets, and even less of the contribution of food biodiversity to nutrition and health outcomes, with much remaining to be learned.

Bending the curve on biodiversity[1] loss

The global food system, with its many harmful subsidies and structural barriers to diversification, largely prioritizes cheap, unhealthy food (see Chapter 4) with a huge environmental footprint and escalating public health budgets, at the cost of diversity on farms and in agricultural landscapes (IPES-Food, 2016; FOLU, 2019). Such a situation is no longer tolerable. We need to urgently move beyond business as usual to achieve a global food transformation that aims to safeguard, creatively use and celebrate the rich diversity of food plants and animals that sustain all of us. By placing diversity at its heart, this new food system would be more sustainable, resilient and equitable, and better equipped to provide healthy and delicious food to all. Achieving such change will require certain key integrative actions and pathways to diversify food production and consumption, while also addressing the drivers of biodiversity loss and ecosystem transformation, including:

- Urgent safeguarding and conservation of remaining globally important genetic diversity including varieties, landraces and breeds and wild relatives of crops and livestock *in situ* in centres of diversity.
- Comprehensively mainstreaming biodiversity into production landscapes and investing in practices that creatively and sustainably use genetic resources to promote and integrate biodiversity-friendly management practices in crop and livestock production. This includes drawing on traditional management practices and knowledge associated with indigenous peoples and local communities, as well as identifying innovations such as diversification of public food procurement and school feeding (see Figure 1.6) and the creation of new market opportunities.
- Promotion of sustainable and healthy diets that empower consumers to improve the healthiness and biodiversity of their diets, and thus create greater demand and desirability for food biodiversity, local varieties and breeds (see Figure 1.4).

FIGURE 1.4 Alaçatı Wild Herb Festival gathers thousands of people, with widespread media coverage, to celebrate the tradition of foraging, cooking and eating nutritious wild edible plants, Turkey, 2015

Source: S. Landersz

The development of the post-2020 strategic plan for the Convention on Biological Diversity (CBD) presents a critical opportunity to contribute to achieving this, and to restore global biodiversity. Key integrative policies and actions, as described here and throughout this book, can contribute to more sustainable food production and consumption with multiple benefits, not just recovery of biodiversity and ecosystems but also benefits to climate and public health.

From words to actions: creating better enabling environments for food biodiversity

There is no shortage of information or guidance on how to change our food systems so that they nourish humanity in a healthy manner, do not harm the environment, reduce public health impacts and are fair and just. The EAT-Lancet Commission Report (Willett et al., 2019) and World Resources Report (Searchinger et al., 2019) are the latest, but there are many reports that suggest key opportunities and recommendations that can inform productive action.

The core of this book is the successful multi-sectoral approach achieved by the Biodiversity for Food and Nutrition (BFN) Project, supported by the Global Environment Facility (GEF). Piloted from 2012–2019 in four mega-diverse countries – Brazil, Kenya, Sri Lanka and Turkey – the initiative explored ways to better mainstream biodiversity for sustainable food systems, healthy diets and improved nutrition. This project arose as a result of increasing interest in the links between ecosystems, biodiversity and nutrition within the Convention on Biological Diversity (CBD), which resulted in the establishment of the CBD Cross-cutting initiative on biodiversity for food and nutrition (www.cbd.int/decision/cop/?id=11037) in 2006 (see Chapter 2). The BFN Project has been the main vehicle for the implementation of the CBD Cross-cutting initiative since its inception.

In its short lifetime, the BFN Project has demonstrated bright spots that can positively contribute to food system transformation, many of which are in line with recommendations and opportunities emanating from an increasing number of global reports on food system change (GloPan, 2016; Bioversity International, 2017; HLPE, 2017; Willett et al., 2019; FOLU, 2019) (see Figure 1.5), as well as leverage points for change identified by the recent International Panel of Experts on Sustainable Food Systems (IPES-Food) report (2016, see Chapter 4). Further opportunities are elaborated upon including urban-rural linkages (Chapter 6), public food procurement and school feeding (Chapter 7), sustainable gastronomy (Chapter 8), culinary tourism (Chapter 9) and food-related innovations with a focus on youth (Chapter 10). What is required now, is the political will, leadership and resources to take these bright spots to scale.

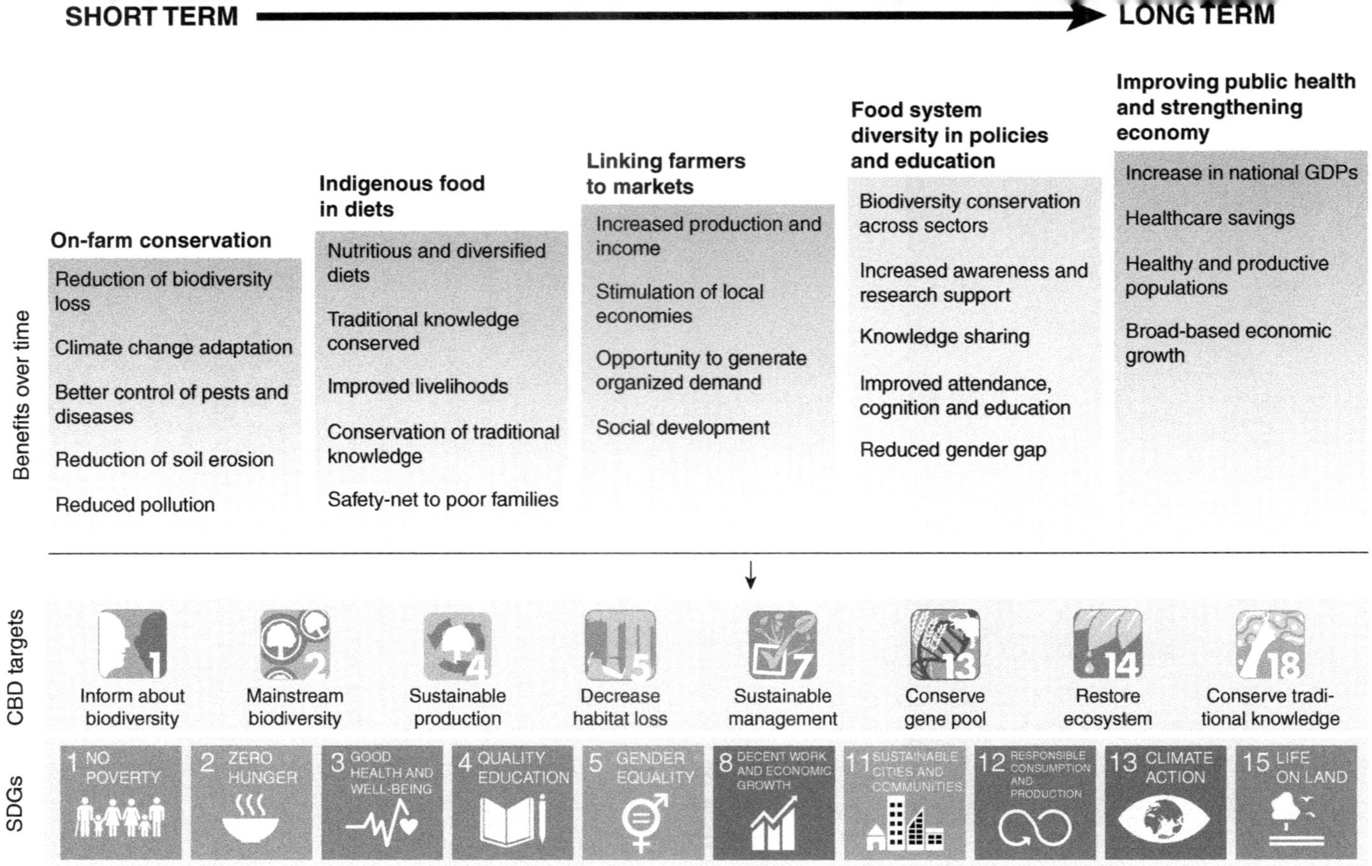

FIGURE 1.5 Objectives of the BFN Project, including how mainstreaming biodiversity for food and nutrition can help achieve multiple SDGs and Aichi Targets

Source: The BFN Project

FIGURE 1.6 School gardens represent an opportunity to put mainstreaming into practice, such as at Mundika school, Kenya, 2018
Source: A. Manjella

Structure of the book

Part I of this book provides a comprehensive overview of the challenges and impacts of poor diets and nutrition from current food systems, and the possible contribution of biodiversity and ecosystem services in addressing these problems. It also reviews the current empirical evidence for the links between biodiversity, diets and nutrition and identifies critical gaps in our knowledge and understanding. Finally, it examines the potential of biodiversity and agroecology to transform food systems and highlights key opportunities where this might happen.

In **Chapter 2**, Burlingame examines the far-reaching negative consequences of unsustainable food systems, poor diets and malnutrition, linking agriculture and environmental issues with human nutrition and health. The chapter discusses the concept of a sustainable diet, with the aim of bringing together the health and agriculture sectors to address the multiple burdens of malnutrition. Sustainable diets, as defined by FAO, are meant to shift the view of malnutrition from a medical challenge with pharmacological interventions, to an ecosystem challenge addressing agriculture and food systems. Such a shift must effectively engage the sectors of health, agriculture and the environment in simultaneously addressing the challenges of malnutrition, from root cause to remediation.

In **Chapter 3**, Hattersley and colleagues shed further light on the relationship between biodiversity and nutrition, reviewing its conceptual basis and critically appraising the existing empirical evidence for these pathway links. Methodological and measurement issues, gaps in the evidence and priorities for improving the evidence base are discussed to ensure a better understanding of the options available for improved health of people and the food system.

In **Chapter 4**, Jacobs and colleagues reflect on how present day farming systems, often dominated by input-intensive crop monocultures and industrial-scale feedlots, supply large volumes of food to global markets, but generate negative outcomes including widespread degradation of land, water and ecosystems, high greenhouse gas emissions, biodiversity loss and persistent malnutrition. The authors state that a fundamental problem of these systems is uniformity and an increasing reliance on chemical fertilizers, pesticides and antibiotics. The alternative model of agriculture is based on diversified farms and farming landscapes, optimizing biodiversity, delivering diversified diets and stimulating interactions between different species, as part of holistic strategies to build long-term fertility, healthy agroecosystems and secure livelihoods. The chapter examines systemic factors keeping countries locked-in to the industrial agriculture model and presents opportunities that exist to facilitate a paradigm shift towards 'diversified agroecological systems'.

Part II of the book is dedicated to the global Biodiversity for Food and Nutrition (BFN) Project and is a single chapter (5) that flows from A: Context, to B: Planning and C: Actions. Section A provides a historical and conceptual overview of the global project, the rationale for the selection of countries involved and the BFN Project's role as a driver of the CBD's Cross-cutting initiative on biodiversity for food and nutrition. Section B highlights the organization of the partner-based project, elaborating on effective multi-sectoral platforms for implementation. Section C dives into the process of bringing the project components – providing evidence, influencing policy and markets, and raising awareness – to life, including best practices, lessons learned and key examples from the countries. This roadmap of the BFN Project is provided to guide other countries and agencies that would like to adopt and adapt a similar approach or integrate it into other ongoing initiatives (see Box 1.1).

BOX 1.1 BFN PROJECT DETAILS COVERED IN PART II

- Identifying multi-sectoral, interdisciplinary partnerships and collaborations
- Identifying project sites, locations and focus
- Prioritizing food biodiversity species and generating nutrition composition data
- Managing data and knowledge
- Influencing policies, institutions and markets
- Developing tools, knowledge and best practices for awareness raising and scaling up
- Innovations and novel approaches employed by the individual BFN countries.

Part III expands on further opportunities for biodiversity and ecosystem services to contribute to healthy and sustainable food systems. The key achievements, lessons learned and good practices of the BFN Project are considered in terms of strategic insights and widely recognized opportunities and leverage points that can contribute to food systems transformation.

In **Chapter 6**, Egal and Forster explore current trends in the intersection of biodiversity, food systems and urban-rural linkages (URLs). In the context of urbanization, biodiversity is often addressed in terms of blue and green corridors or natural habitats within urban and peri-urban spaces; here the authors focus instead on the dynamic interactions of biodiversity and food systems *across* the urban-rural continuum. Due to the dual processes of urbanization and globalization, diets are changing worldwide with increasing consumption of high-calorie, nutrient-poor foods high in fats, sweeteners and salt. The authors stress that dynamically changing food systems, with significant impacts on diets and nutrition, are inseparable from the relationships between rural and urban areas, and deeply connected to the underlying natural resource foundation of food systems. While rural migrants bring their food habits to towns and cities, they often cannot find familiar foods (and biodiversity). Different distribution systems, living conditions and quality of food access result in a subsequent hybridization of urban diets.

The authors also highlight how recent global policy agendas – SDGs (goal 11 and related goals), the Milan Urban Food Policy Pact, New Urban Agenda, ICLEI and UN Habitat's Urban Rural Linkages Guiding Principles – are beginning to address interconnected challenges and opportunities for integrating urban and rural areas, stressing that significant gaps still exist between place-based practices, the lack of research to generate practice-based evidence and related global policy. Pointing out that the formal literature of peer-reviewed, rigorous analysis and synthesis in this area of intersection is weak, they call attention to the promising level of practice in relevant innovations, investments and initiatives in all regions of the world. The chapter concludes with recommendations that could help inform a new consensus and policy narrative based on the intersection of biodiversity, food systems and URLs.

In **Chapter 7**, Swensson and Tartanac cover the increasing attention given to institutional food procurement programmes (IFPP) as an important policy instrument with the potential to deliver multiple benefits for multiple beneficiaries, including food consumers, food producers and local communities. IFPP have the possibility – based on sound policy and regulatory frameworks – to determine not only the way food is produced and procured, but, in particular: what food is purchased (such as local, socio- and bio-diverse, nutritious, healthy, culturally appropriate); from whom food is purchased (for example from local and/ or family farming producers, small and medium food enterprises, women, youth and other vulnerable groups); and from which type of production (for example from agricultural production that ensures environmental sustainability as well as conservation of biodiversity). The authors demonstrate the potential of IFPP to influence both consumption and production patterns, contributing to the transformation of local food systems. The chapter also examines the role of biodiversity and the development of linkages with family farming, agroecological production and the promotion of sociobiodiversity in diversifying procurement, identifying opportunities as well as the barriers and constraints to the better inclusion of food biodiversity.

In **Chapter 8**, Legrand and colleagues trace the role and history of food and gastronomy, arguing that the gastronomic sector is a critical interface in which consumers and chefs negotiate what is eaten and, consequently, what food is sourced. The impact of local and organic food on the gastronomy sector is critically reviewed as well as the concept of 'sustainable gastronomy' with an emphasis on challenges associated with 'sustainable dining options'. The linkages between food biodiversity, gastronomy and local sourcing are explored and critiqued, highlighting that biodiversity plays a vital role in 'what we eat, when we eat it and, how we consume it'. Therefore, a diet rich in biodiversity should be a priority for any sustainable gastronomy. The authors conclude that consumers as well as gastronomic 'producers' have a role to play in the type and location of food grown. However, current gastronomic food trends such as 'local' and 'organic' are unlikely to have a substantial impact on our food production systems without recognition that food choices are not only the concern of restaurateurs and chefs but necessitate significant support from policymakers.

In **Chapter 9**, Berno explores the relationship between food, biodiversity, cuisine and tourism, with a focus on agriculture-tourism linkages, local food systems, community tourism development, and tourism and gastronomy. Examples from the South Pacific highlight the importance of food biodiversity and local food systems to tourism in Small Island Developing States (SIDS). A case study of an agritourism venture in Fiji provides an example of how tourism can be used as a tool to conserve biodiversity, in this case with specific reference to taro, a culturally and nutritionally important staple crop in the Pacific.

In **Chapter 10**, Gee and Lee examine varied perspectives on innovation: innovation as a disrupting force of 'business as usual' in food systems; food innovation as a driver of smart regional growth; youth as drivers of innovation; and culinary innovation such as the New Nordic Movement. Referring to the etymological root of 'innovation' as 'restoration' or 'renewal', the authors focus on practices that can restore natural biodiversity and renew local ecosystems and communities. This conceptual basis is used to explore the incorporation of local food biodiversity into food and agriculture innovations, with attention given to emerging youth-led and focused practices. The authors point out that while innovation is often associated with cutting-edge technology, it can be equally applied to new linkages between pre-existing methods, practices and actors.

The chapter highlights opportunities and challenges of mobile apps, urban agriculture, youth movements and more, which are assessed to suggest how innovative approaches – social and technological – can be best leveraged for fair and effective promotion and protection of local food biodiversity across the producer-consumer spectrum.

Finally, **Chapter 11** offers concluding thoughts on what has been accomplished so far, the growing recognition of food biodiversity's role in transforming food systems, as well as opportunities that exist to carry the work of the BFN Project and similar initiatives farther into the future.

Note

1 Mace, G.M., Barret, M., Burgess, N.D., Cornell, S.E., Freeman, R., Grooten, M., Purvis, A. (2018) 'Aiming higher to bend the curve of biodiversity loss', *Nature Sustainability*, no 1, pp 448–451.

References

Bioversity International. (2017) *Mainstreaming Agrobiodiversity in Sustainable Food Systems: Scientific Foundations for an Agrobiodiversity Index – Summary*, Bioversity International, Rome, Italy.

Blasbalg, T.L., Wispelwey, B., Deckelbaum, R.J. (2011) 'Econutrition and utilization of food-based approaches for nutritional health', *Food and Nutrition Bulletin*, vol 32, no 1, pp S4–13. doi: 10.1177/15648265110321S102.

Diaz, S., Settele, J., Brondízio, E., Ngo, H., Guèze, M., Agard, J., Arneth, A., Balvanera, P., Brauman, K., Butchart, S., Chan, K. (2019) *Summary for Policymakers of the Global Assessment Report on Biodiversity and Ecosystem Services of the Intergovernmental Science-Policy Platform on Biodiversity and Ecosystem Services*. Available at: www.ipbes.net/news/ipbes-global-assessment-summary-policymakers-pdf

Englberger, L., Aalbersberg, W., Ravi, P., Bonnin, E., Marks, G.C., Fitzgerald, M.H., Elymore, J. (2003a) 'Further analyses on Micronesian banana, taro, breadfruit and other foods for provitamin A carotenoids and minerals', *Journal of Food Composition and Analysis*, vol 16, no 2, pp 219–236. doi: 10.1016/S0889-1575(02)00171-0.

Englberger, L., Aalbersberg, W., Fitzgerald, M.H., Marks, G.C., Chand, K. (2003b) 'Provitamin A carotenoid content of different cultivars of edible pandanus fruit tectorius', *Journal of Food Composition and Analysis*, vol 16, no 2, pp 237–247. doi: 10.1016/S0889-1575(02)00169-2.

Englberger, L., Johnson, E. (2013) 'Traditional foods of the Pacific: Go Local, a case study in Pohnpei, Federated States of Micronesia', in Fanzo, J., Hunter, D., Borelli, T., Mattei, F. (eds) *Diversifying Food and Diets: Using Agricultural Biodiversity to Improve Nutrition and Health*, pp 231–241, Routledge, Abingdon.

Fanzo, J., Hunter, D., Borelli, T., Mattei, F. (eds) (2013) *Diversifying Food and Diets: Using Agricultural Biodiversity to Improve Nutrition and Health*, Routledge, Abingdon.

FAO. (2017) *The Future of Food and Agriculture – Trends and Challenges*, Food and agriculture Organization of the United Nations, Rome, Italy.

FAO. (2019) *The State of the World's Biodiversity for Food and Agriculture*, J. Bélanger & D. Pilling (eds.). FAO Commission on Genetic Resources for Food and Agriculture Assessments, Rome. 572 pp. Available at: www.fao.org/3/CA3129EN/CA3129EN.pdf

Fentahun, M.T., Hager, H. (2009) 'Exploiting locally available resources for food and nutritional enhancement: Wild fruits diversity, potential and state of exploitation in the Amahara region of Ethiopia', *Food Security*, vol 1, pp 207–219. doi: 10.1007/s12571-009-0017-z.

FOLU. (2019) *Growing Better: Ten Critical Transitions to Transform Food and Land Use*. The Global Consultation Report of the Food and Land Use Coalition. Available at: www.foodandlandusecoalition.org/wp-content/uploads/2019/09/FOLU-GrowingBetter-GlobalReport.pdf

Forum for the Future. (2014) *Key Challenges for a Sustainable Food Supply*. Available at: www.changemakers.com/sites/default/files/sustainable_food_supply_-_nesta_paper_-_final_version.pdf

Garnett, S.T., Burgess, N.D., Fa, J.E., Fernández-Llamazares, A., Molnár, Z., Robinson, C. J., Watson, J.E.M., Zander, K.K., Austin, B., Brondizio, E.S., Collier, N.F., Duncan, T., Ellis, E., Geyle, H., Jackson, M.V., Jonas, H., Malmer, P., McGowan, B., Sivongxay, A., Leiper, I. (2018) 'A spatial overview of the global importance of Indigenous lands for conservation', *Nature Sustainability*, vol 1, pp 369–374. doi: 10.1038/s41893-018-0100-6.

GBD 2017 Diet Collaborators. (2019) 'Health effects of dietary risks in 195 countries, 1990–2017: A systematic analysis for the Global Burden of Disease Study 2017', *The Lancet*, vol 393, no 10184, pp 1958–1972.

Global Panel on Agriculture and Food Systems for Nutrition. (2016) *Food Systems and Diets: Facing the challenges of the 21st century*, London.

HLPE. (2017) *Nutrition and Food Systems. A Report by the High Level Panel of Experts on Food Security and Nutrition*, High Level Panel of Experts on Food Security and Nutrition, Rome, Italy.

Huang, A.S., Tanudjaja, L., Lum, D. (1999) 'Content of alpha-, beta-carotene, and dietary fiber in 18 sweet potato varieties grown in Hawaii', *Journal of Food Composition and Analysis*, vol 12, no 2, pp 147–151. doi: 10.1006/jfca.1999.0819.

Hunter, D., Borelli, T., Beltrame, D.M.O., Oliveira, C.N.S., Coradin, L., Wasike, V.W., Wasilwa, L., Mwai, J., Manjella, A., Samarasinghe, G.W.L., Madhujith, T., Nadeeshani, H. V.H., Tan, A., Ay, S.T., Güzelsoy, N., Lauridsen, N., Gee, E., Tartanac, F. (2019) 'The potential of neglected and underutilized species for improving diets and nutrition', *Planta*, vol 250, no 3, pp 709–729. doi: 10.1007/s00425-019-03169-4.

IPCC. (2019) *IPCC Special Report on Climate Change, Desertification, Land Degradation, Sustainable Land Management, Food Security, and Greenhouse gas fluxes in Terrestrial Ecosystems Summary for Policymakers*, Intergovernmental Panel on Climate Change. Available at: www.ipcc.ch/site/assets/uploads/2019/08/Edited-SPM_Approved_Microsite_FINAL.pdf

IPES-Food. (2016) *From Uniformity to Diversity: A Paradigm Shift from Industrial Agriculture to Diversified Agroecological Systems*, International Panel of Experts on Sustainable Food systems, Available at: www.ipes-food.org/_img/upload/files/UniformityToDiversity_FULL.pdf

Jones, A.D., Kennedy, G., Raneri, J.E., Borelli, T., Hunter, D., Creed-Kanashiro, M. (2019) 'Agricultural biodiversity and diets: Evidence, indicators and next steps', in Zimmerer, K.S., de Haan, S. (eds) *Agrobiodiversity. Integrating Knowledge for a Sustainable Future*, pp 213–224, The MIT Press, Cambridge, MA.

Kobori, N.C., Rodriguez-Amaya, D.B. (2008) 'Uncultivated Brazilian green leaves are richer sources of carotenoids than are commercially produced leafy vegetables', *Food and Nutrition Bulletin*, vol 29, no 4, pp 320–328. doi: 10.1177/156482650802900408.

Kuhnlein, H., Erasmus, B., Spigelski, D. (2009) *Indigenous Peoples' Food Systems: The Many Dimensions of Culture, Diversity and Environment for Nutrition and Health*, Food and Agriculture Organization of the United Nations, Centre for Indigenous Peoples' Nutrition and Environment, Rome, Italy.

Kuhnlein, H.V. (2017) 'Holding on to agrobiodiversity: Human nutrition and health of Indigenous peoples', in Hunter, D., Guarino, L., Spillane, C., McKeown, P.C. (eds) *Routledge Handbook of Agricultural Biodiversity*, pp 388–400, Routledge, Abingdon.

McGarry, D.K., Shackleton, C.M. (2009) 'Children navigating rural poverty: Rural children's use of wild resources to counteract food insecurity in the Eastern Cape province, South Africa', *Journal of Children and Poverty*, vol 15, no 1, pp 19–37. doi: 10.1080/10796120802677594.

Myers, S.S., Zanobetti, A., Kloog, I., Huybers, P., Leakey, A.D.B., Bloom, A.J., Carlisle, E., Dietterich, L.H., Fitzgerald, G., Hasegawa, T., Holbrook, N.M., Nelson, R.L., Ottman, M.J., Raboy, V., Sakai, H., Sartor, K.A., Schwartz, J., Seneweera, S., Tausz, M., Usui, Y. (2014) 'Increasing CO2 threatens human nutrition', *Nature*, vol 510, pp 139–142. doi: 10.1038/nature13179.

Raneri, J.E., Kennedy, G. (2017) 'Agricultural biodiversity for healthy diets and food systems', in Hunter, D., Guarino, L., Spillane, C., McKeown, P.C. (eds) *Routledge Handbook of Agricultural Biodiversity*, pp 401–416, Routledge, Abingdon.

Rubiang-Yalambing, L., Arcot, J., Greenfield, H., Holford, P. (2014) 'Aibika (Abelmoschus manihot L.): Genetic variation, morphology and relationships to micronutrient composition', *Food Chemistry*, vol 193, pp 62–68. doi: 10.1016/j.foodchem.2014.08.058.

Ruiz, M., Vernooy, R. (2012) *The Custodians of Biodiversity Sharing Access and Benefits to Genetic Resources*, Earthscan, Abingdon.

Searchinger, T., Waite, R., Hanson, C., Ranganathan, J., Matthews, E. (2019) *World Resources Report Creating a Sustainable Food Future: A Menu of Solutions to Feed Nearly 10 Billion People by 2050*, Worlds Resources Institute.

Sthapit, B., Rao, R.R., Lamers, H., Sthapit, S. (2017) 'Uncovering the role of custodian farmers in the on-farm conservation of agricultural biodiversity', in Hunter, D., Guarino, L., Spillane, C., McKeown, P.C. (eds) *Routledge Handbook of Agricultural Biodiversity*, pp 549–562, Routledge, Abingdon.

Toledo, A., Burlingame, B. (2006) 'Biodiversity and nutrition: A common path toward global food security and sustainable development', *Journal of Food Composition and Analysis*, vol 19, pp 477–483.

WHO, CBD. (2015) *Connecting Global Priorities: Biodiversity and Human Health A State of Knowledge Review*, World Health Organization and Convention on Biological Diversity. Available at: www.cbd.int/health/SOK-biodiversity-en.pdf

Willett, W., Rockström, J., Loken, B., Springmann, M., Lang, T., Vermeulen, S., Garnett, T., Tilman, D., DeClerck, F., Wood, A., Jonell, M., Clark, M., Gordon, L.J., Fanzo, J., Hawkes, C., Zurayk, R., Rivera, J.A., De Vries, W., Sibanda, L.M., Afshin, A., Chaudhary, A., Herrero, M., Agustina, R., Branca, F., Lartey, A., Fan, S., Crona, B., Fox, E., Bignet, V., Troell, M., Lindahl, T., Singh, S., Cornell, S.E., Reddy, K.S., Narain, S., Nishtar, S., Murray, C.J.L. (2019) 'Food in the Anthropocene: The EAT-Lancet Commission on healthy diets from sustainable food systems', *The Lancet Commissions*, vol 393, no 10170, pp 447–492. doi: 10.1016/S0140-6736(18)31788-4.

2

CHALLENGES AND IMPACTS OF POOR DIETS AND NUTRITION

Barbara Burlingame

Introduction

The effects of poor diets, and malnutrition as their consequence, are far-reaching, well beyond the realm of public health and disease. Agriculture and environmental issues are inextricably linked to human nutrition, and the impacts of poor diets on sustainable development are equally and profoundly detrimental. The reality of many food systems and their associated diets is devastation – for individuals, communities, nations and the planet. The concept of sustainable diets, as defined by the Food and Agriculture Organization of the United Nations (FAO), is to shift the view of malnutrition from a medical challenge with pharmacological interventions (for example supplements, fortificants, therapeutic formulations), to an ecosystem challenge addressing agriculture and food systems. This shift would effectively engage the sectors of health, agriculture and the environment in simultaneously addressing the challenges of malnutrition, from root cause to remediation. Topics addressed in this chapter include malnutrition in all its forms, highlighting micronutrient deficiencies; sectors and their interventions, causing and alleviating malnutrition; and reframing human nutrition as an ecosystem service.

Malnutrition in all its forms

'Malnutrition', most simply defined as 'bad' nutrition, has many manifestations. The crudest classification is undernutrition (including stunting and underweight) versus overnutrition (including overweight, obesity and diet-related chronic diseases). Common to the entire spectrum of under and overnutrition is micronutrient malnutrition.

Most common examples of malnutrition are defined by anthropometry, i.e. body measurements. For children, age, height and weight are required for stunting, wasting and underweight determinations (de Onis and Branca, 2016). For adults, height and weight plugged into an equation gives body mass index (BMI), the *de facto* standard measurement for defining overweight and obesity. These measurements are easily and inexpensively determined and reported for individuals, households and populations. Clinical manifestations of malnutrition, which require slightly more specialized training, are also commonly determined. Examples include checking eyes for Bitot spots, indicative of vitamin A deficiencies; examining the neck for evidence of enlarged thyroid gland, indicative of iodine deficiencies; and blood pressure checks for evidence of risk for heart and cerebrovascular diseases. Medical and laboratory specialists employ biochemical analyses mainly for determining micronutrient malnutrition and identification of risk factors, symptoms and diseases, for example low serum vitamin A as a micronutrient deficiency marker, or blood lipid profiles for heart disease risk. Table 2.1 provides a list of various forms of malnutrition with definitions, including causes and remedies of relevance to biodiversity for food and nutrition.

TABLE 2.1 Different forms of malnutrition

Form	*Definition*	*Causes*	*Key remedies*
Child stunting	Low height for age; often used as a marker for chronic undernutrition.	Insufficient quantity of food and poor quality of diet over prolonged period; high morbidity, low mortality.	Agriculture and environment sector interventions: sustainable agriculture, agroecology and traditional food systems. Better quality diets, greater intake of highly nutritious food plants and animals.
Child wasting	Low weight for height; often used as a marker for acute undernutrition.	Insufficient quantity of food and poor quality of diet over prolonged period; high mortality.	Health, agriculture and environment sector interventions. More dietary energy and protein, and micronutrient-rich foods, coupled with short-term therapeutic formulations, emergency rations.
Child underweight	Low weight for age	Insufficient energy intake (i.e. quantity of food), often, but not necessarily, related to quality of diet.	Agriculture and environment sector interventions: sustainable agriculture, agroecology and traditional food

(*Continued*)

TABLE 2.1 (Cont.)

Form	*Definition*	*Causes*	*Key remedies*
		Associated with poverty and disrupted food systems.	systems. Better quality diets, greater intake of highly nutritious food plants and animals.
Child obesity	High weight for height	Excessive energy intake, often linked to high fat and simple carbohydrate intakes, and low energy expenditure.	All sectors' interventions. Requires drastic remediation within food systems and along food chains.
Adult underweight	Body Mass Index <16	Insufficient energy intake (i.e. quantity of food), often, but not necessarily, related to quality of diet. Associated with poverty and disrupted food systems.	Agriculture and environment sector interventions: sustainable agriculture, agroecology and traditional food systems. Better quality diets, greater intake of highly nutritious food plants and animals.
Adult obesity	Body Mass Index >30	Excessive energy intake, often linked to high fat and simple carbohydrate intakes, and low energy expenditure.	All sectors' interventions. Requires drastic remediation within food systems and along food chains.
Diet-related non-communicable diseases	Cardiovascular diseases, certain cancers, diabetes.	Usually chronic, i.e. morbidity developing over a prolonged period, from poor quality diets. Often associated with high intake of fats, sugars and ultra-processed foods.	Health sector interventions for treatment. Agriculture and environment sectors' intervention for prevention.
Micronutrient malnutrition	Deficiencies of vitamins & minerals, most notably Iodine, vitamin A and iron.	Poor quality diet. Invariably found in all forms of undernutrition, but also a common feature with overweight, obesity and non-communicable diseases.	Agriculture and environment sectors' interventions to remediate food systems to foster availability and increased intake of fresh fruits and vegetables, whole grains, nuts, seeds and animal-source foods such as eggs, meat, insects, aquatic species.

In FAO's flagship publication, *State of Food Security and Nutrition in the World* (FAO, 2018)[1], the key statement is that rates of hunger continue to rise.

> For the third year in a row, there has been a rise in world hunger. The absolute number of undernourished people, i.e. those facing chronic food deprivation, has increased to nearly 821 million in 2017, from around 804 million in 2016. These are levels from almost a decade ago

The causes are complex, but it is clear that policies, programmes and interventions employed over many years by the agriculture sector are not working, and the brief periods of success have not been sustainable in reversing the prevalence of malnutrition in most of its forms.

Wasting, low weight for height, continues to affect over 50 million children, with death as the frequent endpoint. Stunting, low height for age, is deemed by the World Health Organization (WHO) to be the most prevalent form of child malnutrition. While some progress has been made in reducing child stunting, levels remain unacceptably high. Latest estimates show around 151 million children under five – or over 22% of the world's population in that age group – were affected by stunting in 2017. Neither wasting nor stunting results from a single causative agent but can be the result of any of a number of factors, individually and in combination. Notwithstanding factors such as communicable diseases, poor hygiene and the effects of poverty, it is poor quality diets – i.e. low and/or imbalanced intakes of micronutrients, amino acids, fatty acids – that are invariably at the heart of the problem of child malnutrition.

In addition to the high rates of undernutrition, it is estimated that 38 million children from 0–5 years are overweight. Adult obesity is worsening with each subsequent estimate, with one in eight now classified as obese (FAO, 2018). Complicating the problem, both undernutrition and overweight/obesity coexist in many countries and many households. Unprecedented until recently is the phenomenon of under and overnutrition coexisting in individuals, for example, the cases of concurrent stunting and overweight in children documented in Latin America (Fernald and Neufeld, 2006), Africa (Mamabolo et al., 2005) and Asia (Do et al., 2018).

It is estimated that 1.5 billion people in the world are affected by micronutrient deficiencies (WHO, 2017). Almost invariably, those suffering from stunting, wasting and underweight are also affected by micronutrient malnutrition. However, many people at the high end of the body mass spectrum, and those with diet-related chronic diseases, are also deficient in micronutrients. Diet-related chronic diseases represent the leading cause of premature death and disability around the world (Willett et al., 2019; Swinburn et al., 2019). Examples include heart disease, cerebrovascular disease, high blood pressure, type 2 diabetes, some cancers and poor bone health.

In the history of interventions, few are sustainable

Regardless of one's preferred definition for 'sustainable' – (i) able to be maintained at a certain rate or level, or (ii) conserving an ecological balance by avoiding depletion of natural resources, or (iii) able to be upheld or defended – most of the interventions addressing malnutrition are not/have not been sustainable.

Bad diets are not sustainable

Malnutrition in all its forms – undernutrition, overweight/obesity with its attendant chronic diseases and vitamin/mineral deficiencies – is largely the result of bad diets and poor dietary choices (Willett et al., 2019; Swinburn et al., 2019). The human toll, the economic cost and the environmental damage caused by bad diets is a burden that is not sustainable. The field of nutrition, the obvious discipline to address malnutrition, is largely the domain of the health sector, where effects of malnutrition are seen most clearly in primary health care and public health. Food *per se* was uncoupled from human health as nutrition became a clinical subject, and laboratory-synthesized nutrients became medicine. Typically, nutrition education, nutrition policies and programs and nutrition in the tertiary education system, all emphasize a disease model for treatment and prevention of malnutrition. The health sector has been battling malnutrition with tenacity for decades. Yet, the prevalence of diet-related diseases and deaths in both developed and developing countries is higher now than ever before (WHO, 2017), and micronutrient deficiency rates are stuck at unacceptably high levels. Malnutrition has not arisen because of the health sector, and therefore cannot be effectively addressed by health sector initiatives alone.

Modern agriculture, and food production generally, is not sustainable

The agriculture sector has long been pushing environmental limits throughout the world. The era of industrial agriculture emerged in the 20th century, with the Green Revolution as one of its better-known manifestations. Food and agriculture became focused on dietary energy supply and little else (Barrett, 2010). Attention was directed toward greater intensification of farming systems and monocultures requiring ever-increasing agricultural chemical use. This in turn led to inefficiencies throughout food systems and along food chains, resulting in dramatic levels of food losses and waste; and environmentally-damaging production to meet dietary demands favouring more meat, more dairy and more ultra-processed convenience foods. Globally, intensive crop production systems are largely devoted to cereal grains and oilseeds, mostly consumed in highly refined forms and found in ultra-processed convenient and fast foods, providing cheap and abundant dietary energy but little in the way of quality nutrition. Thus, global agriculture is a major driver of malnutrition, most notably overweight/obesity, micronutrient malnutrition and diet-related chronic diseases. Therefore, the agriculture sector must play a key role in providing the solution to nutrition problems (Story et al., 2009; Bioversity International, 2009).

Indeed, global health, represented by WHO and the United Nations International Children's Emergency Fund (UNICEF), and global agriculture, represented by FAO and the Consultative Group on International Agricultural Research (CGIAR), have been tackling the problems of malnutrition for many decades, but the problems are not abating. Often the policies, programmes and interventions were undertaken independently rather than collaboratively, and the actions of one agency would sometimes undermine the actions of the other (Elinder, 2005); for example, a food-based initiative by the agriculture sector to enhance micronutrient intakes with local fruit and vegetable production, was inadvertently compromised by the widespread introduction of ready to use therapeutic formulations (Bazzano et al., 2017). The author found that:

> using commercially processed nutritious foods for nutritional interventions could result in the deterioration of nutritional status amongst the poor rather than the intended improvement, as these products are likely replacing otherwise healthier, nutritionally adequate local foods, and cost more than real food alternatives.

Environments are not sustainable

The tensions between the environment and agriculture sectors are significant. Environmental degradation and biodiversity loss were early casualties of the Green Revolution, and the pattern continues today (IPES-Food, 2016). There is consensus that agricultural production and food consumption are the most significant causes of environmental pressures, especially habitat change, biodiversity loss, climate change, fish depletion, water use and chemical contamination (IAASTD, 2008; Rockström et al., 2009). Modern agricultural production systems not only destroy landscapes and biodiversity but seed their own demise in the case of monoculture agriculture which can lead to the faster build-up of pests and diseases, and the rapid spread of pathogens. Examples include the kiwifruit infestation by *Pseudomonas syringae* pv. *actinidiae* (Psa) (Gould et al., 2015) and *Fusarium* wilt disease in Cavendish bananas (Ploetz and Churchill, 2011). These intensive systems lead to, and even demand, high levels of agricultural chemical input, which in turn leads to greater contamination of soils, ground water supplies and food (Camargo and Alonso, 2006). In addition, these intensive agricultural systems, along with both intensive and extensive livestock systems, are considered the main drivers of climate change via greenhouse gas emissions (Eurostat, 2018) and of biodiversity loss.

Modern agriculture's 'success' provided ample evidence that dietary energy supply, the common metric for undernourishment, can be satisfied without biodiversity. What became abundantly clear at the same time was that micronutrient supply was not being satisfied. Indeed, many forms of malnutrition, for example obesity and non-communicable diseases, were, and continue to be, exacerbated by modern agriculture and food production. Noting the linkage between biodiversity, food and nutrition and the need to enhance sustainable use of biodiversity to combat hunger and malnutrition, the environment sector took action through the Convention on Biological Diversity (CBD), developing and implementing the '*Cross-cutting initiative on biodiversity for food and nutrition*' (CBD, 2006). The issues addressed were, among other things, conserving food biodiversity through sustainable consumption. The multisectoral approach envisions nutrition-driven agriculture within environmental limits. The key focus was food biodiversity delivering quality diets, over and above the agriculture sector's attention to dietary energy and the health sector's attention to nutrient supplements delivered as medicine and therapeutic formulations. The environment sector – biodiversity conservation through sustainable use – had answers to nutrition challenges that neither health nor agriculture had alone or in combination (Toledo and Burlingame, 2006).

Human nutrition as ecosystem service[2]

The concept of ecosystem services has been an implied, over-arching principle for the relationship between people and their environment for the entirety of human existence. However, ecosystem services only became explicit, defined and made popular as a result of the Millennium Ecosystem Assessment (MA) (IAASTD, 2008) in the early 2000s.[3] The MA grouped ecosystem services under four pillars: provisioning, regulating, supporting and cultural. Food was listed as a provisioning ecosystem service. But with the acceleration of transformations in food supplies brought about by modern agricultural production, and the flow-on effects for human nutrition, a critical re-evaluation of provisioning ecosystem services seemed overdue, a decade after the MA was published. Meanwhile, another group of scientists was busy re-evaluating nutrition based on decades of failures, unintended consequences and collateral damage from policies, programmes and interventions (Latham, 2010). Central to this discussion was a fundamental question: what is the basic unit of nutrition (Burlingame, 2001, 2012): nutrients (as described in all basic nutrition textbooks), foods (Jacobs and Tapsell, 2007) or diets (Burlingame and Dernini, 2012)?

The health sector used individual micronutrients as that basic unit. Malnutrition was a disease, and the treatment called for nutrition to be administered as vitamin A capsules, iron tablets, iodized salt and ready to use therapeutic formulations. The flaws in this model are manifold, the two most important being that a measured, single micronutrient deficiency almost invariably signals complete micronutrient inadequacy, not simply a deficiency of that measured nutrient; and the approach is largely unsustainable.

The agriculture sector used food, *per se*, as the basic unit, often expressed as the quantity of foods by weight and their dietary energy equivalent in kilocalories. Nutrition was administered, largely, as starchy staples coming from intensive, industrial monoculture agriculture. In many places, as much as 80% of the diet (dietary energy) came from not much more than refined rice, wheat or maize. Again, the flaws in this model, which has its origins in the Green Revolution, are manifold (Conway and Barbie, 1988; Smith, 1998; IAASTD, 2008). For nutrition, poor dietary quality in terms of micronutrients was one result, along with ever-rising levels of overweight and obesity. Furthermore, there was grave concern for the environment with this model of agriculture, with increasing damage to ecosystems and significant and ongoing biodiversity loss.

Considering the many decades of actions from both health and agriculture sectors to solve the problems of malnutrition, and the lack of sustainable, consistent success, a new model was needed. Perhaps 'diet' should be that basic unit of nutrition, but not simply diet, rather sustainable diet (Gussow and Clancy, 1986; Reddy et al., 2009; Burlingame and Dernini, 2019). The rationale was that this approach, under an environmental sustainability banner, would bring together the health and agriculture sectors to address the multiple burdens of malnutrition, i.e. undernutrition, overweight/obesity, diet-related chronic diseases and most importantly, micronutrient deficiencies (Burlingame and Dernini, 2012).

This model necessitated different sectors coming together to explore common problems and issues. Interestingly, we can credit the environment sector for taking the initiative in formalising this model. Informally and incidentally, nutritionists have included environment sector issues in their research and practice. For example, food composition databases present nutrient data on food biodiversity (FAO, INFOODS, 2008, 2017). Food-based dietary guidelines sometimes include a recommendation to decrease environmental footprints by eating locally and consuming less meat, the Mediterranean Diet Pyramid being one such example (Dernini et al., 2013). But it was the environment sector, with leadership from the CBD, that was first to acknowledge and operationalize the idea that multisectoral solutions were required in order to avert collateral damage in one sector caused by policies and programmes in another. In 2006, the Conference of the Parties, which is the governing body of the CBD, developed and endorsed the *Cross-cutting initiative on biodiversity for food and nutrition* (CBD, 2006). Table 2.2 shows the timeline.

The *Cross-cutting initiative on biodiversity for food and nutrition* included a set of operational objectives, which continue to be developed and promoted, and have found their way into the Sustainable Development Goals and other global commitments:

TABLE 2.2 Timeline for development and adoption of the Convention on Biological Diversity's initiatives related to Biodiversity for Food and Nutrition

Event	*Report*	*URL*
COP 7 Decision VII/32: The programme of work of the Convention and the Millennium Development Goals (Kuala Lumpur, Malaysia, February 2004)	• Noting the linkage between biodiversity, food and nutrition, and the need to enhance sustainable use of biodiversity to combat hunger and malnutrition … • Requests … a *Cross-cutting initiative on biodiversity for food and nutrition* to work together with relevant organizations, in order to strengthen existing initiatives on food and nutrition, enhance synergies and fully integrate biodiversity concerns into their work, with a view to the achievement of … relevant Millennium Development Goals.	www.cbd.int/decision/cop/default.shtml?id=7769
Commission on Genetic Resources for Food and Agriculture 10th Regular Session (Rome, Italy, November 2004)	• … requested how it could best support countries, on request, to generate, compile and disseminate cultivar-specific nutrient composition data, as well as indicate the relative priority of obtaining cultivar-specific dietary consumption data, in order to demonstrate the role of biodiversity in nutrition and food security.	www.fao.org/tempref/docrep/fao/meeting/014/j3951e.pdf

(*Continued*)

TABLE 2.2 (Cont.)

Event	*Report*	*URL*
Subsidiary Body on Scientific, Technical and Technological Advice (SBSTTA) Recommendation X/9 (Bangkok, Thailand, 7–11 February 2005)	• Presented options for a *Cross-cutting initiative on biodiversity for food and nutrition* • Recommended elements for an international initiative on biodiversity for food and nutrition.	www.cbd.int/recommendation/sbstta/?id=10689
Intergovernmental Technical Working Group on Plant Genetic Resources 3rd Session (Rome, Italy, October 2005)	• to undertake the necessary consultations and bring forward options for consideration by the Conference of Parties at its Eighth Meeting for a *Cross-cutting initiative on biodiversity for food and nutrition* within the CBD's existing programme of work on agricultural biodiversity. • 8 high priority actions	www.fao.org/fileadmin/templates/agphome/documents/PGR/ITWG/ITWG3/p3w5E.pdf
COP 8 Decision VIII/23A Cross-cutting initiative on biodiversity for food and nutrition (Curitiba, Brazil, March 2006)	• Adopts the framework for a *Cross-cutting initiative on biodiversity for food and nutrition.*	www.cbd.int/decision/cop/default.shtml?id=11037
COP 12 Decision XII/21: Biodiversity and human health (Pyeongchang, Republic of Korea, 6–17 October 2014)	• *Encourages* parties and other governments to promote cooperation between sectors and agencies responsible for biodiversity and those responsible for human health;	www.cbd.int/convention/results/?id=13384&kw=nutrition&t0=nutrition

	• *Recognizes* the relevance of the *Cross-cutting initiative on biodiversity for food and nutrition* for the linkages among biodiversity, food, nutrition and human health.	
COP 14 Draft decision submitted Health and biodiversity (Sharm El-Sheikh, Egypt, 17–29 November 2018)	• Acknowledging that consideration of health-biodiversity linkages can contribute to improving several aspects of human health and well-being, including through the prevention and reduction of both infectious and non-communicable diseases, and by supporting nutrition and healthy diets. • To compile information … on the design, management and implementation of production systems based on the conservation and sustainable use of biological diversity and traditional knowledge and the corresponding benefits to nutrition and healthy diets, particularly, but not restricted to, vulnerable and marginalized sectors.	www.cbd.int/doc/c/831c/4f46/a600bbe7338826cde2a93d8b/cop-14-l-04-en.pdf

- To substantiate the links between biodiversity, food and nutrition, in particular clarifying the relationship between biodiversity, dietary diversity and food preferences and the relevant links between human health and ecosystem health.
- To mainstream the conservation and sustainable use of biodiversity into agendas, programmes and policies related to nutrition, health, agriculture and hunger and poverty reduction.
- To counter the loss of diversity in human diets, and in ecosystems, by conserving and promoting the wider use of biodiversity for food and nutrition.
- To raise awareness of the links between biodiversity, food and nutrition and the importance of biodiversity conservation to meeting health and development objectives, including the elimination of hunger.

Key partners in the *Cross-cutting initiative on biodiversity for food and nutrition* at the global level were FAO, WHO and Bioversity International from the CGIAR. FAO and Bioversity International immediately took up the challenge with enthusiasm and energy and with many initiatives and outputs. These included the work on biodiversity indicators for human nutrition (FAO/INFOODS/Bioversity International, 2008, 2010), biodiversity food composition databases (Charrondiere et al., 2012), the sustainable diets initiative (Burlingame and Dernini, 2012), conferences and conference sessions, projects under the 'Sustainable Food Systems Programme' in the One Planet Network[4] and the ongoing Global Environment Facility (GEF) supported 'Biodiversity for Food and Nutrition' (BFN) Project[5] represented in this book and many other resources (Fanzo et al., 2013).

In an important decision-making forum, the Commission on Genetic Resources for Food and Agriculture (CGRFA), engaged with the Nutrition Division of FAO, requesting a paper addressing key issues in nutrition (FAO, 2013). The report from the CGRFA states the following:

> The Commission highlighted the importance of biodiversity for food and nutrition and noted that its potential role in nutrition is underexplored and undervalued. It welcomed the progress FAO had made in awareness raising and requested FAO to continue its leading role in the Cross-Cutting Initiative on Biodiversity for Food and Nutrition. The Commission appreciated that food biodiversity, in the context of the Initiative, regarded genetic resources, including neglected and underutilized species, and noted that improved information on their nutrient contents could facilitate new market opportunities. The Commission requested FAO to further develop its work on biodiversity and nutrition, recognizing the importance of linking food biodiversity and the environment sector to human nutrition and healthy diets, and of the *concept that nutrients in food and whole diets, as well as food, should be explicitly regarded as ecosystem services.*
>
> *CGRFA, 2013*

Later, at its 15th Regular Session in January 2015, the CGRFA endorsed the Voluntary Guidelines for Mainstreaming Biodiversity into Policies, Programmes and National and Regional Plans of Action on Nutrition, with the aim of assisting countries to make the best use of biodiversity for food and agriculture in their nutrition programmes. The guidelines provide examples of how mainstreaming could be implemented, depending on countries' needs and capabilities, as appropriate. The Commission stressed that implementation should be based on scientific evidence and consistent with relevant international obligations and that 'governments and stakeholders are encouraged, where appropriate, to implement these guidelines' (CGRFA, 2015).

As new as it all seems, references and inferences for human nutrition as an ecosystem service can be found going back millennia in the writings and teachings of doctors and philosophers. If we look back nearly 150 years, we find one of the earliest university-level nutrition programmes, developed by Ellen Swallow at the Massachusetts Institute of Technology in the late 19th century. A nutrition pioneer, Swallow called her subject Human Ecology. Fundamental to this course was the principle that human health and environmental health were linked, with food and nutrition being the connecting forces (Merchant, 2007).

Key features of ecosystems are their diversity. Efforts to define 'healthy' diets independently of their ecosystem can lead to the unintended consequence of exacerbating malnutrition (HLPE, 2017). Clear examples have been presented in case studies of indigenous food systems (Kuhnlein et al., 2009, 2013). The report of the 2nd 'International Conference on Sustainable Diets' held in Mongolia in 2013 made the point explicitly in the context of human nutrition and livestock. Meat and dairy, largely from a local breed of horse feeding on local pasture species, are fundamental to diets of many Mongolians, and livestock are fundamental to the sustainability of the ecosystems. Thus, were Mongolia to heed the global calls to decrease or make more efficient livestock systems for dairy and meat, micronutrient malnutrition and disruption to the ecosystems would be a consequence. The conference unanimously endorsed the principle that for understanding and characterising sustainable diets, the context is the ecosystem; the ecosystem must provide the guidance, and that the essence of sustainable diets is an ecosystem approach (Burlingame, 2014b).

Transformational change

Transformational change to food systems and diets has been a rallying cry for many decades, and continues today (Willett et al., 2019; Burlingame and Dernini, 2019). Since the 1950s, a huge body of scientific research developed showing that food systems as products of industrial agriculture were not sustainable, contributing to both malnutrition and environmental damage (Gannon et al., 1959; Chichester, 1965). The resource-intensive Green Revolution, requiring high chemical inputs for ever-increasing production yields, was successful in increasing the supply of starchy staple foods and contributing to improved food security as measured by dietary energy supply but was also deemed unsustainable, contributing to significant environmental degradation, biodiversity loss and indirectly to several forms of malnutrition. Nevertheless, the intensification of agriculture and the focus on production yields in terms of mass and dietary energy was supported by policies and programmes of governments and the International Agricultural Research Organizations and the FAO (Harwood, 2019).

More recently, as the harmful consequences of these actions became overwhelmingly obvious, agriculture joined the voices of civil society and researchers in calling for 'transformational change'. The term is now prominent in vision statements of UN agencies and policy documents of government departments, along with mission statements of civil society organizations, and in the titles of research papers in the scientific literature. Yet, the progressive degradation of diets and food systems continues at an ever-increasing pace, with only incremental changes evident at the periphery. Notwithstanding the consequences already realized from inaction, a transformational change is still required (UN, 2014; GPAFSN, 2016; HLPE, 2017). The rates of malnutrition and associated morbidity and mortality provide the striking justification (Swinburn et al., 2019).

Until recently, the calls for transformational change have been largely unheeded. However, as food systems and diets have been progressively worsening, the calls for dramatic change have become increasingly loud and compelling. In many of these calls, biodiversity for food and nutrition features prominently.

The Sustainable Development Goals (SDGs), endorsed by UN member nations in 2015, require transformational changes in order to be achieved. SDG 2 was constructed as a plea for health, agriculture and the environment sectors to work together. The short title for this goal is simply 'End hunger', while the long title is 'End hunger, achieve food security and improved nutrition, and promote sustainable agriculture', with biodiversity explicitly mentioned in its targets. The UN specialized agencies, WHO, FAO, the United Nations Environment Programme (UNEP), are trying to break down sectoral silos and encourage member nations to do the same. Margaret Chan, ex-Director General of the WHO, said in her October 2016 address to the Regional Committee for the Western Pacific:

> We hear many calls for multisectoral action. But we seldom see these calls acted upon in practical arrangements in your countries. The region enters the era of sustainable development with an exceptionally refined and comprehensive action agenda. *It calls for nothing less than a transformational change.*

In 2010, a large working group of scientists and stakeholders was charged with negotiating a definition for a term that requires collaborative efforts from many sectors and disciplines in order to be realized; that term is sustainable diets. It was fitting that the working group was held as a breakout session during the 'International Conference on Sustainable Diets and Biodiversity', convened by FAO and Bioversity International, and attended by researchers in nutrition, agriculture and environmental sciences and representatives from UN agencies, civil society, academia and the private sector (FAO, 2010). The now widely used definition is:

> Sustainable diets are diets with low environmental impacts that contribute to food and nutrition security and to healthy life for present and future generations. They are protective and respectful of biodiversity and ecosystems, culturally acceptable, accessible, economically fair and affordable, nutritionally adequate, safe and healthy, while optimizing natural and human resources.
>
> *FAO, 2010, 2012*

Despite some progress on a range of fronts, the challenges still remain, and more needs to be done. Initiatives such as the GEF-supported Biodiversity for Food and Nutrition (BFN) Project (described extensively in Part II of this book) are working to connect environmental and human health. The 2030 Agenda for Sustainable Development along with the United Nations Decade of Action on Nutrition 2016–2025, and scores of other initiatives (Burlingame, 2019), aim to achieve transformational change to address the challenges and impacts of poor diets and nutrition. These goals and plans of action provide roadmaps and guidance for intensifying policies, programmes and interventions to improve nutrition, one fundamental requirement of which is to transform food systems. Progress is being documented but the pace is far from 'transformational'. The future of people and the planet is at stake.

Notes

1 See also: FAO, IFAD, UNICEF, WFP and WHO (2019). The State of Food Security and Nutrition in the World 2019. Safeguarding against economic slowdowns and downturns. Rome, FAO.

2 This section is excerpted and modified from a short essay, Double Dividend, by Barbara Burlingame, appearing in the 2017 Massey University Press *New Zealand Land and Food Annual, No Free Lunch: Can NZ food the world sustainably?* and is reproduced with permission.

3 The Millennium Ecosystem Assessment (MA) was initiated by the United Nations Secretary-General Kofi Annan in 2000, with the objective of assessing the consequences of ecosystem change for human well-being and the scientific basis for action needed to enhance the conservation and sustainable use of those systems and their contribution to human well-being. The MA involved more than 1,360 experts worldwide. The findings were published in five technical volumes and six synthesis reports, providing a scientific appraisal of the condition and trends in ecosystem services and the options to restore, conserve or enhance the sustainable use of ecosystems. www.millenniumassessment.org/en/index.html.

4 The Sustainable Food Systems (SFS) Programme of the One Planet Network 10 Year Framework Programme on Sustainable Consumption and Productions, is a multi-stakeholder partnership focused on catalyzing more sustainable food consumption and production patterns. The vision enables partners to collaborate on joint initiatives, which range from normative, advocacy and policy support activities, to research and development projects as well as on-the-ground implementation activities that address our food systems challenges. The programme promotes a holistic approach, taking into account the interconnections and trade-offs between all elements and actors in food systems.

5 http://www.b4fn.org/.

References

Barrett, C.B. (2010) 'Measuring food insecurity', *Science*, vol 327, pp825–828, doi: 10.1126/science.1182768.

Bazzano, A.N., Potts, K.S., Bazzano, L.A., Mason, J.B. (2017) 'The life course implications of ready to use therapeutic food for children in low-income countries', *International Journal of Environmental Research and Public Health*, vol 14, no 4, article 403, doi: 10.3390/ijerph14040403.

Bioversity International. (2009) *Mainstreaming Agrobiodiversity in Sustainable Food Systems Scientific Foundations for an Agrobiodiversity Index*, Bioversity International, Rome, Italy.

Burlingame, B. (2001) 'What is nutrient?', *Journal of Food Composition and Analysis*, vol 14, no 1, p1. doi: 10.1016/j.jfca.2009.05.003.

Burlingame, B. (2014a) 'Grand challenges in nutrition and environmental sustainability', *Frontiers in Nutrition*, vol 1, article 3, doi: 10.3389/fnut.2014.00003.

Burlingame, B. (2014b) 'Chapter 2: Defining relationships: Synergies and trade-offs between health and environmental impacts', in *Sustainable Diets: Food for Healthy People and a Healthy Planet: Workshop Summary*, pp11–34, The National Academies Press, Washington, DC. Available at: www.ncbi.nlm.nih.gov/books/NBK189800/

Burlingame, B. (2019) 'Towards a code of conduct for sustainable diets', in Burlingame, B., Dernini, S. (eds), *Sustainable Diets: Linking Nutrition and Food Systems*, Centre for Agriculture and Bioscience International (CABI), Oxfordshire.

Burlingame, B., Dernini, S. (eds) (2012) *Sustainable Diets and Biodiversity: Directions and Solutions for Policy, Research and Action*, Food and Agriculture Organization of the United Nations, Rome, Italy.

Burlingame, B., Dernini, S. (eds) (2019) *Sustainable Diets: Linking Nutrition and Food Systems*, Centre for Agriculture and Bioscience International (CABI), Oxfordshire.

Camargo, J.A., Alonso, A. (2006) 'Ecological and toxicological effects of inorganic nitrogen pollution in aquatic ecosystems: a global assessment', *Environment International*, vol 32, pp831–849.

CBD. (2006) 'Decision adopted by the Conference of the Parties to the Convention on Biological Diversity at its eighth meeting VIII/23', Agricultural biodiversity, Conference of the Parties to the Convention on Biological Diversity Eighth meeting Curitiba, Brazil, 20–31 March 2006, Convention on Biological Diversity. Available at: www.cbd.int/doc/decisions/cop-08/cop-08-dec-23-en.pdf

CGRFA. (2013) *Report of the Commission on Genetic Resources for Food and Agriculture Fourteenth Regular Session Rome, 15–19 April 2013*, Food and Agriculture Organization of the United Nations, Rome, Italy. Available at: www.fao.org/3/mg538e/mg538e.pdf

CGRFA. (2015) *Report of the Commission on Genetic Resources for Food and Agriculture Fifteenth Regular Session Rome, 19–23 January 2015*, Food and Agriculture Organization of the United Nations, Rome. Available at: www.fao.org/3/a-mm660e.pdf

Charrondiere, U.R., Stadlmayr, B., Rittenschober, D., Nowak, V., Nilsson, E., Burlingame, B. (2012) *FAO/INFOODS Food Composition Database for Biodiversity Version 2.0 – BioFoodComp2.0*, Food and Agriculture Organization of the United Nations, Rome, Italy.

Chichester, C.O. (ed) (1965) 'Research in Pesticides', Proceedings of the Conference on Research Needs and Approaches to the Use of Agricultural Chemicals form a Public Health Viewpoint Held at the University of California Davis, California, October 1–3, 1964, Academic Press, New York.

Conway, G.R., Barbie, E.B. (1988) 'After the green revolution: Sustainable and equitable agricultural development', *Futures*, vol 20, no 6, pp651–670. doi: 10.4324/9781315066820.

de Onis, M., Branca, F. (2016) 'Childhood stunting: A global perspective', *Maternal & Child Nutrition*, vol 12, no S1, pp12–26. doi: 10.1111/mcn.12231.

Dernini, S., Meybeck, A., Burlingame, B., Gitz, V., Lacirignola, C., Debs, P., Capone, R., El Bilali, H. (2013) 'Developing a methodological approach for assessing the sustainability of diets: The Mediterranean diet as a case study', *New Medit*, vol 12, no 3, pp28–36.

Do, L.M., Lissner, L., Ascher, H. (2018) 'Overweight, stunting, and concurrent overweight and stunting observed over 3 years in Vietnamese children', *Global Health Action*, vol 11, no 1. doi: 10.1080/16549716.2018.1517932.

Elinder, L.S. (2005) 'Obesity, hunger, and agriculture: The damaging role of subsidies', *British Medical Journal*, vol 331, pp1333–1336, doi: 10.1136/bmj.331.7528.1333.

Eurostat. (2018) 'Greenhouse gas emissions by source sector', Available at: http://appsso.eurostat.ec.europa.eu/nui/show.do?dataset=env_air_gge&lang=en

Fanzo, J., Hunter, D., Borelli, T., Mattei, F. (2013) *Diversifying Food and Diets: Improving Agricultural Biodiversity to Improve Nutrition and Health*, Routledge, Oxfordshire.

FAO. (2010) *Final Document: International Scientific Symposium Biodiversity And Sustainable Diets United Against Hunger, 3–5 November 2010 FAO Headquarters, Rome*, Food and Agriculture Organization of the United Nations, Rome. Available at: www.fao.org/ag/humannutrition/28507-0e8d8dc364ee46865d5841c48976e9980.pdf

FAO. (2012) *Sustainable Diets and Biodiversity: Directions and Solutions for Policy, Research and Action*, Burlingame, B., Dernini, S. (eds), The Food and Agriculture Organization of the United Nations, Rome. Available at: www.fao.org/3/i3004e/i3004e.pdf

FAO. (2013) *Commission on Genetic Resources for Food and Agriculture Item 2.5 of the Provisional Agenda Fourteenth Regular Session Rome, 15–19 April 2013 Review of Key Issues on Biodiversity and Nutrition, CGRFA-14/13/8*, Food and Agriculture Organization of the United Nations.

FAO, IFAD, UNICEF, WFP, WHO. (2018) *The State of Food Security and Nutrition in the World 2018. Building Climate Resilience for Food Security and Nutrition*, Food and Agriculture Organization of the United Nations, Rome, Italy. Available at: www.fao.org/3/i9553en/i9553en.pdf

FAO, INFOODS. (2017) *Food Composition Database for Biodiversity Version 4.0 – BioFoodComp4.0*, Food and Agriculture Organization of the United Nations, Rome, Italy. Available at: www.fao.org/3/a-i7364e.pdf

FAO, INFOODS, Bioversity International. (2008) *Expert Consultation on Nutrition Indicators for Biodiversity 1. Food Composition*, Food and Agriculture Organization of the United Nations, Rome, Italy. Available at: www.fao.org/3/a-a1582e.pdf

FAO, INFOODS, Bioversity International. (2010) *Expert Consultation on Nutrition Indicators for Biodiversity 2. Food Consumption*, Food and Agriculture Organization of the United Nations, Rome, Italy. Available at: www.fao.org/docrep/014/i1951e/i1951e.pdf

Fernald, L.C., Neufeld, L.M. (2006) 'Overweight with concurrent stunting in very young children from rural Mexico: Prevalence and associated factors', *European Journal of Clinical Nutrition*, vol 61, pp623–632. doi: 10.1038/sj.ejcn.1602558.

Gannon, N., Link, R.P., Decker, G.C. (1959) 'Pesticide residues in meat and milk, storage of dieldrin in tissues and its excretion in milk of dairy cows fed dieldrin in their diets', *Journal of Agricultural and Food Chemistry*, vol 7, no 12, pp824–826. doi: 10.1021/jf60106a001.

Gould, E.M., Black, M.Z., Clark, G., Tanner, D.J., Benge, J. (2015) 'Tools for managing the kiwifruit bacterial canker disease *Pseudomonas syringae pv actinidiae (Psa)*', *Acta Hortic*, vol 105, pp39–46. doi: 10.17660/ActaHortic.2015.1105.6.

GPAFSN. (2016) *Food Systems and Diets: Facing the Challenges of the 21st Century*, Global Panel on Agriculture and Food Systems for Nutrition, London, UK. Available at: http://ebrary.ifpri.org/utils/getfile/collection/p15738coll5/id/5516/filename/5517.pdf

Gussow, J.D., Clancy, K.L. (1986) 'Dietary guidelines for sustainability', *Journal of Nutrition Education and Behavior*, vol 18, no 1, pp1–5. doi: 10.1016/S0022-3182(86)80255-2.

Harwood, J. (2019) 'Was the green revolution intended to maximise food production?' *International Journal of Agricultural Sustainability*, vol 17, no 4, pp312–325.

HLPE. (2017) *Nutrition and Food Systems*, A report by the High Level Panel of Experts on Food Security and Nutrition of the Committee on World Food Security, Rome. Available at: www.fao.org/3/a-i7846e.pdf

IAASTD. (2008) *Agriculture at a Crossroads Synthesis Report: A Synthesis of the Global and Sub-Global IAASTD Reports*, International Assessment of Agricultural Knowledge, Science and Technology for Development, Washington, DC. Available at: http://wedocs.unep.org/bitstream/handle/20.500.11822/7862/-Agriculture%20at%20a%20crossroads%20-%20Synthesis%20report-2009Agriculture_at_Crossroads_Synthesis_Report.pdf?sequence=3&isAllowed=y

IPES-Food. (2016) *From Uniformity to Diversity: A Paradigm Shift from Industrial Agriculture to Diversified Agroecological Systems*, International Panel of Experts on Sustainable Food systems. Available at: www.ipes-food.org/_img/upload/files/UniformityToDiversity_FULL.pdf

Jacobs, D.R., Tapsell, L.C. (2007) 'Food, not nutrients, is the fundamental unit in nutrition', *Nutrition Reviews*, vol 65, no 10, pp439–450. doi: 10.1111/j.1753-4887.2007.tb00269.x.

Kuhnlein, H., Erasmus, B., Spigelski, D. (2009) *Indigenous Peoples' Food Systems: The many Dimensions of Culture, Diversity and Environment for Nutrition and Health*, Food and Agriculture Organization of the United Nations and Centre for Indigenous Peoples' Nutrition and Environment, Rome, Italy. Available at: www.fao.org/3/i0370e/i0370e.pdf

Kuhnlein, H., Erasmus, B., Spigelski, D., Burlingame, B. (2013) *Indigenous Peoples' Food Systems & Well-Being Interventions & Policies for Healthy Communities*, Food and Agriculture Ogranization of the United Nations and Centre for Indigenous Peoples' Nutrition and Environment, Rome, Italy. Available at: www.fao.org/3/i3144e/i3144e.pdf

Latham, M. (2010) 'The great vitamin A fiasco', *World Nutrition*, vol 1, no 1, pp12–45.

Mamabolo, R.L., Alberts, M., Steyn, N.P., Delemarre-van de Waal, H.A., Levitt, N.S. (2005) 'Prevalence and determinants of stunting and overweight in 3-year-old black South African children residing in the Central Region of Limpopo Province, South Africa', *Public Health Nutrition*, vol 8, no 5, pp501–508. doi: 10.1079/PHN2005786.

Merchant, C. (2007) *American Environmental History: An Introduction*, Colombia University Press, New York.

Ploetz, R.C., Churchill, A.C.L. (2011) 'Fusarium wilt: The banana disease that refuses to go away', in Van den Bergh, I., Smith, M., Swennen, R., Hermanto, C. (eds) Proceedings of the International ISHS-ProMusa Symposium on Global Perspectives on Asian Challenges: Guangzhou, China, September 14–18, 2009, Acta horticulturae, Vol. 897, pp519–526.

Reddy, S., Lang, T., Dibb, S. (2009) *Setting the Table: Advice to Government on Priority Elements of Sustainable Diets*, Sustainable Development Commission, London, UK.

Rockström, J., Steffen, W., Noone, K., Persson, A., Chapin, F.S., Lambin, E., Lenton, T.M., Scheffer, M., Folke, C, Schellnhuber, H.J., Nykvist, B., de Wit, C.A., Hughes, T., van der Leeuw, S., Rodhe, H., Sörlin, S., Snyder, P.K., Costanza, R., Svedin, U., Falkenmark, M., Karlberg, L., Corell, R.W., Fabry, V.J., Hansen, J., Walker, B., Liverman, D., Katherine, R., Crutzen, P., Foley, J. (2009) 'Planetary boundaries: Exploring the safe operating space for humanity', *Ecology and Society*, vol 14, no 2, article 32. Available at: www.ecolo gyandsociety.org/vol14/iss2/art32/

Smith, L. (1998) 'Can FAO's measure of chronic undernourishment be strengthened?', *Food Policy*, vol 23, no 5, pp425–445. doi: 10.1016/S0306-9192(98)00049-9.

Story, M., Hamm, M.W., Wallinga, D. (2009) 'Food systems and public health: linkages to achieve healthier diets and healthier communities', *Journal of Hunger & Environmental Nutrition*, vol 4, no 3-4, pp219–224. doi: 10.1080/19320240903351463.

Swinburn, B.A., Kraak, V.I., Allender, S., Atkins, V.J., Baker, P.I., Bogard, J.R., Brinsden, H., Calvillo, A., Schutter, O.D., Devarajan, R., Ezzati, M., Friel, S., Goenka, S., Hammond, R.A., Hastings, G., Hawkes, C., Herrero, M., Hovmand, P.S., Howden, M., Jaacks, L.M., Kapetanaki, A.B., Kasman, M., Kuhnlein, H.V., Kumanyika, S.K., Larijani, B., Lobstein, T., Long, M.W., Matsudo, V.K.R., Mills, S.D. H., Morgan, G., Morshed, A., Nece, P.M., Pan, A., Patterson, D.W., Sacks, G., Shekar, M., Simmons, G.L., Smit, W., Tootee, A., Vandevijvere, S., Waterlander, W. E., Wolfenden, L., Dietz, W.H. (2019) 'The global syndemic of obesity, undernutrition, and climate change: The Lancet Commission report', *The Lancet Commissions*, vol 393, no 10173, pp791–846. doi: 10.1016/S0140-6736(18)32822-8.

Toledo, A., Burlingame, B. (2006) 'Biodiversity and nutrition: A common path toward global food security and sustainable development', *Journal of Food Composition and Analysis*, vol 19, no 6–7, pp477–483. doi: 10.1016/j.jfca.2006.05.001.

UN. (2014) 'United Nations General Assembly, Human Rights Council Twenty-fifth Session'. *Special Rapporteur on the Right to Food, Olivier De Schutter; Final report: The transformative potential of the right to food*, The United Nations. Available at: www.srfood.org/images/stories/pdf/officialreports/20140310_finalreport_en.pdf

WHO. (2017) *The Double Burden of Malnutrition*, World Health Organization, Geneva.

Willett, W., Rockström, J., Loken, B., Springmann, M., Lang, T., Vermeulen, S., Garnett, T., Tilman, D., DeClerck, F., Wood, A., Jonell, M., Clark, M., Gordon, L.J., Fanzo, J., Hawkes, C., Zurayk, R., Rivera, J.A., De Vries, W., Sibanda, L.M., Afshin, A., Chaudhary, A., Herrero, M., Agustina, R., Branca, F., Lartey, A., Fan, S., Crona, B., Fox, E., Bignet, V., Troell, M., Lindahl, T., Singh, S., Cornell, S.E., Reddy, K.S., Narain, S., Nishtar, S., Murray, C.J.L. (2019) 'Food in the Anthropocene: The EAT–Lancet Commission on healthy diets from sustainable food systems', *The Lancet Commissions*, vol 393, no 10170, pp447–492. doi: 10.1016/S0140-6736(18)31788-4.

3

EVIDENCE FOR THE ROLE OF BIODIVERSITY IN SUPPORTING HEALTHY, DIVERSE DIETS AND NUTRITION

Libby Hattersley, Bruce Cogill, Danny Hunter and Gina Kennedy

BIOVERSITY INTERNATIONAL

1 Introduction

Biological diversity and dietary diversity are considered essential components of healthy and sustainable food systems and diets (IAASTD, 2008; FAO, 2011; Burlingame and Dernini, 2012; Fanzo et al., 2013; FAO and WHO, 2014; WHO and CBD, 2015; IPES-Food, 2016; GLOPAN, 2016; Bioversity International, 2017; HLPE, 2017; Bélanger and Pilling, 2019; Fanzo, 2019; Swinburn et al., 2019; Willett et al., 2019). From the production perspective, there is strong evidence showing that biological diversity (biodiversity) supports the provisioning of a range of ecosystem services essential to food production and contributes to resilience within food systems, including against threats posed by climate change (Millennium Ecosystem Assessment, 2005a; FAO, 2011, 2019; Allen et al., 2014; IPBES, 2019). From the consumption perspective, the principle of dietary diversity has been embedded in national food-based dietary guidelines (FBDGs) for decades, with most including a recommendation to eat a variety of foods every day for optimal health. Many FBDGs also include a recommendation to eat a variety of fruits and vegetables, and some a diversity of foods within each food group (Fischer and Garnett, 2016; Willett et al., 2019). Eating a range of nutrient-rich foods from different food groups is considered important to ensure adequate intakes of all of the macro and micronutrients known to be essential to human health (WHO, 2003; FAO, 2015a). Synergistic interactions between nutrients and other bioactive compounds (including soluble and insoluble fibre, flavonoids, phenols, sterols, carotenoids and isothiocyanates) in different foods are thought to convey additional health benefits (Liu, 2003; Jacobs and Tapsell, 2007).

There is general consensus on the conceptual linkages between biodiversity and human health (Millennium Ecosystem Assessment, 2005b), and between biodiversity, diets and nutrition (Remans and Smukler, 2013, Figure 7.2). However, remarkably little is actually known about how these complex interactions and linkages operate in practice (Hough, 2014). This chapter reviews available evidence on biodiversity's contribution to diets and nutrition, highlights key limitations in this evidence, and identifies several priorities for future research. It argues for greater collaboration across disciplines to improve study designs and quality; consistent use of validated measurement tools and indicators; and more attention to the role of contextual factors, as well as to measuring impacts on all aspects of diet quality, and on nutrition and health outcomes.

The review of the evidence in this chapter is organized into two main sections. Evidence for the role of food biodiversity in supporting healthy, diverse diets is reviewed in Section 2 below. Section 3 reviews available evidence for the contribution of food biodiversity and dietary diversity to nutrition and related health outcomes. Food biodiversity is broadly defined as: 'the diversity of plants, animals and other organisms used for food, covering the genetic resources within species, between species, and provided by ecosystems' (FAO and Bioversity International, 2017). It is closely linked to (although is broader than) the concept of agricultural biodiversity (or agrobiodiversity) which includes all components of biological diversity of relevance to food and agriculture, and all components of biological diversity that constitute agricultural ecosystems, also termed agroecosystems (FAO, 1997; Fanzo, 2019). Food biodiversity can be cultivated or reared (cultivated biodiversity), accessed through markets or food transfer programs, or sourced from the wild (wild food biodiversity).

2 Evidence that food biodiversity supports healthy, diverse diets

The multiple pathways linking biodiversity and nutrition are long, complex and dynamic, and may be influenced by multiple potential confounding factors. This intricacy presents considerable analytical and empirical challenges. As a result, much of the empirical research specifically investigating these links to-date has focused on dietary intake measures as an intermediate nutrition outcome. Experimental research provides the most powerful design to establish a causal relationship between food biodiversity and diet quality. However, few intervention studies focused specifically on this link have been conducted. Nonetheless, some insights can be drawn from evaluations of nutrition-sensitive agricultural intervention programs that have included strategies to diversify food production (Section 2.1).

By far the most common approach used to investigate the biodiversity-diet link has been observational research using cross-sectional data to document associations between different aspects of food biodiversity and dietary intake. While unable to establish causation, observational studies can be useful for exploring and unravelling hypothesized associations and impact pathways, and for generating new hypotheses for intervention research. Observational evidence for an association between food biodiversity and dietary diversity/quality is reviewed in Section 2.2. Although yet to be tested for impacts on dietary intakes, use of diet optimization models to demonstrate how incorporation of food biodiversity can support healthier, more diverse diets is a nascent area of research that offers promise. Section 2.3 overviews current research in this area.

The evidence reviewed in this section is focused overwhelmingly on locally available food biodiversity within food systems categorized as traditional (HLPE, 2017), where most of the foods available are locally produced and seasonal, production is dominated by smallholder farmers who consume much of what they produce, and retail is dominated by informal vending and wet markets. In these contexts, food biodiversity has the potential to directly improve the variety and quality of diets, and to reduce peaks and troughs in household food availability. This includes playing an important role in smoothing seasonal food shortages and contributing to dietary diversity in rural communities during 'hungry' or 'lean' seasons between harvests, or during periods of unexpected household shocks such as crop failure or drought (Savy et al., 2006; Kennedy et al., 2017). There has been comparatively little research undertaken on the links between food biodiversity and diets in contexts in which food supply chains are longer, post-harvest segments (including processing, packaging and promotion) play a greater role in influencing diets, and populations engage with a wider range of retail formats.

2.1 Evidence from intervention studies

There has been a proliferation of studies evaluating the impacts of nutrition-sensitive agricultural programs (NSAPs) over the last few decades. Early NSAPs focused on addressing deficiencies in a handful of priority micronutrients (particularly vitamin A and iron), and evaluation approaches were developed around a need to demonstrate impact in terms of these priorities. Successive reviews of these studies have found little evidence that agricultural interventions impact nutritional status, or maternal or child anthropometric measures (Berti et al., 2004; World Bank, 2007; Bhutta et al., 2008; Arimond et al., 2011; Masset et al., 2011, 2012; Girard et al., 2012; Meeker and Haddad, 2013; Webb, 2013; Ruel and Alderman, 2013; Webb and Kennedy, 2014; Ruel et al., 2018; Bird et al., 2019). There is consensus in the reviews that this lack of evidence can be at least partly attributed to widespread methodological weaknesses in the studies, rather than a lack of effectiveness of the interventions themselves.

While the strongest approach to elucidating the impact of agriculture on nutrition would be to conduct a randomized controlled trial (RCT), the practical and cost considerations of using RCTs in this type of research are typically prohibitive. As a result, most NSAP studies have involved quasi-experimental designs. Common methodological limitations in this literature include lack of statistical power, lack of an appropriate control or comparison group, lack of control for confounding or effect modification, and poor evaluation designs (Ruel et al., 2018). High levels of heterogeneity have also made it difficult to make comparisons across studies and conduct meta-analyses.

There has been greater attention to methodological quality in more recent studies, and evaluations of good-quality RCTs are emerging (see Ruel et al., 2018; Bird et al., 2019). There has also been greater attention in recent NSAPs on strategies to diversify food production at the household or farm level, through homestead food production systems, home vegetable gardens, and small animal and livestock rearing (Ruel et al., 2018). There has also been a shift in focus towards using measures of dietary diversity (at the food group level) to assess impacts on food security and nutrition (Herforth and Ballard, 2016; Verger et al., 2019).

Evaluations of these more recent NSAPs suggest that promoting production diversity can lead to increases in dietary diversity at the household, and sometimes individual level, particularly when combined with well-designed behaviour change communication (BCC) components and a strong focus on women´s empowerment (Ruel et al., 2018; Bird et al., 2019). This includes findings from the first cluster-RCT to investigate the impact of a homestead food production programme (with strong BCC and women´s empowerment components) on child nutrition outcomes, which documented significant improvements in maternal dietary diversity and consumption of micronutrient-rich foods, in addition to positive impacts on maternal and child health outcomes (Olney et al., 2015, 2016).

However, despite the greater attention to dietary diversity in NSAP studies, it is difficult to elucidate the specific contribution of improved food biodiversity (at the farm and/or household level) to dietary intakes because few NSAPs have explicitly focused on increasing production diversity, and measures of production diversity are still not consistently used in this literature. When both production and consumption diversity have been measured (generally at the food group level), use of different categories of food groups for production and consumption has complicated interpretation of findings (Sibhatu et al., 2015; Koppmair et al., 2017). Inconsistent use and interpretation of these indicators also makes it difficult to compare across studies (Verger et al., 2019).

2.2 Evidence from observational studies

A large number of small-scale observational studies assessing the links between agriculture and nutrition have been conducted in recent years, with many including a focus on cultivated (and reared) biodiversity at the household or farm level, and/or dietary diversity (Section 2.2.1 below). Some of these have also sought to assess the role of access to, and participation in, markets in mediating the relationship (Section 2.2.4). There has been significantly less attention to the role of wild food biodiversity (Section 2.2.2), and even less on the contribution of forests and agro-forestry systems to dietary intakes (Section 2.2.3), despite their potential to make a significant contribution to dietary diversity and nutrition in specific contexts.

2.2.1 Cultivated food biodiversity

Recent reviews of studies that have analysed associations between cultivated biodiversity (measured at the crop group or species level) and dietary diversity scores (as proxy indicators of diet quality, measured at the food group or food type level) have found evidence of a small, positive association, at least among poor, isolated households and communities with limited market access or imperfect market structures (Powell et al., 2015; Jones, 2017; Ruel et al., 2018; Sibhatu and Qaim, 2018a). There is also limited evidence of an association between dietary species richness (in addition to dietary diversity) and adequate intake of multiple micronutrients where, for every additional species consumed, dietary nutrient adequacy increases (Lachat et al., 2018).

Given the small effect sizes observed, the nutritional significance of the association, even in settings of very low agricultural diversity, is unclear. Jones (2017) cautioned that potentially large and unrealistic increases in cultivated biodiversity may be required to have a nutritionally meaningful impact on dietary diversity.

The relationship between cultivated biodiversity and dietary diversity also appears to be highly context-specific (Jones, 2017; Ruel et al., 2018). At very low levels of biodiversity, a marginal increase in cultivated biodiversity has a much greater effect on diet diversity than on farms with moderate to high levels of production diversity. In contexts where on-farm biodiversity is already high, agricultural diversification may have minimal-to-no impact, and even a negative impact, on dietary diversity (Sibhatu et al., 2015; Jones, 2017; Ruel et al., 2018). Market access and participation also appear to be key factors that modify the relationship (see 2.2.2 below).

In general, this literature suffers from many of the same methodological weaknesses as the NSAP literature, including: a lack of appropriate controls or comparison groups, as well as accounting for potential mediating and confounding factors; and significant heterogeneity in measurement approaches and indicators used, making comparisons between studies difficult (Ruel et al., 2018).

2.2.2 Market access and participation

Even among subsistence-oriented households and within the context of traditional food systems, there is interplay between own production, sale to and purchase from markets and diversified consumption, with both own production and market participation being important to diets of smallholder producer families

(HLPE, 2017). Market participation refers to the market-based sale, purchase or acquisition of foods through retail markets (ranging from small informal local markets and small neighbourhood shops/kiosks to supermarkets, hypermarkets and ready-to-eat food retail), food transfer programs, bartering or food sharing.

Access to, and integration into, markets has the potential to improve food security, income, dietary diversity and nutrition and health outcomes (Darrouzet-Nardi and Masters, 2015; Reardon et al., 2015; Bellon et al., 2016; Sibhatu and Qaim, 2018b). On the other hand, greater market access has also been associated with the nutrition transition, which is characterized by greater consumption of processed, energy-dense foods, and the gradual shift from a high burden of undernutrition to rising rates of overnutrition, overweight and diet-related chronic diseases (Baker and Friel, 2016).

The small, positive association between cultivated biodiversity and dietary diversity appears to operate independently of market access and relative market-orientation of a household/farm. However, market access and participation (in addition to the extent of existing on-farm diversification), appear to be important effect modifiers (Jones, 2017; Ruel et al., 2018). There is relatively consistent (albeit, again, limited) evidence that market access is independently, positively associated with dietary diversity, at least among poor, rural communities in low and middle-income countries (Remans et al., 2011; Sibhatu et al., 2015; Bellon et al., 2016; Jones, 2017; Koppmair et al., 2017). Evidence on the role of market access among other population groups and other settings is very limited, although it appears that the influence of market access is likely to be highly context-specific (Jones, 2017).

Market access and participation have largely been measured in this literature using simple proxy indicators of market access, such as travel time or distance to closest market, road or population centre. Relative market-orientation (or commercialization) of a household's agricultural production may be a more direct proxy indicator of market participation, with some limited evidence that households with farms that are at least partly market-oriented (typically measured as proportion of cultivated area devoted to non-food cash crops) have more diverse diets than less market-oriented farms (Jones, 2017). However, the evidence is too limited to determine the nature, and even direction, of this relationship. A mixture of cultivated biodiversity along with enhanced infrastructure for improved market access is likely to be the optimum combination for improving dietary diversity (Ickowitz et al., 2019).

2.2.3 Wild food biodiversity

Food biodiversity can also be gathered from forests, water bodies and uncultivated areas within diverse agricultural landscapes. Wild food biodiversity includes plants, other forest products (for example fruit, nuts, mushrooms and bush meats), insects and aquatic species. Many wild foods are collected from within, or at the margins of, agricultural production systems, and some foods are considered semi-wild, or semi-domesticated (Halwart, 2006; Bharucha and Pretty, 2010).

Wild foods can be much richer in nutrients and food components than many domesticated species (Burlingame et al., 2009; Bharucha and Pretty, 2010; Termote et al., 2014; WHO and CBD, 2015). They tend to be resilient to harsh conditions and have significant potential to contribute to the nutrient-density of diets year-round, provided they are available, safe, consumed in sufficient quantities and the nutrients are bioavailable. Wild foods play an important role in contributing to food security and livelihoods for millions of people worldwide (Bioversity International, 2017; Ickowitz et al., 2019), although their actual consumption, contribution to diets relative to other foods, and nutritional significance within diets is elusive and varies widely between populations, settings and seasons. Globally, consumption of wild foods appears to be declining due to a combination of urbanization, costs, habitat loss and changing dietary preferences (Termote et al., 2012; WHO and CBD, 2015).

Evidence on the contribution of wild food biodiversity to diets and nutrition is limited by a severe lack of food composition data for many neglected and underutilized cultivated and wild foods (Termote et al., 2014; de Bruyn et al., 2016), as well as practical challenges in measuring wild food consumption and contribution to the diet relative to other foods (Bharucha and Pretty, 2010).

There is some evidence that wild fish in particular make an important contribution to dietary diversity and nutrition among poor populations in low- and middle-income countries for whom seafood products have traditionally been by far the most frequently consumed animal-source foods. In these contexts, the expansion of aquaculture (which is typically characterized by a limited number of large commercial species) has been associated with reduced diversity of fish species consumed, along with reduced intakes of essential micronutrients (including zinc, iron and calcium) and fatty acids (Belton and Thilsted, 2014; Belton et al., 2014). Bogard et al. (2017) similarly found that despite an increasing trend in fish consumption in Bangladesh, which has been attributed to commercial species (mainly carp and tilapia) replacing non-farmed species, there was a decline in intakes of iron, zinc and calcium from fish.

2.2.4 Forests and agro-forestry systems

Despite their diminishing significance in terms of land area and populations, the majority of the world's wild biodiversity is found in and around forests, particularly tropical forests. In addition to wild foods, forests and trees provide essential ecosystem services that support agricultural production (Millennium Ecosystem Assessment, 2005b) and potentially, therefore, cultivated food biodiversity.

Several studies that have paired satellite imaging with dietary data to assess the association between forest and tree cover and various metrics of diet diversity and quality, have identified positive associations between forest and tree cover, and dietary diversity in children (Johnson et al., 2013; Ickowitz et al., 2014). In 15 Demographic and Health surveys from sub-Saharan Africa, Jones et al. (2017), found an association between deforestation and lower diet diversity of children. In Indonesia, mixed mosaic/agro-forestry landscapes have been associated with more frequent consumption of the largest number of micronutrient-rich food groups (Ickowitz et al., 2016). However, the authors recognize that the associations are complex and were not evident until analyses were undertaken on provincial as compared to aggregate national level. There are also challenges in distinguishing between wild and cultivated food biodiversity from forests.

The hunting, harvesting and use of certain wild foods, particularly bush meats, forest products and fish, raises issues relating to sustainable natural resource management and conservation (Bharucha and Pretty, 2010; Powell et al., 2015). However, wild animal-source foods can be an essential source of micronutrients that might otherwise be lacking in local diets (Golden et al., 2011). Harnessing traditional knowledge and cultural practices relating to wild foods and landscape management has been proposed as a strategy to simultaneously support sustainable land management and food systems and improve nutrition (Powell et al., 2015), although there has been limited research in this area and questions remain as to how scalable this approach is to improve food security and nutrition in increasingly urbanized populations (Cogill, 2015).

2.3 Evidence from modelling studies

Diet modelling with linear programming has been used to identify optimal diets in a range of contexts and for a range of nutrition purposes, including: complementary feeding for infants and young children; emergency and therapeutic feeding and food aid; national food programs; dietary guidelines; promotion of the use of locally available nutritious foods, including wild foods; and modelling of population and individual diets that meet specific nutrient, environmental and cost constraints (Maillot et al., 2008; Maillot and Drewnowski, 2011; WWF, 2011; Macdiarmid et al., 2012; Horgan et al., 2016; van Dooren, 2018; Willett et al., 2019; Swinburn et al., 2019).

Food composition studies have demonstrated the significant variation in nutrient content both between and within species (for example between different types of fruits, such as mango, banana and apples), as well as between different cultivars and traditional varieties of a single species (for example bananas) (de Bruyn et al., 2016; Bioversity International, 2017). There is emerging work using optimization models to demonstrate how incorporation of this food biodiversity can improve micronutrient density of the diet (Levesque, unpublished; Ekesa et al., 2019; Wessells et al., 2019). To date, this evidence has been used to develop nutrition education materials. In Benin, for example, a recipe book has been developed to demonstrate the preparation and added nutritional value of complementary foods that incorporate diverse, locally available species such as niébé (*Vigna unguiculata*), a pulse common in Benin, and a wide range of locally available dark green leafy vegetables such as Corète/Crincrin (*Corchorus olitorius*), Moringa, (*Moringa oleifera*), *Bidens pilosa* or Grande morelle (*Solanum macrocarpum*) (Bodjrènou et al., 2018). However, impacts of these approaches, including on micronutrient adequacy of diets, are yet to be tested.

3 Evidence that food biodiversity and dietary diversity improve nutrition and health outcomes

3.1 Evidence from intervention studies

Successive reviews of NSAP studies have found little evidence of an impact of production strategies, including production diversification strategies, on micronutrient intakes, micronutrient status or anthropometric measures. It is likely that methodological limitations in these studies have hampered their ability to demonstrate impacts on nutrition and health outcomes. Programs that have included production diversification strategies and have successfully improved micronutrient intakes have generally included well-designed behaviour change communication components and a strong focus on women´s status and empowerment through agriculture (Ruel, 2001; Girard et al., 2012; Ruel et al., 2018).

3.2 Evidence from observational studies

Evidence from observational studies on the relationship between food biodiversity and nutrition and health outcomes is similarly weak and inconsistent. The strongest evidence is for a weak association between cultivated biodiversity and child linear growth (using height-for-age Z scores), with the greatest effect size among older and severely stunted children (Kumar et al., 2015; Jones, 2017). However, the magnitude of observed associations is small. No consistent associations have been observed between cultivated biodiversity and other indices of child anthropometric status (weight-for-age and weight-for-height Z scores) (Jones, 2017).

Nonetheless, the importance of dietary diversity to nutrition and health outcomes is supported by evidence from large-scale nutrition epidemiological studies, most of which have been conducted in high-income countries. The most robust evidence available is for a protective effect of fruit and vegetable diversity in the diet against cardiovascular diseases and certain cancers (Jansen et al., 2004; Wirt and Collins, 2009; Buchner et al., 2010; Isa, 2013). There is limited evidence for the protective effect of diversity within other food groups, including cereals and animal-source foods, and for other health outcomes.

There has been growing emphasis on healthy dietary patterns containing a variety of healthy foods, particularly fruits and vegetables, and other plant-based foods (as well as lowering intakes of red and processed meats, other highly-processed foods, and refined grains), rather than on individual foods and nutrients. The most recent exemplar of this is the Planetary Health Diet developed by the EAT-Lancet Commission on Food, Planet and Health (Willett et al., 2019). Dietary patterns that have received the most attention in the literature to-date include the Mediterranean-style diet and Dietary Approaches to Stop Hypertension (DASH), as well as general Dietary Guidelines-related patterns. There is strong evidence for the protective effect of adherence to one of these healthy dietary patterns against cardiovascular diseases including coronary heart disease and stroke (USDA, 2014; Katz and Meller, 2014; Tapsell et al., 2016), and moderate evidence of an association with more favourable outcomes relating to body weight and obesity risk (USDA, 2014). There is some limited evidence of a reduced risk of type II diabetes (USDA, 2014), all-cause mortality (Sofi et al., 2008; Wirt and Collins, 2009), overall or specific cancer risk (McCullough et al., 2000a, 2000b; Harnack et al., 2002; Benetou et al., 2008; Wirt and Collins, 2009; Couto et al., 2011), and development of metabolic syndrome (Meydani, 2005; Serra-Majem et al., 2006). However, it is not possible to distinguish the specific role of dietary diversity to these associations. Overall, given the long-standing emphasis in dietary guidelines on the importance of consuming a diverse diet, it is surprising, and unfortunate, that there has not been greater explicit attention to dietary diversity in nutrition epidemiological studies.

4 Knowledge gaps and priorities for future research

The body of evidence linking food biodiversity to diets and nutrition continues to grow. However, available studies provide only limited evidence for the role of biodiversity in supporting healthy, diverse diets, and even less for the contribution of food biodiversity to nutrition and health outcomes. Studies that have specifically investigated these links have predominantly involved observational study designs using cross-sectional data; a short-coming that has already been highlighted by Jones (2017) and others. While offering the potential to uncover or confirm hypothesized linkages and associations, these studies do not enable causal inferences to be drawn. The large number of small-scale, highly context-specific studies also makes it difficult to extrapolate or generalize findings to other settings and population groups.

Lack of accounting for potential mediating and contextual factors is another key limitation, although attention to these factors is growing. The most evidence available is for the role of market access, which appears to be an important effect modifier of the relationship between biodiversity and dietary diversity. Women's empowerment – as defined by female control of household resources, women's workload and women's group membership – may also play a positive role in buffering the effects of low production diversity on dietary diversity and child health outcomes (Jones et al., 2014; Malapit et al., 2015). Other sociocultural, socioeconomic and contextual factors known to influence the choice of foods produced, purchased and consumed include household income/wealth, parental (particularly maternal) education, social and gender norms influencing women´s status within the household, control of economic and food resources, time demands and allocation, intra-household food allocation, child-feeding and caregiving practices, food preparation, storage and cooking techniques and individual health/disease status (Masset et al., 2012). Household income/wealth has been positively associated with dietary diversity (Hatloy et al., 2000; Hoddinott and Yohannes, 2002; Ruel, 2003; Thorne-Lyman et al., 2010; Jones et al., 2014), although it is usually insufficient alone to bring about sustained improvements in nutrition and health outcomes (Arimond et al., 2011; Webb and Kennedy, 2014). These and other contextual factors influence, interact with and impact each other in ways that are often synergistic, non-linear and highly context-specific.

Dietary choices are also influenced by the broad sociopolitical, socioeconomic and sociocultural context, including policy and institutional environments. In the case of nutrient-dense foods, including indigenous fruits and vegetables for example, their supply and consumption is often limited by economic, agricultural and cultural factors placing limits on their potential to contribute to diets and nutrition (Cogill, 2015). In an enabling environment for food biodiversity, an understanding of the variability in nutrient composition would be used to select and promote the most nutrient-dense species, varieties and breeds for production systems (de Bruyn et al., 2016). These production systems would be directly linked to processors, markets and consumers, with information transmitted to consumers to enable them to make informed food choices.

Weaknesses in the measurement tools and indicators used are another key limitation in the evidence. Most studies that have measured dietary diversity have used simple counts of the number of crop species grown on farm (crop count) to measure farm production diversity, with some studies also including a count of livestock species. Given that food composition studies have identified potentially nutritionally-meaningful differences in nutrient composition at the species and varietal levels, this focus on diversity between (and, to a lesser extent, within) food groups does not account for the full potential of food biodiversity (Lachat et al., 2018). To address this constraint, FAO and Bioversity International (2017) have released guidelines on assessing biodiversity in dietary intake surveys.

A similar criticism can be made of market-based measures that do not collect information on food group and within food group diversity available for purchase. In those studies that have assessed the role of markets as an effect modifier, various but not very precise measures of market access have been used, including presence of a market in the local community, distance to a market (expressed in kilometres or walking/travel time) and proximity to an urban area.

In terms of dietary assessment, the vast majority of studies in this literature have measured diversity in diets using simple counts of the number of different food groups consumed by an individual or household over various reference time periods (typically ranging from one day to two weeks). Dietary diversity scores are considered relatively simple, practical tools for assessing diets, particularly in low-resource settings and in situations in which more in-depth dietary assessment is not feasible (Ruel, 2003). When measured using the score-based approach, and after controlling for household income and a variety of other potential confounding factors at the individual and household levels, dietary diversity has been consistently associated with micronutrient adequacy of the diet in infants and children aged 6–23 months (FANTA, 2006; Steyn et al., 2006; Kennedy et al., 2007) and women of reproductive age (Arimond et al., 2010), and with child growth (Arimond and Ruel, 2004; Sawadogo et al., 2006; Kennedy et al., 2007; Rah et al., 2010) in low- and middle-income countries. There is less consistent evidence that dietary diversity scores are an appropriate indicator of dietary quality, or predictor of nutrition and health outcomes, in other age and population groups.

Further, although using simple (food-group-based) dietary diversity scores as a proxy for diet quality does, in an aggregate form, represent consumption of diverse species, it does not reflect the full nutritional potential of food biodiversity (Bioversity International, 2017). Using a food group approach makes it impossible to gain intra-species information on the breeds, traditional varieties or cultivars consumed. For example one can obtain information on the percentage of the target population that consumed 'fruit' in the previous 24 hours, but one would not necessarily know the inter-species group diversity consumed (for example banana, apple or orange) or how many species or varieties contribute to the food group and during which time periods of the year (Raneri and Kennedy, 2017).

Dietary diversity scores are also insufficient to capture the totality of dietary patterns and dietary quality. Diversity is an essential, but not the sole, component of dietary quality. Other key dimensions of a healthy diet are adequacy (of consumption of dietary energy, macronutrients and essential micronutrients), balance and moderation. A diet being more diversified does not necessarily translate into better diet quality and nutritional outcomes if, for example, diversity is not nutritionally significant or even detrimental, foods produced/collected are not nutrient-rich, culturally-appropriate or consumed by the target population in sufficient quantities or are limited by low bioavailability or interactions with anti-nutrients, toxins and absorption inhibitors (Ruel, 2001; Webb, 2013). Individual health/disease status is also likely to play an important role, including the role of the gut microbiome (Bernstein, 2014; Sandifer et al., 2015).

The foods and food groups included in biodiversity and dietary diversity scores, and how the scores are calculated and compared, strongly influences outcomes. Using even a simple metric of species count (for example cassava, maize, rice) as a measure of production diversity, for example, will likely not have the same impact on dietary diversity as when food group diversity (for example cereals, fruits, vegetables) is measured. When processed food and other discretionary food items are included in a dietary diversity score, a diet may receive a higher score, but also be associated with higher energy and fat density of the diet, along with higher risk of overweight and a range of chronic disease outcomes (Swinburn et al., 2019). While this may not be a concern in very resource-constrained settings, particularly subsistence-oriented rural farming communities, it is a growing concern worldwide given rapid urbanization, rapidly changing diets associated with the nutrition transition (away from traditional diets to higher intakes of sugars, fats and animal-source foods) and the associated rising burden of overweight, obesity and diet-related chronic diseases.

The methods used in some studies have also been criticized when they do not match on-farm production diversity with the concept of increasing dietary diversity as expressed in individual dietary diversity scores (Verger et al., 2017). Two standardized indicators for measurement of individual diet – one for children 6–23 months of age and the other for women of reproductive age – are based upon nutritionally meaningful food groups and are internationally-recognized proxies of the micronutrient adequacy of the diet for their target population groups (WHO, 2008; FAO and FHI 360, 2016). Consistent use of these two indicators when researching these complex relationships can overcome at least some of the methodological weaknesses raised here.

Overall, the main conclusions that can be drawn from the evidence available are that food biodiversity can make an important contribution to dietary diversity, and potentially nutritional status, although the nature and strength of the relationship is likely to vary in different contexts. There is consistent evidence from observational studies to suggest a small, positive relationship between cultivated biodiversity and dietary diversity, at least among poor, isolated households and in areas with limited market access. Intervention strategies to increase cultivated biodiversity have been shown to provide modest positive impacts on dietary diversity and are more likely to succeed when combined with other strategies, including well-designed behaviour change communication (BCC) components and a strong focus on women's empowerment. However, there is little evidence available to indicate that these strategies lead to sustained improvements in nutrition and health outcomes. There is also limited evidence from observational studies of an association between food biodiversity and nutrition outcomes. Wild edible plants, insects, aquatic species and bush meats may contribute to food security and diets in specific populations and settings, although more research is needed on the role of wild foods, and forests and agro-forestry systems, in contributing to diets and nutrition. Evidence from nutritional epidemiological studies is generally supportive of the role of dietary diversity (particularly fruit and vegetable diversity) and healthy, diverse dietary patterns in supporting human health, although there needs to be more explicit attention to measures of diet diversity and diet quality in this literature.

In short, much remains to be learned about biodiversity's contribution to diets, nutrition and health. Given current public, private sector and policy interest in this area, more robust evidence is urgently needed. There is a need for collaboration across disciplines; development of research capacities in under-studied agrobiodiversity hot spots and under studied populations; greater attention to high-quality experimental study designs, mixed methods approaches, and use of appropriate statistical modelling tools, as well as to opportunities to monitor policies and programs that link biodiversity and nutrition (Allen et al., 2014; Prosperi et al., 2014; Fanzo, 2019). Another area requiring urgent attention is the need to generate more detailed information on the nutrient and non-nutrient components of diverse foods, including neglected and underutilized cultivated and wild foods. Finally, new studies need a greater focus on methodological rigour, including consistent use of validated definitions and measures of food biodiversity, market access and participation and dietary diversity; and greater attention to measuring the contribution of food biodiversity to other aspects of diet quality and to nutrition and health outcomes.

The effect of food biodiversity on nutrition and health outcomes represents a wicked problem. Recent research has employed the best aggregate transdisciplinary data and methods to tackle broad issues of diet, health, environment and challenges such as climate change, urbanization and changing food systems (Swinburn et al., 2019; Willett et al., 2019). Many gaps in our understanding remain including how best to understand complex systems. Describing a complex system is challenging and can be approached both tactically and strategically. We have tackled many of the tactical challenges in using complex systems in this chapter by focusing on tools, indicators, methods, terminology and the specifics of the study of systems. However, to be able to influence or engender a particular outcome, a strategic approach is critical to understanding how the various drivers or determinants interact and influence one another. Studies that tackle the role of gender or inequity on diet are needed. A combination of a strategic and tactical approach will ensure a better understanding of the options available for improved health of populations and the food system.

Funding source

This work was supported by the CGIAR´s Independent Science and Partnership Council and Bioversity International.

References

Allen, T., Prosperi, P., Cogill, B., Flichman, G. (2014) 'Agricultural biodiversity, social-ecological systems and sustainable diets', *Proceedings of the Nutrition Society*, vol 73, no 4, pp 498–508. doi: 10.1017/S002966511400069X.

Arimond, M., Hawkes, C., Ruel, M., Sifri, Z., Berti, P., LeRoy, J., Low, J., Brown, L., Frongillo, E. (2011) 'Agricultural interventions and nutrition: Lessons from the past and new evidence', in Thompson, B., Amoroso, L. (eds) *Combating Micronutrient Deficiencies: Food-Based Approaches*, Food and Agriculture Organization of the United Nations and CABI International, Rome, Italy, pp 41–75. Available at: www.fao.org/3/a-am027e.pdf

Arimond, M., Ruel, M.T. (2004) 'Dietary diversity is associated with child nutritional status: Evidence from 11 demographic and health surveys', *Journal of Nutrition*, vol 134, no 10, pp 2579–2585. doi: 10.1093/jn/134.10.2579.

Arimond, M., Wiesmann, D., Becquey, E., Carriquiry, A., Daniels, M., Deitchler, M., Fanou-Fogny, N., Joseph, M., Kennedy, G., Martin-Prevel, Y., Torheim, L.E. (2010) 'Simple food group diversity indicators predict micronutrient adequacy of women's diets in 5 diverse, resource-poor settings', *Journal of Nutrition*, vol 140, no 11, pp 2059S–2069S. doi: 10.3945/jn.110.123414.

Baker, P., Friel, S. (2016) 'Food systems transformations, ultra-processed food markets and the nutrition transition in Asia', *Globalization and Health*, vol 12, no article, pp 80. doi: 10.1186/s12992-016-0223-3.

Bélanger, J., Pilling, D. (eds) (2019) *The State of the World's Biodiversity for Food and Agriculture*, Commission on Genetic Resources for Food and Agriculture and Food and Agriculture Organization of the United Nations, Rome, Italy. Available at: www.fao.org/3/CA3129EN/CA3129EN.pdf

Bellon, M.R., Ntandou-Bouzitou, G.D., Caracciolo, F. (2016) 'On-farm diversity and market participation are positively associated with dietary diversity of rural mothers in Southern Benin, West Africa', *PLoS ONE*, vol 11, no 9, pp e0162535. doi: 10.1371/journal.pone.0162535.

Belton, B., Thilsted, S.H. (2014) 'Fisheries in transition: Food and nutrition security implications for the global South', *Global Food Security*, vol 3, no 1, pp 59–66. doi: 10.1016/j.gfs.2013.10.001.

Belton, B., van Asseldonk, I.J.M., Thilsted, S.H. (2014) 'Faltering fisheries and ascendant aquaculture: Implications for food and nutrition security in Bangladesh', *Food Policy*, vol 44, pp 77–87. doi: 10.1016/j.foodpol.2013.11.003.

Benetou, V., Trichopoulou, A., Orfanos, P., Naska, A., Lagiou, P., Boffetta, P., Trichopoulos, D. (2008) 'Conformity to traditional Mediterranean diet and cancer incidence: The Greek EPIC cohort', *British Journal of Cancer*, vol 99, pp 191–195. doi: 10.1038/sj.bjc.6604418.

Bernstein, A.S. (2014) 'Biological diversity and public health', *Annual Review of Public Health*, vol 35, pp 153–167. doi: 10.1146/annurev-publhealth-032013-182348.

Berti, P.R., Krasevec, J., Fitzgerald, S. (2004) 'A review of the effectiveness of agriculture interventions in improving nutrition outcomes', *Public Health Nutrition*, vol 7, no 5, pp 599–609. doi: 10.1079/PHN2003595.

Bharucha, Z., Pretty, J. (2010) 'The roles and values of wild foods in agricultural systems', *Philosophical Transactions of The Royal Society B Biological Sciences*, vol 365, no 1554, pp 2913–2926. doi: 10.1098/rstb.2010.0123.

Bhutta, Z.A., Ahmed, T., Black, R.E., Cousens, S., Dewey, K., Giugliani, E., Haider, B. A., Kirkwood, B., Morris, S.S., Sachdev, H.P.S., Shekar, M. (2008) 'What works?

Interventions for maternal and child undernutrition and survival', *The Lancet*, vol 371, no 9610, pp 417–440. doi: 10.1016/S0140-6736(07)61693-6.

Bioversity International. (2017) *Mainstreaming Agrobiodiversity in Sustainable Food Systems: Scientific Foundations for an Agrobiodiversity Index*, Bioversity International, Rome, Italy. Available at: www.bioversityinternational.org/mainstreaming-agrobiodiversity/

Bird, F.A., Pradhan, A., Bhavani, R.V., Dangour, A.D. (2019) 'Interventions in agriculture for nutrition outcomes: A systematic review focused on South Asia', *Food Policy*, vol 82, pp 39–49. doi: 10.1016/j.foodpol.2018.10.015.

Bodjrènou, F.S.U., Termote, C, Honnougan, F., Mitchodigni, I., Amoussa Hounkpatin, W., Ntandou Bouzitou, G., Kennedy, G. (2018) *Livret de recettes des aliments de complément pour les enfants âgés de 6-23 mois au Bénin*, Bioversity International, Bureau de l'Afrique de l'Ouest et du Centre, Dakar.

Bogard, J.R., Farook, S., Marks, G.C., Waid, J., Belton, B., Ali, M., Toufique, K., Mamun, A., Thilsted, S.H. (2017) 'Higher fish but lower micronutrient intakes: Temporal changes in fish consumption from capture fisheries and aquaculture in Bangladesh', *PLoS ONE*, vol 12, no 4, pe0175098. doi: 10.1371/journal.pone.0175098.

Buchner, F.L., Bueno-de-Mesquita, H.B., Ros, M.M., Overvad, K., Dahm, C.C., Hansen, L., Tjønneland, A., Clavel-Chapelon, F., Boutron-Ruault, M.C., Touillaud, M., Kaaks, R., Rohrmann, S., Boeing, H., Nöthlings, U., Trichopoulou, A., Zylis, D., Dilis, V., Palli, D., Sieri, S., Vineis, P., Tumino, R., Panico, S., Peeters, P.H., van Gils, C.H., Lund, E., Gram, I.T., Braaten, T., Sánchez, M.J., Agudo, A., Larrañaga, N., Ardanaz, E., Navarro, C., Argüelles, M.V., Manjer, J., Wirfält, E., Hallmans, G., Rasmuson, T., Key, T.J., Khaw, K. T., Wareham, N., Slimani, N., Vergnaud, A.C., Xun, W.W., Kiemeney, L.A., Riboli, E. (2010) 'Variety in fruit and vegetable consumption and the risk of lung cancer in the European prospective investigation into cancer and nutrition', *Cancer Epidemiology, Biomarkers and Prevention*, vol 19, no 9, pp 2278–2286. doi: 10.1158/1055-9965.EPI-10-0489.

Burlingame, B., Charrondiere, R., Mouille, B. (2009) 'Food composition is fundamental to the cross-cutting initiative on biodiversity for food and nutrition', *Journal of Food Composition and Analysis*, vol 22, no 5, pp 361–365. doi: 10.1016/j.jfca.2009.05.003.

Burlingame, B., Dernini, S. (eds) (2012) *Sustainable Diets and Biodiversity: Directions and Solutions for Policy, Research and Action*, Food and Agriculture Organization of the United Nations, Rome, Italy. Available at: www.fao.org/3/a-i3004e.pdf

Cogill, B. (2015) 'Contributions of indigenous vegetables and fruits to dietary diversity and quality', *Acta Hortic*, vol 1102, pp 213–228. doi: 10.17660/ActaHortic.2015.1102.27.

Couto, E., Boffetta, P., Lagiou, P., Ferrari, P., Buckland, G., Overvad, K., Dahm, C.C., Tjønneland, A., Olsen, A., Clavel-Chapelon, F., Boutron-Ruault, M.C., Cottet, V., Trichopoulos, D., Naska, A., Benetou, V., Kaaks, R., Rohrmann, S., Boeing, H., von Ruesten, A., Panico, S., Pala, V., Vineis, P., Palli, D., Tumino, R., May, A., Peeters, P.H., Bueno-de-Mesquita, H.B., Büchner, F.L., Lund, E., Skeie, G., Engeset, D., Gonzalez, C. A., Navarro, C., Rodríguez, L., Sánchez, M.J., Amiano, P., Barricarte, A., Hallmans, G., Johansson, I., Manjer, J., Wirfärt, E., Allen, N.E., Crowe, F., Khaw, K.T., Wareham, N., Moskal, A., Slimani, N., Jenab, M., Romaguera, D., Mouw, T., Norat, T., Riboli, E., Trichopoulou, A. (2011) 'Mediterranean dietary pattern and cancer risk in the EPIC cohort', *British Journal of Cancer*, vol 104, no 9, pp 1493–1499. doi: 10.1038/bjc.2011.106.

Darrouzet-Nardi, A.F., Masters, W.A. (2015) 'Urbanization, market development and malnutrition in farm households: Evidence from the Demographic and Health Surveys, 1986–2011', *Food Security*, vol 7, no 3, pp 521–533. doi: 10.1007/s12571-015-0470-9.

de Bruyn, J., Ferguson, E., Allman-Farinelli, M., Darnton-Hill, I., Maulaga, W., Msuya, J., Alders, R. (2016) 'Food composition tables in resource-poor settings: Exploring current limitations and opportunities, with a focus on animal-source foods in sub-Saharan Africa', *British Journal of Nutrition*, vol 116, no 10, pp 1709–1719. doi: 10.1017/S0007114516003706.

Ekesa, B., Nabuma, D., Kennedy, G. (2019) 'Content of iron and Vitamin A in common foods given to children 12–59 months old from North Western Tanzania and Central Uganda', *Nutrients*, vol 11, no 3, p 484. doi: 10.3390/nu11030484.

FANTA. (2006) *Developing and Validating Simple Indicators of Dietary Quality and Energy Intake of Infants and Young Children in Developing Countries: Summary of findings from analysis of 10 data sets August 2006 Working Group on Infant and Young Child Feeding Indicators*, Food and Nutrition Technical Assistance Project, Washington, DC. Available at: www.fantaproject.org/research/indicators-dietary-quality-intake-children

Fanzo, J. (2019) 'Biodiversity: an essential natural resource for improving diets and nutrition', in Fan, S., Yosef, S., Pandya-Lorch, S. (eds) *Agriculture for Improved Nutrition: Seizing the Momentum*, International Food Policy Research Institute and CABI. Available at: http://ebrary.ifpri.org/cdm/ref/collection/p15738coll2/id/133090

Fanzo, J., Hunter, D., Borelli, T., Matei, F. (eds) (2013) *Diversifying Food and Diets: Using Agricultural Biodiversity to Improve Nutrition and Health*, Biodiversity International, Rome, Italy. Available at: www.bioversityinternational.org/fileadmin/user_upload/online_library/publications/pdfs/Diversifying_food_and_diets_1688_01.pdf

FAO. (1997) *The State of the World's Plant Genetic Resources for Food and Agriculture*, Food and Agriculture Organization of the United Nations, Rome, Italy. Available at: www.fao.org/3/a-w7324e.pdf

FAO. (2011) *Biodiversity for Food and Agriculture Contributing to Food Security and Sustainability in a Changing World Outcomes of an Expert Workshop held by FAO and the Platform on Agrobiodiversity Research from 14–16 April 2010 in Rome, Italy*, Food and Agriculture Organization of the United Nations and Platform for Agrobiodiversity Research, Rome, Italy. Available at: www.fao.org/fileadmin/templates/biodiversity_paia/PAR-FAO-book_lr.pdf

FAO. (2015a) *Food Based Dietary Guidelines*, Food and Agriculture Organization of the United Nations, Rome, Italy. Available at: www.fao.org/nutrition/education/food-dietary-guidelines/regions/en/

FAO. (2019) *The State of the World's Biodiversity for Food and Agriculture*, J. Bélanger & D. Pilling (eds.). FAO Commission on Genetic Resources for Food and Agriculture Assessments, Rome, 572 pp. Available at: www.fao.org/3/CA3129EN/CA3129EN.pdf

FAO and Bioversity International. (2017) *Guidelines on Assessing Biodiverse Foods in Dietary Intake Surveys*, Food and Agriculture Organization of the United Nations, Rome, Italy. Available at: www.fao.org/3/a-i6717e.pdf

FAO and FHI 360. (2016) *Minimum Dietary Diversity for Women: A Guide to Measurement*, Food and Agriculture Organization of the United Nations and USAID's Food and Nutrition Technical Assistance III Project, Rome, Italy. Available at: www.fao.org/3/a-i5486e.pdf

FAO and WHO. (2014) *Second International Conference on Nutrition Rome, 19–21 November 2014 Conference Outcome Document: Framework for Action*, Food and Agriculture Organization of the United Nations and World Health Organization, Rome, Italy. Available at: www.fao.org/3/a-mm215e.pdf

Fischer, C.G., Garnett, T. (2016) *Plates, Pyramids and Planets: Developments in National Healthy and Sustainable Dietary Guidelines: A State of Play Assessment*, Food and Agriculture Organization of the United Nations and Food Climate Research Network, University of Oxford. Available at: www.fao.org/documents/card/en/c/d8dfeaf1-f859-4191-954f-e8e1388cd0b7/

Girard, A.W., Self, J.L., McAuliffe, C., Olude, O. (2012) 'The effects of household food production strategies on the health and nutrition outcomes of women and young children: A systematic review', *Paediatric Perinatal Epidemiology*, vol 26, no Suppl 1, pp 205–222. doi: 10.1111/j.1365-3016.2012.01282.x.

GLOPAN. (2016) *Food Systems and Diets: Facing the Challenges of the 21st Century*, Global Panel on Agriculture and Food Systems for Nutrition, London, UK. Available at: www.glopan.org/foresight

Golden, C.D., Fernald, L.C.H., Brashares, J.S., Rasolofoniaina, B.J.R., Kremen, C. (2011) 'Benefits of wildlife consumption to child nutrition in a biodiversity hotspot', *Proceedings of the National Academy of Sciences USA*, vol 108, no 49, pp 19653–19656. doi: 10.1073/pnas.1112586108.

Halwart, M. (2006) 'Biodiversity and nutrition in rice-based aquatic ecosystems', *Journal of Food Composition and Analysis*, vol 19, no 6–7, pp 747–751. doi: 10.1016/j.jfca.2006.03.012.

Harnack, L., Nicodemus, K., Jacobs, D.R., Folsom, A. (2002) 'An evaluation of the Dietary Guidelines for Americans in relation to cancer occurrence', *American Journal of Clinical Nutrition*, vol 76, no 4, pp 889–896. doi: 10.1093/ajcn/76.4.889.

Hatloy, A., Halland, J., Diarra, M., Oshaug, A. (2000) 'Food variety, socioeconomic status and nutritional status in urban and rural areas in Koutiala (Mali)', *Public Health Nutrition*, vol 3, pp 57–65.

Herforth, A., Ballard, T.J. (2016) 'Nutrition indicators in agriculture projects: Current measurement, priorities, and gaps', *Global Food Security*, vol 10, pp 1–10. doi: 10.1016/j.gfs.2016.07.004.

HLPE. (2017) *Nutrition and Food Systems: A Report by The High Level Panel of Experts on Food Security and Nutrition September 2017*, Committee on World Food Security, High Level Panel of Experts. Available at: www.fao.org/fileadmin/user_upload/hlpe/hlpe_documents/HLPE_Reports/HLPE-Report-12_EN.pdf

Hoddinott, J., Yohannes, Y. (2002) *Discussion Paper 136: Dietary Diversity as a Food Security Indicator*, International Food Policy Research Institute, Washington DC. Available at: http://ebrary.ifpri.org/cdm/ref/collection/p15738coll2/id/47966

Horgan, G.W., Perrin, A., Whybrow, S., Macdiarmid, J.I. (2016) 'Achieving dietary recommendations and reducing greenhouse gas emissions: Modelling diets to minimise the change from current intakes', *International Journal of Behavioral Nutrition and Physical Activity*, vol 13. article, 46. doi: 10.1186/s12966-016-0370-1.

Hough, R.L. (2014) 'Biodiversity and human health: Evidence for causality?', *Biodiversity and Conservation*, vol 23, no 2, pp 267–288. doi: 10.1007/s10531-013-0614-1.

IAASTD. (2008) *Agriculture at a Crossroads: Synthesis Report*, International Assessment of Agricultural Knowledge, Science and Technology for Development, London, UK. Available at: www.fao.org/fileadmin/templates/est/Investment/Agriculture_at_a_Crossroads_Global_Report_IAASTD.pdf

Ickowitz, A., Powell, B., Rowland, D., Jones, A., Sunderland, T. (2019) 'Agricultural intensification, dietary diversity, and markets in the global food security narrative', *Global Food Security*, vol 20, pp 9–16. doi: 10.1016/j.gfs.2018.11.002.

Ickowitz, A., Powell, B., Salim, M.A., Sunderland, T.C.H. (2014) 'Dietary quality and tree cover in Africa', *Global Environmental Change*, vol 24, pp 287–294. doi: 10.1016/j.gloenvcha.2013.12.001.

Ickowitz, A., Rowland, D., Powell, B., Salim, M.A.T. (2016) 'Forests, trees, and micronutrient-rich food consumption in Indonesia', *PLoS ONE*, vol 11, no 5, pe0154139. doi: 10.1371/journal.pone.0154139.

IPBES. (2019) *Summary for Policymakers of the Global Assessment Report on Biodiversity and Ecosystem Services*, Intergovernmental Science-Policy Panel on Biodiversity and Ecosystem Services, Bonn, Germany.

IPES-Food. (2016) *From Uniformity to Diversity: A Paradigm Shift from Industrial Agriculture to Diversified Agroecological Systems*, International Panel of Experts on Sustainable Food systems. Available at: https://cgspace.cgiar.org/bitstream/handle/10568/75659/UniformityToDiversity_FullReport.pdf?sequence=1&isAllowed=y

Isa, F. (2013) 'Dietary consumption and diet diversity and risk of developing bladder cancer: Results from the South and East China case–control study', *Cancer Causes and Control*, vol 24, no 5, pp 885–895. doi: 10.1007/s10552-013-0165-5.

Jacobs, D.R., Tapsell, L.C. (2007) 'Food, not nutrients, is the fundamental unit in nutrition', *Nutrition Reviews*, vol 65, no 10, pp 439–450. doi: 10.1111/j.1753-4887.2007.tb00269.x.

Jansen, M., Bueno-de-Mesquita, H., Feskens, E., Streppel, M.T., Kok, F.J., Kromhout, D. (2004) 'Reports: Quantity and variety of fruit and vegetable consumption and cancer risk', *Nutrition and Cancer*, vol 48, no 2, pp 142–148. doi: 10.1207/s15327914nc4802_3.

Johnson, K.B., Jacob, A., Brown, M.E. (2013) 'Forest cover associated with improved child health and nutrition: Evidence from the Malawi Demographic and Health Survey and satellite data', *Global Health: Science and Practice*, vol 1, no 2, pp 237–248. doi: 10.9745/GHSP-D-13-00055.

Jones, A. (2017) 'Critical review of the emerging research evidence on agricultural biodiversity, diet diversity, and nutritional status in low- and middle-income countries', *Nutrition Reviews*, vol 75, no 10, pp 769–782. doi: 10.1093/nutrit/nux040.

Jones, A.D., Acharya, Y., Galway, L.P. (2017) 'Deforestation and child diet diversity: A geospatial analysis of 15 sub-Saharan African countries', *The Lancet*, vol *389*, pp S11. doi: 10.1016/S0140-6736(17)31123-6.

Jones, A.D., Shrinivas, A., Bezner-Kerr, R. (2014) 'Farm production diversity is associated with greater household dietary diversity in Malawi: Findings from nationally representative data', *Food Policy*, vol 46, pp 1–12. doi: 10.1016/j.foodpol.2014.02.001.

Katz, D.L., Meller, S. (2014) 'Can we say what diet is best for health?', *Annual Review of Public Health*, vol 35, pp 83–103. doi: 10.1146/annurev-publhealth-032013-182351.

Kennedy, G., Stoian, D., Hunter, D., Kikulwe, E. (2017) 'Agricultural biodiversity for healthy, diverse diets', in Bailey, A. (ed) *Mainstreaming Agrobiodiversity in Sustainable Food Systems: Scientific Foundations for an Agrobiodiversity Index*, Bioversity International, Rome, Italy, pp 10–13. Available at: www.bioversityinternational.org/mainstreaming-agrobiodiversity/

Kennedy, G.L., Pedro, M.R., Seghieri, C., Nantel, G., Brouwer, I. (2007) 'Dietary diversity score is a useful indicator of micronutrient intake in non-breast-feeding Filipino children', *The Journal of Nutrition*, vol 137, no 2, pp 472–477. doi: 10.1093/jn/137.2.472.

Koppmair, S., Kassie, M., Qaim, M. (2017) 'Farm production, market access and dietary diversity in Malawi', *Public Health Nutrition*, vol 20, no 2, pp 325–335. doi: 10.1017/S1368980016002135.

Kumar, N., Harris, J., Rawat, R. (2015) 'If they grow it, will they eat and grow? Evidence from Zambia on agricultural diversity and child undernutrition', *The Journal of Development Studies*, vol 51, no 8, pp 1060–1077. doi: 10.1080/00220388.2015.1018901.

Lachat, C., Raneri, J. E., Walker Smith, K., Kolsteren, P., Van Damme, P., Verzelen, K., Penafiel, D., Vanhove, W., Kennedy, G., Hunter, D., Odhiambo, F.O., Ntandou-Bouzitou, G., De Baets, B., Ratnasekera, D., Ky, H.T., Remans, R., Termote, C.

(2018) 'Dietary species richness as a measure of food biodiversity and nutritional quality of diets', *PNAS*, vol 115, no 1, pp 127–132. doi: 10.1073/pnas.1709194115.
Liu, R.H. (2003) 'Health benefits of fruit and vegetables are from additive and synergistic combinations of phytochemicals', *American Journal of Clinical Nutrition*, vol 78, no 3, pp 517S–520S. doi: 10.1093/ajcn/78.3.517S.
Macdiarmid, J.I., Kyle, J., Horgan, G.W., Loe, J., Fyfe, C., Johnstone, A., McNeill, G. (2012) 'Sustainable diets for the future: Can we contribute to reducing greenhouse gas emissions by eating a healthy diet?', *American Journal of Clinical Nutrition*, vol 96, no 3, pp 632–639. doi: 10.3945/ajcn.112.038729.
Maillot, M., Drewnowski, A. (2011) 'Energy allowances for solid fats and added sugars in nutritionally adequate U.S. diets estimated at 17–33% by a linear programming model', *Journal of Nutrition*, vol 141, no 2, pp 333–340. doi: 10.3945/jn.110.131920.
Maillot, M., Ferguson, E. L., Drewnowski, A., Darmon, N. (2008) 'Nutrient profiling can help identify foods of good nutritional quality for their price: A validation study with linear programming', *Journal of Nutrition*, vol 138, no 6, pp 1107–1113. doi: 10.1093/jn/138.6.1107.
Malapit, H.J.L., Kadiyala, S., Quisumbing, A.R., Cunningham, K., Tyagi, P. (2015) 'Women's empowerment mitigates the negative effects of low production diversity on maternal and child nutrition in Nepal', *The Journal of Develoment Studies*, vol 51, no 8, pp 1097–1123. doi: 10.1080/00220388.2015.1018904.
Masset, E., Haddad, L., Cornelius, A., Isaza-Castro, J. (2011) *A Systematic Review of Agricultural Interventions that Aim to Improve Nutritional Status of Children*, EPPI-Centre, Social Science Research Unit, Institute of Education, University of London, London.
Masset, E., Haddad, L., Cornelius, A., Isaza-Castro, J. (2012) 'Effectiveness of agricultural interventions that aim to improve nutritional status of children: Systematic review', *British Medical Journal*, vol 344, pd8222. doi: 10.1136/bmj.d8222.
McCullough, J., Feskanich, D., Stampfer, M., Rosner, B.A., Hu, F.B., Hunter, D.J., Variyam, J.N., Colditz, G.A., Willett, W.C. (2000b) 'Adherence to the *dietary guidelines for Americans* and risk of major chronic disease in women', *The American Journal of Clinical Nutrition*, vol 72, no 5, pp 1214–1222. doi: 10.1093/ajcn/72.5.1214.
McCullough, M., Feskanich, D., Rimm, E., Giovannucci, E.L., Ascherio, A., Variyam, J. N., Spiegelman, D., Stampfer, M.J., Willett, W.C. (2000a) 'Adherence to the *Dietary Guidelines for Americans* and risk of major chronic disease in men', *The American Journal of Clinical Nutrition*, vol 72, no 5, pp 1223–1231. doi: 10.1093/ajcn/72.5.1223.
Meeker, J., Haddad, L. (2013) *A State of the Art Review of Agriculture-Nutrition Linkages an AgriDiet Position Paper*, University College Cork, Cork, Ireland. Available at: https://opendocs.ids.ac.uk/opendocs/bitstream/handle/123456789/3035/AgiDiet%20Global%20Review%20FINAL.pdf?sequence=1&isAllowed=y
Meydani, M. (2005) 'A Mediterranean-style diet and metabolic syndrome', *Nutrition Reviews*, vol 63, no 9, pp 312–314. doi: 10.1111/j.1753-4887.2005.tb00146.x.
Millennium Ecosystem Assessment. (2005a) 'Biodiversity regulation of ecosystem services', in Hassan, R., Scholes, R., Ash, N. (eds) *Ecosystems and Human Well-being: Current State and Trends*, Island Press, Washington, DC. Available at: www.millenniumassessment.org/documents/document.280.aspx.pdf
Millennium Ecosystem Assessment. (2005b) *Ecosystems and Human Well-being: Biodiversity Synthesis*, World Resources Institute, Washington, DC. Available at: www.millenniumassessment.org/documents/document.354.aspx.pdf
Olney, D. K., Bliznashka, L., Pedehombga, A., Dillon, A., Ruel, M. T., Heckert, J. (2016) 'A 2-year integrated agriculture and nutrition program targeted to mothers of young

children in Burkina Faso reduces underweight among mothers and increases their empowerment: A cluster-randomized controlled trial', *Journal of Nutrition*, vol 146, no 5, pp 1109–1117. doi: 10.3945/jn.115.224261.

Olney, D. K., Pedehombga, A., Ruel, M.T., Dillon, A. (2015) 'A 2-year integrated agriculture and Nutrition and Health Behavior Change Communication Program targeted to women in Burkina Faso reduces anemia, wasting, and diarrhea in children 3–12.9 months of age at baseline: A cluster-randomized controlled trial', *Journal of Nutrition*, vol 145, no 6, pp 1317–1324. doi: 10.3945/jn.114.203539.

Powell, B., Thilsted, S.H., Ickowitz, A., Termote, C., Sunderland, T., Herforth, A. (2015) 'Improving diets with wild and cultivated biodiversity from across the landscape', *Food Security*, vol 7, no 3, pp 535–554. doi: 10.1007/s12571-015-0466-5.

Prosperi, P., Allen, T., Padilla, M., Peri, I., Cogill, B. (2014) 'Sustainability and food and nutrition security: A vulnerability assessment framework for the Mediterranean region', *Sage Open*, vol 4, pp 1–15. doi: 10.1177/2158244014539169.

Rah, J.H., Akhter, N., Semba, R.D., de Pee, S., Bloem, M.W., Campbell, A.A., Moench-Pfanner, R., Sun, K., Badham, J., Kraemer, K. (2010) 'Low dietary diversity is a predictor of child stunting in rural Bangladesh', *European Journal of Clinical Nutrition*, vol 64, pp 1393–1398.

Raneri, J., Kennedy, G. (2017) 'Agricultural biodiversity for healthy diets and healthy food systems', in Hunter, D., Guarino, L., Spillane, C., McKeown, P. (eds) *Routledge Handbook of Agriculture Biodiversity*, Routledge, Abingdon, pp 401–416.

Reardon, T., Boughton, D., Tschirley, D., Haggblade, S., Dolishlager, M., Minten, B., Hernandez, R. (2015) 'Urbanization, diet change, and transformation of the downstream and midstream of the agrifood system: Effects on the poor in Africa and Asia', *Faith and Economics*, vol 66, pp 43–63.

Remans, R., Flynn, D. F. B., DeClerck, F., Diru, W., Fanzo, J., Gaynor, K., Lambrecht, I., Mudiope, J., Mutuo, P.K., Nkhoma, P., Siriri, D., Sullivan, C., Palm, C. A. (2011) 'Assessing nutritional diversity of cropping systems in African villages', *PloS One*, vol 6, no 6, pe21235. doi: 10.1371/journal.pone.0021235.

Remans, R., Smukler, S. (2013) 'Linking biodiversity and nutrition: Research methodologies', in Fanzo, J., Hunter, D., Borelli, T., Matei, F. (eds) *Diversifying Food and Diets: Using Agricultural Biodiversity to Improve Nutrition and Health*, Biodiversity International, Rome, Italy, pp 140–163. Available at: www.bioversityinternational.org/e-library/publications/detail/diversifying-food-and-diets/

Ruel, M. T. (2001) *Can Food-based Strategies Help Reduce Vitamin A and Iron Deficiencies? A Review of Recent Evidence, IFPRI Food Policy Review 5*, International Food Policy Research Institute, Washington, DC. Available at: http://ebrary.ifpri.org/cdm/ref/collection/p15738coll2/id/79381

Ruel, M.T. (2003) 'Operationalizing dietary diversity: A review of measurement issues and research priorities', *Journal of Nutrition*, vol 133, no 11 Suppl 2, pp 3911S–26S. doi: 10.1093/jn/133.11.3911S.

Ruel, M.T., Alderman, H. (2013) 'Nutrition-sensitive interventions and programmes: How can they help to accelerate progress in improving maternal and child nutrition?' *The Lancet*, vol 382, no 9891, pp 536–551. doi: 10.1016/S0140-6736(13)60843-0.

Ruel, M.T., Quisumbing, A.R., Balagamwala, M. (2018) 'Nutrition-sensitive agriculture: What have we learned so far?' *Global Food Security*, vol 17, pp 128–153. doi: 10.1016/j.gfs.2018.01.002.

Sandifer, P.A., Sutton-Grier, A.E., Ward, B.P. (2015) 'Exploring connections among nature, biodiversity, ecosystem services, and human health and wellbeing: Opportunities to enhance health and biodiversity conservation', *Ecosystem services*, vol 12, pp 1–15. doi: 10.1016/j.ecoser.2014.12.007.

Savy, M., Martin-Prével, Y., Traissac, P., Eymard- Duvernay, S., Delpeuch, F. (2006) 'Dietary diversity scores and nutritional status of women change during the seasonal food shortage in rural Burkina Faso', *Journal of Nutrition*, vol 136, no 10, pp 2625–2632. doi: 10.1093/jn/136.10.2625.

Sawadogo, P.S., Martin-Prevel, Y., Savy, M., Kameli, Y., Traissac, P., Traore, A.S., Delpeuch, F. (2006) 'An infant and child feeding index is associated with the nutritional status of 6- to 23-month-old children in rural Burkina Faso', *Journal of Nutrition*, vol 136, pp 656–663.

Serra-Majem, L., Roman, B., Estruch, R. (2006) 'Scientific evidence of interventions using the Mediterranean diet: A systematic review', *Nutrition Reviews*, vol 64, Supplement 1, pp S27–S47. doi: 10.1111/j.1753-4887.2006.tb00232.x.

Sibhatu, K.T., Krishna, V.V., Qaim, M. (2015) 'Production diversity and dietary diversity in smallholder farm households', *Proceedings of the National Academy of Sciences of the United States of America*, vol 112, no 34, pp 10657–10662. doi: 10.1073/pnas.1510982112.

Sibhatu, K. T., Qaim, M. (2018a) 'Review: Meta-analysis of the association between production diversity, diets, and nutrition in smallholder farm households', *Food Policy*, vol 77, pp 1–18. doi: 10.1016/j.foodpol.2018.04.013.

Sibhatu, K.T., Qaim, M. (2018b) 'Farm production diversity and dietary quality: Linkages and measurement issues', *Food Security*, vol 10, no 1, pp 47–59. doi: 10.1007/s12571-017-0762-3.

Sofi, F., Cesari, F., Abbate, R., Gensini, G.F., Casini, A. (2008) 'Adherence to Mediterranean diet and health status: Meta-analysis', *British Medical Journal*, vol 337, pa1344. doi: 10.1136/bmj.a1344.

Steyn, N., Nel, J., Nantel, G., Kennedy, G., Labadarios, D. (2006) 'Food variety and dietary diversity scores in children: Are they good indicators of dietary adequacy?' *Public Health Nutrition*, vol 9, no 5, pp 644–650. doi: 10.1079/PHN2005912.

Swinburn, B.A., Kraak, V.I., Allender, S., Atkins, V.J., Baker, P.I., Bogard, J.R., Brinsden, H., Calvillo, A., De Schutter, O., Devarajan, R., Ezzati, M., Friel, S., Goenka, S., Hammond, R.A., Hastings, G., Hawkes, C., Herrero, M., Hovmand, P.S., Howden, M., Jaacks, L.M., Kapetanaki, A.B., Kasman, M., Kuhnlein, H.V., Kumanyika, S.K., Larijani, B., Lobstein, T., Long, M.W., Matsudo, V.K.R., Mills, S.D. H., Morgan, G., Morshed, A., Nece, P.M., Pan, A., Patterson, D.W., Sacks, G., Shekar, M., Simmons, G.L., Smit, W., Tootee, A., Vandevijvere, S., Waterlander, W. E., Wolfenden, L., Dietz, W.H. (2019) 'The global syndemic of obesity, undernutrition, and climate change: The lancet commission report', *The Lancet Comissions*, vol 393, no 10173, pp 791–846. doi: 10.1016/S0140-6736(18)32822-8.

Tapsell, L.C., Neale, E.P., Satija, A., Hu, F.B. (2016) 'Foods, nutrients, and dietary patterns: Interconnections and implications for dietary guidelines', *Advances in Nutrition*, vol 7, no 3, pp 445–454. doi: 10.3945/an.115.011718.

Termote, C., Bwama Meyi, M., Djailo, B.D., Huybregts, L., Lachat, C., Kolsteren, P., Van Damme, P. (2012) 'A biodiverse rich environment does not contribute to a better diet: A case study from DR Congo', *PloS one*, vol 7, no 1, pp e30533. doi: 10.1371/journal.pone.0030533.

Termote, C., Raneri, J., Deptford, A., Cogill, B. (2014) 'Assessing the potential of wild foods to reduce the cost of a nutritionally adequate diet: An example from eastern Baringo District, Kenya', *Food and Nutrition Bulletin*, vol 35, no 4, pp 458–479. doi: 10.1177/156482651403500408.

Thorne-Lyman, A. L., Valpiani, N., Sun, K., Semba, R.D., Klotz, C. L., Kraemer, K., Akhter, N., de Pee, S., Moench-Pfanner, R., Sari, M., Bloem, M.W. (2010) 'Household dietary diversity and food expenditures are closely linked in rural Bangladesh, increasing the risk of malnutrition due to the financial crisis', *Journal of Nutrition*, vol 140, no 1, pp 182S–188S. doi: 10.3945/jn.109.110809.

USDA. (2014) *A Series of Systematic Reviews on the Relationship between Dietary Patterns and Health Outcomes*, United States Department of Agriculture. Available at: https://pdfs.semanticscholar.org/9dc0/ebed6c942f2be042fe359c12a801f1dbc690.pdf?_ga=2.25536826.662661264.1569243781-5220422.1566290231

van Dooren, C. (2018) 'A review of the use of linear programming to optimize diets, nutritiously, economically and environmentally', *Frontiers in Nutrition*, vol 5, p48. doi: 10.3389/fnut.2018.00048.

Verger, E.O., Ballard, T.J., Dop, M.C., Martin-Prevel, Y. (2019) 'Systematic review of use and interpretation of dietary diversity indicators in nutrition-sensitive agriculture literature', *Global Food Security*, vol 20, pp 156–169. doi: 10.1016/j.gfs.2019.02.004.

Verger, E.O., Jones, A.D., Dop, M.C., Martin-Prével, Y. (2017) 'On the appropriate use and interpretation of dietary diversity scores'. Response to: 'Farm production diversity and individual-level dietary diversity' by Koppmair and Qaim', *Public Health Nutrition*, vol 20, no 11, pp 2073–2074. doi: 10.1017/S1368980017001343.

Webb, P. (2013) *Impact Pathways from Agricultural Research to Improved Nutrition and Health: Literature Analysis and Research Priorities*, Food and Agriculture Organization of the United Nations and World Health Organization. Available at: www.fao.org/fileadmin/user_upload/agn/pdf/Webb_FAO_paper__Webb_June_26_2013_.pdf

Webb, P., Kennedy, E. (2014) 'Impacts of agriculture on nutrition: nature of the evidence and research gaps', *Food and Nutrition Bulletin*, vol 35, no 1, pp 126–132. doi: 10.1177/156482651403500113.

Wessells, K.R., Young, R.R., Ferguson, E.L., Ouédraogo, C.T., Faye, M.T., Hess, S.Y. (2019) 'Assessment of dietary intake and nutrient gaps, and development of food-based recommendations, among pregnant and lactating women in Zinder, Niger: an optifood linear programming analysis', *Nutrients*, vol 11, no 1, article 72. doi: 10.3390/nu11010072.

WHO. (2003) *WHO Technical Report Series 916: Diet, Nutrition and the Prevention of Chronic Diseases, Report of a Joint WHO/FAO Expert Consultation*, World Health Organization, Geneva. Available at: http://health.euroafrica.org/books/dietnutritionwho.pdf

WHO. (2008) *Indicators for Assessing Infant and Young Child Feeding Practices: Conclusions of a Consensus Meeting Held 6–8 November 2007 in Washington, DC*, World Health Organization, Geneva, Switzerland. Available at: www.who.int/nutrition/publications/iycf_indicators_for_peer_review.pdf

WHO and CBD. (2015) *Connecting Global Priorities: Biodiversity and Human Health. A State of Knowledge Review*, World Health Organization Biological Diversity. Available at: www.cbd.int/health/SOK-biodiversity-en.pdf

Willett, W., Rockström, J., Loken, B., Springmann, M., Lang, T., Vermeulen, S., Garnett, T., Tilman, D., DeClerck, F., Wood, A., Jonell, M., Clark, M., Gordon, L.J., Fanzo, J., Hawkes, C., Zurayk, R., Rivera, J.A., De Vries, W., Sibanda, L.M., Afshin, A., Chaudhary, A., Herrero, M., Agustina, R., Branca, F., Lartey, A., Fan, S., Crona, B., Fox, E., Bignet, V., Troell, M., Lindahl, T., Singh, S., Cornell, S.E., Reddy, K.S., Narain, S., Nishtar, S., Murray, C.J.L. (2019) 'Food in the Anthropocene: the EAT–Lancet Commission on healthy diets from sustainable food systems', *The Lancet Commissions*, vol 393, no 10170, pp 447–492. doi: 10.1016/S0140-6736(18)31788-4.

Wirt, A., Collins, C.E. (2009) 'Diet quality what is it and does it matter?', *Public Health Nutrition*, vol 12, no 12, pp 2473–2492. doi: 10.1017/S136898000900531X.

World Bank. (2007) *From Agriculture to Nutrition: Pathways, Synergies and Outcomes*, The World Bank, Washington, DC. Available at: http://documents.worldbank.org/curated/en/241231468201835433/From-agriculture-to-nutrition-pathways-synergies-and-outcomes

WWF. (2011) 'Livewell: A balance of healthy and sustainable food choices. World Wide Fund for Nature'.

4

FROM UNIFORMITY TO DIVERSITY

The potential of agroecology to transform food systems

Nick Jacobs, Chantal Wei-Ying Clément and Emile Frison

Introduction

Today's food and farming systems have succeeded in supplying large volumes of foods to global markets, but are generating negative outcomes on multiple fronts: widespread degradation of land, water and ecosystems; high greenhouse gas (GHG) emissions; biodiversity losses; persistent hunger, and micronutrient deficiencies alongside the rapid rise of obesity and diet-related diseases; and livelihood stresses for farmers around the world. Many of these problems are linked specifically to the input-intensive crop monocultures and industrial-scale feedlots that now dominate farming landscapes. The uniformity at the heart of these systems, and their reliance on chemical fertilizers, pesticides and preventive use of antibiotics, leads systematically to negative outcomes and vulnerabilities. What is required is a fundamentally different model of agriculture based on diversifying farms and farming landscapes, replacing chemical inputs, optimizing biodiversity and stimulating interactions between different species, as part of holistic strategies to build long-term fertility, healthy agroecosystems and secure livelihoods, i.e. 'diversified agroecological systems'. This chapter will examine the systemic factors that are keeping the industrial agriculture model in place while highlighting the many opportunities that exist to shift the balance in favour of more diversified, agroecological and healthier food systems.

The multiple, severe and interconnected crises in industrial food and farming systems

Many of the deepest problems in today's food and farming systems trace back to a specific mode of production that has taken root around the world: highly-specialized, chemical-intensive, large-scale, industrial agriculture. Chemical inputs, at the heart of this model, are driving unprecedented impacts on plant and insect life. A study of protected areas in Germany found that more than 75% of flying insects have disappeared in 25 years (Hallmann et al., 2017). In addition, biodiversity loss costs some 3% of global GDP each year (Kumar, 2010); global pollinator loss, which severely jeopardizes crop yields, has been estimated to cost EUR €190–€310 billion per year (Gallai et al., 2009).

Meanwhile, food and farming systems contribute up to 30% of global GHG emissions (Vermeulen et al., 2012), largely as a result of intensive cropping, animal farming and land use change to make way for agricultural expansion. Industrial agriculture remains the largest contributor to land degradation, which continues at an alarming rate of 12 million hectares/year, equivalent to the total agricultural land of the Philippines (ELD, 2015). Meanwhile, 70% of all water withdrawn from aquifers, streams and lakes is used for agriculture – often at unsustainable rates (UN Water, 2013). The division of labour in highly-specialized industrial systems means that these impacts are particularly acute in export commodity regions, while wealthier regions are able to externalize their footprint. For example, as much as 31% of the land required to meet EU food demand is located outside Europe (European Commission, 2013); the EU imports some 22 million tons of soya-based animal feed every year (Laaninen, 2015), including from South American countries where illegal deforestation, evictions, pesticide poisoning and rights abuses have been alleged in intensive export cropping zones (Ezquerro-Cañete, 2016).

The global supply chains that have developed hand in hand with the spread of industrial commodity production are also squeezing the livelihoods of farmers and food-workers around the world. Seventy percent of the global agrochemical industry is now in the hands of only three companies (IPES-Food, 2017b), and up to 90% of the global grain trade is controlled by four multinationals (Murphy et al., 2012). In 2011, the five largest food retailers in 13 EU Member States had a combined market share of over 60% (European Commission, 2017); the top four US food retailers accounted for just under 40% of national grocery sales in 2015 – double the four-firm concentration ratio from the early 1990s (USDA, 2016). In this context, dominant players have been able to drive down prices and working conditions in supply chains – affecting seasonal migrant labourers, food retail staff and self-employed delivery workers alike, as well as driving a spate of recent strikes (Chapman, 2018). Labour conditions are systematically poor for hired and migrant farm labourers (ILO, 2015).

Many small farmers, especially women, struggle to emerge above subsistence level, often lacking access to credit, technical support and markets – or facing the uncertainties of volatile prices on global commodity markets (FAO, 2004). Even in wealthier regions, farmers' incomes are being squeezed to breaking point. The share of EU food chain value going to agriculture dropped from 31% in 1995 to 24% in 2005 (European Parliament, 2009), and has more recently been estimated at around 21% (European Parliament, 2015). In the US, net farm income has declined from USD$120 billion in 2013 to an estimated $70 billion in 2019 (or just over $60 billion in 2018) (USDA, 2019).

Despite the production of large volumes of staple commodities, industrial food and farming systems are failing to provide food and nutrition security for all. Despite decreases in the percentage of the global population going hungry over recent decades, 820 million people still suffered from hunger in 2017 (FAO, 2018b). The world faces the irony of small-scale farming communities making up about 50% of the hungry (Arsenault, 2015).

The emergence of industrial agriculture has also had an impact on the nutritional quality of foods and dietary patterns around the world. In the last 100 years, some 75% of global crop genetic diversity has disappeared from farmers' fields, as a consequence of abandoning multiple local crops in favour of genetically uniform, high-yielding varieties (FAO, 1999). 'Underutilized' or minor crops (for example indigenous leafy vegetables, small-grained cereals, legumes, pulses, wild fruits and tree crops) have been abandoned in favour of industrially-produced varieties of rice, maize and wheat (Hawkes, 2007; Jacobsen et al., 2013). This has contributed to a situation in which 2 billion people around the world are afflicted by the 'hidden hunger' of micronutrient deficiencies (Frison et al., 2011; Bioversity International, 2014), while over 1.9 billion people are now obese or overweight (WHO, 2015).

In addition to the impacts of poor nutrition, industrial agriculture affects health through multiple interconnected pathways, including dangerous working conditions; chronic exposure to contaminants in the water, soil and air; and consumption of unsafe or contaminated foods (IPES-Food, 2017a). In the Vietnamese Mekong Delta, studies have found all types of drinking water sources to be contaminated with pesticides at dangerous concentrations (Van Toan et al., 2013; Chau et al., 2015). In the US, exposure to Endocrine Disrupting Chemicals (EDCs) via pesticides has been estimated to cost $42 billion per year (Attina et al., 2016). The very high human and economic costs generated by current food and farming systems are not reflected in the price of food (Hamm et al., 2018), however the total costs to society of the 'cheap' food produced by industrial food systems is enormous (TEEB, 2018).

Even the high yields that industrial agriculture was designed to deliver are now under threat. A meta-analysis of yield developments around the world from 1961–2008 found that only slightly more than half of all global rice and wheat areas are still experiencing yield increases (Ray et al., 2012). Climate change, to which industrial agriculture is a major contributor, is likely to provoke crop losses due to increasingly severe floods and droughts, while increased atmospheric carbon dioxide levels threaten the nutritional value of food crops (Watts et al., 2015; Ziska et al., 2016; Niles et al., 2017). Moreover, farmers' options for adapting to changing environments are limited by the reduction in practical applications of genetic diversity that has gone hand in hand with the spread of industrial agriculture (Vigouroux et al., 2011).

The growing promise of agroecological alternatives

Alternatives to industrial agriculture are, however, emerging fast. These alternatives take a range of names (for example organic, biodynamic or regenerative agriculture, permaculture, eco-agriculture), but can be broadly grouped together under the umbrella of *diversified agroecological systems*, i.e. systems based on diversifying farms and farming landscapes, replacing chemical inputs with organic inputs, optimizing biodiversity and stimulating interactions between different species, as part of holistic strategies to build long-term fertility, healthy agroecosystems and secure livelihoods (IPES-Food, 2016).

For years now, the environmental benefits of alternative production models have been increasingly recognized (Folke et al., 2002; Tengö and Belfrage, 2004; IAASTD, 2009; Tirado and Cotter, 2010; Mijatović et al., 2013; Prieto et al., 2015; Pretty et al., 2018). However, there is now growing awareness that these are not just eco-friendly niche solutions but can in fact succeed in reconciling concerns such as food security, environmental protection, climate change adaptation, nutritional adequacy and social equity. Agroecology has been recognized by the International Assessment of Agricultural Knowledge, Science and Technology for Development's (IAASTD, 2009) global agriculture assessment and by the FAO (2018a) as key to building sustainable food systems, in light of the systemic approach it employs, with equal emphasis on economic, environmental and social dimensions of sustainability.

Evidence is particularly strong on the ability of diversified agroecological systems to deliver strong and stable outputs on the basis of building environmental resilience and highly functioning agroecosystems that generate abundant ecosystem services (for example pollination, water purification, soil water retention, pest and disease control) and to limit losses and enable recovery in the face of environmental stresses and shocks (Holt-Giménez, 2002; Badgley et al., 2007; Cardinale et al., 2007; Gliessman, 2007; Picasso et al., 2008; IAASTD, 2009; Alonso and Guzmán, 2010; Tirado and Cotter, 2010; Pretty et al., 2011; Mijatović et al., 2013; Aguilera et al., 2014; Altieri et al., 2015; Prieto et al., 2015; Rodale Institute, 2015).

New studies from a range of contexts are rapidly building our understanding of these effects. A 6.5-year comparison study in Argentina found major increases in soil organic matter in agroecological production systems compared to industrial, alongside yield and income benefits for agroecological farmers (Virginia et al., 2018). Direct yield comparisons continue to show slightly higher productivity in conventional/industrial systems compared to organic agriculture in industrialized countries (see IPES-Food, 2016). However, yields in organic agriculture do not necessarily reflect the full potential of agroecology. For example, Ponisio et al. (2015) found that organic yields are 19.2% lower than conventional yields, but that two agricultural diversification practices – multi-cropping and crop rotations – substantially reduce the yield gap (to 9% and 8%, respectively) when applied in organic systems. Outcomes can be expected to improve further if commensurate support and investment is made available to develop and spread agroecological knowledge (Pretty, 2006; De Schutter, 2010).

Furthermore, diversified agroecological systems have shown the capacity to increase production in the places where additional food is desperately needed (Badgley et al., 2007; Pretty et al., 2011). In West Africa, the application of the System of Rice Intensification (SRI), an agroecological, low-GHG and low-input methodology for increasing rice productivity, has increased yields on 50,000 farms by 56% for irrigated rice and 86% for lowland rain-fed rice, while using 90% less seed, 30–50% less water and less agrochemicals. The average income for farmers using SRI was 41% higher than for those using conventional practices (Styger and Traoré, 2018).

There is also growing evidence of positive linkages between agricultural diversity and nutritional diversity at the household and local level, through the increased availability of nutrient-rich diverse foods throughout the year (Herforth, 2010; Remans et al., 2011; Oyarzun et al., 2013; Jones et al., 2014; Carletto et al., 2015; Kumar et al., 2015; Shively and Sununtnasuk, 2015; KC et al., 2016; IPES-Food, 2018). Furthermore, crop and livestock diversification acts as a form of self-insurance for farmers, allowing income to be stabilized in the face of crop failure/livestock loss (Johnston et al., 1995; Papademetriou and Dent, 2001; Gliessman, 2007), acting as a crucial stabilizer in hurricane and flood-risk areas (FAO, 2013), and allowing production to continue year-round (Powell et al., 2015). A study covering five continents over 40 years found that despite lower yields, organic agriculture was significantly more profitable (by 22–35%) than conventional agriculture (Crowder and Reganold, 2015; Reganold and Wachter, 2016). An eight-country study found that the number of crops that a given farm produces is positively correlated to household income, as well as dietary diversity (Pellegrini and Tasciotti, 2014). Data recently gathered in France has shown that organic farms on average produced 2.4 annual work units compared to 1.5 on conventional farms and are more likely to create permanent salaried jobs (Massis and Hild, 2016; Sautereau and Benoit, 2016).

The diverse nature of agroecology means that these experiments are geographically-dispersed, context-specific, and small-scale, often lending themselves to qualitative documentation in the shape of case studies (see for example IPES-Food, 2018). A picture is nonetheless starting to emerge in terms of how agroecology could scale out, and with what implications for global food security. A recent modelling exercise found that a 100% conversion to organic agriculture by 2050 would be viable in terms of global food production and land use, providing that it is accompanied by significant reductions in livestock and animal feed production, and reduced food waste (Muller et al., 2017). A study looking specifically at the European region found that wholesale transition to agroecology over the coming decades would not undermine food security or entail major dependency on imports, providing that livestock production is deintensified and that food and feed competition is reduced – among other shifts (Poux and Aubert, 2018).

Real-world experimentation is now taking place on new scales and will yield crucial insights as it evolves. For example, in the Indian state of Andhra Pradesh, the local government is aiming to scale out 'zero-budget natural farming' to six million farmers by 2027 (Government of Andhra Pradesh, 2017). A 2018 meta-study estimated that some 163 million farms – comprising 29% of all farms worldwide and 9% of agricultural land – are now engaging in various forms of redesign of their production systems (for example agroforestry, integrated pest management) (Pretty et al., 2018).

Why is a major paradigm shift yet to occur? The eight lock-ins of industrial food systems

The shift towards agroecology has not, however, been commensurate to the promise shown by these alternative systems, or the severe crisis facing industrial agriculture. A major paradigm shift is yet to occur among a critical mass of farmers, food businesses and policymakers. A holistic *food system* perspective helps to explain this inertia, shedding light on the way in which policies, regulatory frameworks, economic incentives, technological pathways, infrastructures and lifestyles have co-evolved and reinforced one another over time. Analysing global food systems through this lens, IPES-Food (2016) identified eight factors locking industrial food and farming systems in place:

Lock-in 1: path dependency

Industrial agriculture requires significant up-front investments, in terms of equipment, training, networks and retail relationships. To see a return on these investments, farmers are often required to scale up to deliver sufficiently high volumes of (low-value) uniform commodity crops. Once these structural shifts have been made, it is difficult for farmers to change course.

Lock-in 2: export orientation

Over decades, production subsidies, energy subsidies, trade liberalization, development aid and a range of other measures have been put in place with a view to producing large volumes of cheap commodity crops for global markets (Leach, 1992; Lines, 2008). The foreign currency reserves generated by these exports has become crucial to afford the imported goods on which countries have become reliant and to service debts (UNCTAD, 2008). Supporting export chains has often been prioritized over other interests (for example ensuring resources for local food production) and in spite of the risks this entails for many developing countries (for example price volatility, declining terms of trade, environmental degradation, competition for land).

Lock-in 3: the expectation of cheap food

Industrial agriculture and shifting consumer habits have helped to facilitate the emergence of mass food retailing, characterized by the abundance of relatively cheap highly-processed foods (Reardon et al., 2003; Gómez and Ricketts, 2013), undifferentiated commodity ingredients (Popkin et al., 2012) and the year-round availability of a wide variety of foods. In many countries, consumers have become accustomed to spending less on food – and are increasingly detached from the realities of how that food is produced and what its impact is on the environment and public health (Bricas et al., 2013). The food industry has therefore become increasingly reliant on the cheap and flexible supply of uniform commodities that industrial agriculture is uniquely positioned to provide.

Lock-in 4: compartmentalized thinking

Highly compartmentalized structures continue to govern the setting of priorities in politics, education, research and business, allowing the solutions offered by industrial agriculture to remain at centre stage. Agricultural ministries, committees and lobbies retain a privileged position relative to other constituencies (for example the environment and health sectors) in determining the policies that shape food systems. Increasingly privatized agricultural research and development programmes remain focused on the handful of commodities for which there is a large enough market to secure significant returns. Classical agricultural research and education systems remain siloed (O'Brien et al., 2013), and ill-adapted to capture the complex social-ecological interactions in food systems (Francis et al., 2003).

Lock-in 5: short-term thinking

Time is required to rebuild soil health and fertility, to increase biodiversity, and to reap the benefits of enhanced resilience in agroecological systems. However, key players in food systems are often required to deliver rapid results. Politicians are locked into short-term electoral cycles that reward policies delivering immediate returns. Meanwhile, publicly traded agribusiness firms are required to deliver rapid returns to shareholders.

Lock-in 6: 'feed the world' narratives

Food security continues to be framed by many prominent actors as a question of how to 'feed the world', or in other words, how to produce sufficient net calories at the global level – particularly in the wake of the 2007–2008 food price spikes. These narratives sideline the questions of nutritional quality, poverty, access, power and equity that have been recognized as essential pieces of the food security puzzle (Sen, 1981; Popay et al., 2008; Martin, 2010; FAO, 2018a). By narrowing the debate to a question of net calorie production, such narratives allow industrial agriculture to be reinvented as the solution.

Lock-in 7: measures of success

Diversified agroecological systems are by definition geared towards producing diverse outputs, while delivering a range of other benefits (for example resource efficiency, reduced GHG emissions, multiple environmental services and social benefits on and off the farm). Narrowly defined indicators of agricultural performance (for example yields of specific crops or productivity per worker) reward large-scale industrial monocultures while failing to capture the benefits of alternative systems; the classic cost-benefit analysis applied to agricultural systems tends to ignore ecological, social and cultural variables (Flores and Sarandón, 2004).

Lock-in 8: concentration of power

The way food systems are currently structured allows value to accrue mainly to a limited number of actors, reinforcing their economic and political dominance, and thus their ability to influence the governance of food systems. For example, in 2015, agribusiness firms spent more than $130 million lobbying US Congress, a figure that exceeds the lobbying efforts of the defence industry (OpenSecrets, 2016).

From systemic lock-ins to systemic leverage points for change

Sparking a transition is therefore a major challenge, requiring a range of imperatives across food systems to be realigned and a series of cycles to be broken. In other words, the solutions must be as systemic and interconnected as the current problems. In its 2016 report, IPES-Food identified a series of steps that can spark a paradigm shift from industrial agriculture to diversified agroecological systems. Many of these steps can build on innovative practices and policies that are already taking root around the world.

Recommendation 1: develop new indicators for sustainable food systems

It is essential to adopt a broader range of indicators, covering long-term ecosystem health; total resource flows; the sustainability of outputs; nutrition and health outcomes; livelihood resilience; and the economic resilience and viability of farms with respect to debt, climate shocks, and so on.

Recommendation 2: shift public support towards diversified agroecological production systems

Governments must shift public support (for example agricultural subsidies) away from industrial production systems, while rewarding the array of positive outcomes in diversified agroecological systems. Governments must implement measures that allow farms to diversify and transition towards agroecology. In particular, they must support young people to enter agriculture and adopt agroecology – before they are locked into the cycles of industrial agriculture.

Recommendation 3: support short supply chains and alternative retail infrastructures

Governments should support and promote short food supply chain initiatives in order to make them a viable, accessible and affordable alternative to mass retail outlets, for example by repurposing infrastructure in cities to favour farmers' markets. In the US, the number of farmers' markets grew by 76% between 2008 and 2014 (USDA, 2014). Community Supported Agriculture (CSA) schemes are already spreading fast in the US, Japan and a range of European countries (Lagane, 2011); in France, some 250,000 people, almost 1% of the working-age population, participate in CSA schemes (Assemblée nationale, 2015; INSEE, 2016).

Recommendation 4: use public procurement to support diverse local agroecological produce

Public procurement should be used with increasing ambition in order to ensure sales outlets for diversified agroecological farms, while providing fresh, nutritious food and diversified diets for the users of public canteens, particularly schoolchildren. In a growing number of municipalities, cities and countries, public procurement programmes have been reformed in order to source local, sustainable, ethical and/or healthy food for public canteens (De Schutter, 2014; Chandler et al., 2015). For example, the city of Copenhagen set incremental targets aiming for 90% organic procurement by 2016 (Hultberg and Madsen, 2012). Many countries in the global South have put home-grown school feeding programmes in place to source food from smallholders and strengthen local agricultural development (Drake et al., 2016).

Recommendation 5: strengthen movements that unify diverse constituencies around agroecology

Platforms bringing together farmers' groups, community-based organizations and social movements with a shared interest in promoting agroecology are essential and must be supported. To become effective, alliances must reach across various divides (for example between producers and consumers, between farmers and researchers, between advocates of food sovereignty, family farming, community-supported agriculture and agroecology), and strong and unified messages must emerge to counter the 'feed the world' narratives which currently hold sway.

Recommendation 6: mainstream diversification, agroecology and holistic food systems approaches into education and research agendas

Public research investments must be redirected towards equipping farmers to shift towards agroecology, in line with a 2015 statement signed by over 300 US scientists and experts (Union of Concerned Scientists, 2015). Building on the 2nd FAO International Symposium on Agroecology in April 2018 (FAO, 2018a), the FAO and other international agencies should mainstream agroecology across their work. Research conducted by the CGIAR Centres should be refocused around diversified agroecological systems and farmer participatory research.

Recommendation 7: develop joined-up food policies at multiple levels

Integrated food policies are required to overcome the traditional biases in sectoral policies (for example export orientation in agricultural policy), to avoid costly contradictions between different policies (for example anti-obesity strategies versus agri-trade policies that make junk food cheap and abundant), and to align various policies with the objective of delivering environmentally, socially and economically sustainable food systems. Integrated food policies must be co-developed by a range of food system stakeholders working on equal footing, in order to break away from existing policy silos, path dependencies and power imbalances. The 2016–2019 process conducted by IPES-Food, 'Towards a Common Food Policy for the EU', offers a precedent for developing a food policy vision based on the collective intelligence of farmers, food businesses, civil society groups, social movements, scientists and policymakers themselves (IPES-Food, 2019).

None of these solutions offer a silver bullet. Collectively, however, they could shift the centre of gravity in food systems, allowing harmful dependencies to be cut, the agents of change to be empowered, alliances to be forged in favour of change, and fundamental shifts to occur in the way that policies are made and priorities are set in food systems.

References

Aguilera, E., Guzmán, G., Alonso, A. (2014) 'Greenhouse gas emissions from conventional and organic cropping systems in Spain. II. Fruit tree orchards', *Agronomy for Sustainable Development*, vol 35, no 2, pp 725–737. doi: 10.1007/s13593-014-0265-y.

Alonso, A.M., Guzmán, G.J. (2010) 'Comparison of the efficiency and use of energy in organic and conventional farming in Spanish agricultural systems', *Journal of Sustainable Agriculture*, vol 34, no 3, pp 312–338. doi: 10.1080/10440041003613362.

Altieri, M., Nicholls, C., Henao, A., Lana, M. (2015) 'Agroecology and the design of climate change-resilient farming systems', *Agronomy for Sustainable Development*, vol 35, no 3, pp 869–890. doi: 10.1007/s13593-015-0285-2.

Arsenault, C. (2015) 'Why are most of the world's hungry people farmers?' World Economic Forum. Available at: www.weforum.org/agenda/2015/05/why-are-most-of-the-worlds-hungry-people-farmers/

Assemblée nationale. (2015) 'Rapport d'information de Mme Brigitte Allain et M. Jean-Charles Taugourdeau déposé en application de l'article 145 du règlement, par la commission des a aires économiques sur les circuits courts et la relocalisation des filières agricoles et alimentaires'. Available at: www.assemblee-nationale.fr/14/rap-info/i2942.asp#P436_71915

Attina, T.M., Hauser, R., Sathyanarayana, S., Hunt, P.A., Bourguignon, J.-P., Myers, J.P., DiGangi, J., Zoeller, R.T., Trasande, L. (2016) 'Exposure to endocrine-disrupting chemicals in the USA: A population-based disease burden and cost analysis', *The Lancet Diabetes & Endocrinology*, vol 4, no 12, pp 996–1003. doi: 10.1016/S2213-8587(16)30275-3.

Badgley, C., Moghtader, J., Quintero, E., Zakem, E., Chappell, M.J., Avilés-Vázquez, K., Samulon, A., Perfecto, I. (2007) 'Organic agriculture and the global food supply', *Renewable Agriculture and Food Systems*, vol 22, no 2, pp 86–108. doi: 10.1017/S1742170507001640.

KC, K.B., Pant, L.P., Fraser, E.D., Shrestha, P.K., Shrestha, D., Lama, A. (2016) 'Assessing links between crop diversity and food self-sufficiency in three agroecological regions of Nepal', *Regional environmental change*, vol 16, no 5, pp 1239–1251. doi: 10.1007/s10113-015-0851-9.

Bioversity International. (2014) *Bioversity International's 10-year Strategy 2014–2024: Agricultural Bio-Diversity Nourishes People and Sustains the Planet*, Bioversity International, Rome, Italy.

Bricas, N., Lamine, C., Casabianca, F. (2013) 'Agricultures et alimentations: des relations à repenser?', *Natures Sciences Sociétés*, vol 21, no 1, pp 66–70. doi: 10.1051/nss/2013084.

Cardinale, B.J., Wright, J.P., Cadotte, M.W., Carroll, I.T., Hector, A., Srivastava, D.S., Loreau, M., Weis, J. J. (2007) 'Impacts of plant diversity on biomass production increase through time because of species complementarity', *Proceedings of the National Academy of Sciences USA*, vol 104, no 46, pp 18123–18128. doi: 10.1073/pnas.0709069104.

Carletto, G., Ruel, M., Winters, P., Zezza, A. (2015) 'Farm-level pathways to improved nutritional status: Introduction to the special issue', *The Journal of Development Studies*, vol 51, no 8, pp 945–957. doi: 10.1080/00220388.2015.1018908.

Chandler, C., Franklin, A., Ochoa, A., Clement, S. (2015) 'Sustainable public procurement of school catering services – A good practice report', INNOCAT. Available at: www.sustainable-catering.eu/fileadmin/user_upload/enewsletter/Documents/INNOCAT_Catering_Report_FINAL.pdf

Chapman, B. (2018) 'Uber Eats and Deliveroo riders to strike alongside McDonald's, Wetherspoons and TGI Fridays employees', *Independent*, 3 October 2018. Available at: www.independent.co.uk/news/business/news/uber-eats-deliveroo-strike-mcdonalds-wetherspoons-mcstrike-industrial-action-a8567286.html

Chau, N.D.G., Sebesvari, Z., Amelung, W., Renaud, F.G. (2015) 'Pesticide pollution of multiple drinking water sources in the Mekong Delta, Vietnam: Evidence from two

provinces', *Environmental Science and Pollution Research*, vol 22, no 12, pp 9042–9058. doi: 10.1007/s11356-014-4034-x.
Crowder, D.W., Reganold, J.P. (2015) 'Financial competitiveness of organic agriculture on a global scale', *Proceedings of the National Academy of Sciences USA*, vol 112, no 24, pp 7611–7616. doi: 10.1073/pnas.1423674112.
De Schutter, O. (2010) *Report submitted by the Special Rapporteur on the right to food, Olivier De Schutter*, United Nations General Assembly, Human Rights Council, Sixteenth Session. Available at: www2.ohchr.org/english/issues/food/docs/A-HRC-16-49.pdf
De Schutter, O. (2014) *The power of procurement: Public purchasing in the service of realizing the right to food*, United Nations Special Rapporteur on the Right to Food, Briefing Note 08 April 2014. Available at: www.srfood.org/images/stories/pdf/otherdocuments/20140514_procurement_en.pdf
Drake, L.J., Woolnough, A., Bundy, D., Burbano, C. (eds) (2016) *Global School Feeding Sourcebook: Lessons from 14 Countries*, World Bank Group, Washington, DC.
ELD. (2015) *Report for policy and decision makers: Reaping economic and environmental benefits from sustainable land management.* Economics of Land Degradation Initiative. Available at: www.eld-initiative.org/fileadmin/pdf/ELD-pm-report_05_web_300dpi.pdf
European Commission. (2013) 'Science for Environment Policy, thematic issue: Global Environmental Impacts of EU Trade in Commodities', European Commission. Available at: https://ec.europa.eu/environment/integration/research/newsalert/pdf/44si_en.pdf
European Commission. (2017) 'The economic impact of modern retail on choice and innovation in the EU food sector: Final Report', European Union, Luxembourg. doi: 10.2763/77405.
European Parliament. (2009) 'Report on fair revenues for farmers: A better functioning food supply chain in Europe (2009/2237(INI))', European Parliament Committee on Agriculture and Rural Development. Available at: www.europarl.europa.eu/sides/getDoc.do?type=REPORT&reference=A7-2010-0225&language=EN#
European Parliament. (2015) 'Parliamentary questions – Answer given by Mr. Hogan on behalf of the Commission', European Parliament. Available at: www.europarl.europa.eu/doceo/document/E-8-2015-000521-ASW_EN.html?redirect
Ezquerro-Cañete, A. (2016) 'Poisoned, dispossessed and excluded: A critique of the neoliberal soy regime in Paraguay', *Journal of Agrarian Change*, vol 16, no 4, pp 702–710. doi: 10.1111/joac.12164.
FAO. (1999) *Women: Users, Preservers and Managers of Agrobiodiversity*, Food and Agriculture Organization of the United Nations, Rome.
FAO. (2004) *The State of Agricultural Commodity Markets: 2004*, Food and Agriculture Organization of the United Nations, Rome.
FAO. (2013) *Resilient Livelihoods: Disaster Risk Reduction for Food and Nutrition Security*, Emergency and Rehabilitation Division, Food and Agriculture Organization of the United Nations.
FAO. (2018a) *Scaling Up Agroecology Initiative Transforming Food and Agricultural Systems in Support of the SDGs: A Proposal Prepared for the International Symposium on Agroecology 3–5 April 2018*, Food and Agriculture Organization of the United Nations, Rome. Available at: www.fao.org/3/I9049EN/i9049en.pdf
FAO. (2018b) *The State of Food Security and Nutrition in the World: Building Climate Resilience for Food Security and Nutrtition*, Food and Agriculture Organization of the United Nations, Rome, Italy.

Flores, C.C., Sarandón, S.J. (2004) 'Limitations of neo-classical economics for evaluating sustainability of agricultural systems: Comparing organic and conventional systems', *Journal of Sustainable Agriculture*, vol 24, no 2, pp 77–91. doi: 10.1300/J064v24n02_08.

Folke, C., Carpenter, S., Elmqvist, T., Gunderson, L., Holling, C.S., Walker, B. (2002) 'Resilience and sustainable development: Building adaptive capacity in a world of transformations', *AMBIO: A journal of the human environment*, vol 31, no 5, pp 437–440. doi: 10.1579/0044-7447-31.5.437.

Francis, C., Lieblein, G., Gliessman, S., Breland, T.A., Creamer, N., Harwood, R., Salomonsson, L., Helenius, J., Rickerl, D., Salvador, R., Wiedenhoeft, M., Simmons, S., Allen, P., Altieri, M., Flora, C., Poincelot, R. (2003) 'Agroecology: the ecology of food systems', *Journal of Sustainable Agriculture*, vol 22, no 3, pp 99–118. doi: 10.1300/J064v22n03_10.

Frison, E.A., Cherfas, J., Hodgkin, T. (2011) 'Agricultural biodiversity is essential for a sustainable improvement in food and nutrition security', *Sustainability*, vol 3, no 1, pp 238–253. doi: 10.3390/su3010238.

Gallai, N., Salles, J.M., Settele, J., Vaissière, B.E. (2009) 'Economic valuation of the vulnerability of world agriculture confronted with pollinator decline', *Ecological economics*, vol 68, no 3, pp 810–821. doi: 10.1016/j.ecolecon.2008.06.014.

Gliessman, S.R. (2007) *Agroecology: The Ecology of Sustainable Food Systems*, CRC Press, Boca Raton, FL.

Gómez, M.I., Ricketts, K.D. (2013) 'Food value chain transformations in developing countries: Selected hypotheses on nutritional implications', *Food Policy*, vol 42, pp 139–150. doi: 10.1016/j.foodpol.2013.06.010.

Government of Andhra Pradesh. (2017) 'Zero-Budget Natural Farming'. Available at: http://apzbnf.in

Hallmann, C.A., Sorg, M., Jongejans, E., Siepel, H., Hofland, N., Schwan, H., Goulson, D., de Kroon, H. (2017) 'More than 75 percent decline over 27 years in total flying insect biomass in protected areas', *PloS One*, vol 12, no 10, doi: 10.1371/journal.pone.0185809.

Hamm, M.W., Frison, E., von der Pahlen, M.C.T. (2018) 'Human health, diets and nutrition: missing links in eco-agri-food systems', *TEEB for Agriculture & Food: Scientific and Economic Foundations Report*, UN Environment.

Hawkes, C. (2007) 'Promoting healthy diets and tackling obesity and diet-related chronic diseases: What are the agricultural policy levers?', *Food and Nutrition Bulletin*, vol 28, 2 suppl, pp S312–S322. doi: 10.1177/15648265070282S210.

Herforth, A. (2010) 'Promotion of traditional African vegetables in Kenya and Tanzania: A case study of an intervention representing emerging imperatives in global nutrition', PhD thesis, Cornell University, New York.

Holt-Giménez, E. (2002) 'Measuring farmers' agroecological resistance after Hurricane Mitch in Nicaragua: A case study in participatory, sustainable land management impact monitoring', *Agriculture, Ecosystems and Environment*, vol 93, no 1–3, pp 87–105. doi: 10.1016/S0167-8809(02)00006-3.

Hultberg, A., Madsen, B.B. (2012) 'Public food in Copenhagen; organic conversion on the road towards sustainable food supply', Copenhagen House of Food and Municipality of Copenhagen. Available at: https://ec.europa.eu/agriculture/sites/agriculture/files/events/2012/food-sustainabilty/hultberg-madsen_en.pdf

ILO. (2015) 'Combating Forced Labour: A Handbook for Employers and Business: Special Action Programme to Combat Forced Labour', International Labour Organization.

Available at: www.ilo.org/wcmsp5/groups/public/—ed_norm/—declaration/documents/publication/wcms_101171.pdf
INSEE. (2016) 'INSEE Travail-Emploi – Population active', Institut national de la statistique et des études économiques. Available at: www.insee.fr/fr/themes/document.asp?ref_id=T14F041
Rodale Institute. (2015) 'The farming systems trial'. Available at: https://rodaleinstitute.org/wp-content/uploads/fst-30-year-report.pdf
International Assessment of Agricultural Knowledge, Science and Technology for Development (IAASTD). (2009) *Synthesis report: A synthesis of the global and sub-global IAASTD reports*, Island Press, Washington DC.
IPES-Food. (2016) *From Uniformity to Diversity: A Paradigm Shift from Industrial Agriculture to Diversified Agroecological Systems*, International Panel of Experts on Sustainable Food Systems.
IPES-Food. (2017a) *Unravelling the Food-Health Nexus: Addressing Practices, Political Economy and Power Relations to Build Healthier Food Systems*, The Global Alliance for the Future of Food and IPES-Food.
IPES-Food. (2017b) *Too Big to Feed: Exploring the impacts of mega-mergers, concentration, concentration of power in the agri-food sector*, International Panel of Experts on Sustainable Food Systems.
IPES-Food. (2018) *Breaking away from industrial food and farming systems: Seven case studies of agroecological transition*, International Panel of Experts on Sustainable Food Systems.
IPES-Food. (2019) *Towards a Common Food Policy for the European Union: The policy reform and realignment that is required to build sustainable food systems in Europe*, The Global Alliance for the Future of Food and IPES-Food.
Jacobsen, S.E., Sørensen, M., Pedersen, S.M., Weiner, J. (2013) 'Feeding the world: Genetically modified crops versus agricultural biodiversity', *Agronomy for Sustainable Development*, vol 33, no 4, pp 651–662. doi: 10.1007/s13593-013-0138-9.
Johnston, G., Vaupel, S., Kegel, F., Cadet, M. (1995) 'Crop and farm diversification provide social benefits', *California Agriculture*, vol 49, no 1, pp 10–16. doi: 10.3733/ca.v049n01p10.
Jones, A.D., Shrinivas, A., Bezner-Kerr, R. (2014) 'Farm production diversity is associated with greater household dietary diversity in Malawi: Findings from nationally representative data', *Food Policy*, vol 46, pp 1–12. doi: 10.1016/j.foodpol.2014.02.001.
Kumar, N., Harris, J., Rawat, R. (2015) 'If they grow it, will they eat and grow? Evidence from Zambia on agricultural diversity and child undernutrition', *The Journal of Development Studies*, vol 51, no 8, pp 1060–1077. doi: 10.1080/00220388.2015.1018901.
Kumar, P. (ed) (2010) *The Economics of Ecosystems and Biodiversity Ecological and Economic Foundations*, Earthscan, London.
Laaninen, T. (2015) 'Imports of GM food and feed Right of Member States to opt out', European Parliamentary Research Service. Available at: www.europarl.europa.eu/EPRS/EPRS-Briefing-559479-Imports-GM-food-and-feed-FINAL.pdf
Lagane, J. (2011) 'Du teikei à l'AMAP, un modèle acculturé', *Développement durable et territoires*, vol 2, no 2, doi: 10.4000/developpementdurable.9013.
Leach, G. (1992) 'The energy transition', *Energy Policy*, vol 20, no 2, pp 116–123. doi: 10.1016/0301-4215(92)90105-B.
Lines, T. (2008) *Making Poverty: A History*, Zed Books, London.
Martin, W. (2010) 'Food security and poverty—a precarious balance', World Bank Blogs: Let's Talk Development. Available at: http://blogs.worldbank.org/developmenttalk/food-security-and-poverty-a-precarious-balance

Massis, D., Hild, F. (2016) 'La pratique de l'agriculture biologique créatrice d'emploi? Une évaluation de l'impact du bio sur la quantité de travail agricole', *Agreste Les Dossiers*, vol 35. Available at: www.eauetbio.org/publications/agreste-juil16-bio-emplois/

Mijatović, D., Van Oudenhoven, F., Eyzaguirre, P., Hodgkin, T. (2013) 'The role of agricultural biodiversity in strengthening resilience to climate change: Towards an analytical framework', *International Journal of Agricultural Sustainability*, vol 11, no 2, pp 95–107. doi: 10.1080/14735903.2012.691221.

Muller, A., Schader, C., Scialabba, N.E.H., Brüggemann, J., Isensee, A., Erb, K.H., Smith, P., Klocke, P., Leiber, F., Stolze, M., Niggli, U. (2017) 'Strategies for feeding the world more sustainably with organic agriculture', *Nature communications*, vol 8, no 1, article 1290, doi: 10.1038/s41467-017-01410-w.

Murphy, S., Burch, D., Clapp, J. (2012) *Cereal Secrets: The World's Largest Grain Traders and Global Agriculture*, Oxfam Research Reports.

Niles, M., Esquivel, J., Ahuja, R., Mango, N. (2017) *Climate Change & Food Systems: Assessing Impacts and Opportunities*, Meridian Institute, Washington, DC.

O'Brien, K., Reams, J., Caspari, A., Dugmore, A., Faghihimani, M., Fazey, I., Hackmann, H., Manuel-Navarette, D., Marks, J., Miller, R., Raivio, K., Romero-Lankao, P., Virji, H., Vogel, C., Winiwarter, V. (2013) 'You say you want a revolution? Transforming education and capacity building in response to global change', *Environmental Science & Policy*, vol 28, pp 48–59. doi: 10.1016/j.envsci.2012.11.011.

OpenSecrets. (2016) 'Lobbying Top Industries'. Available at: www.opensecrets.org/lobby/top.php?showYear=2015&indexType=i

Oyarzun, P.J., Borja, R.M., Sherwood, S., Parra, V. (2013) 'Making sense of agrobiodiversity, diet, and intensification of smallholder family farming in the Highland Andes of Ecuador', *Ecology of Food and Nutrition*, vol 52, no 6, pp 515–541. doi: 10.1080/03670244.2013.769099.

Papademetriou, M.K., Dent, F.J. (eds) (2001) *Crop Diversification in the Asia Pacific Region*, Food and Agriculture Organization of the United Nations Regional Office for Asia and the Pacific, Bangkok. Available at: https://coin.fao.org/coin-static/cms/media/9/13171763115260/2001_03_high.pdf

Pellegrini, L., Tasciotti, L. (2014) 'Crop diversification, dietary diversity and agricultural income: Empirical evidence from eight developing countries', *Canadian Journal of Development Studies/Revue canadienne d'études du développement*, vol 35, no 2, pp 211–227. doi: 10.1080/02255189.2014.898580.

Picasso, V.D., Brummer, E.C., Liebman, M., Dixon, P.M., Wilsey, B.J. (2008) 'Crop species diversity affects productivity and weed suppression in perennial polycultures under two management strategies', *Crop Science*, vol 48, no 1, pp 331–342. doi: 10.2135/cropsci2007.04.0225.

Ponisio, L., M'Gonigle, L.K., Mace, K.C., Palomino, J., de Valpine, P., Kremen, C. (2015) 'Diversification practices reduce organic to conventional yield gap', *Proceedings of the Royal Society B: Biological Sciences*, vol 282, no 1799, pp 20141396–20201413. doi: 10.1098/rspb.2014.1396.

Popay, J., Escorel, S., Hernández, M., Johnston, H., Mathieson, J., Rispel, L. (2008) 'Understanding and Tackling Social Exclusion Final Report to the WHO Commission on Social Determinants of Health From the Social Exclusion Knowledge Network February 2008', World Health Organization Social Exclusion Knowledge Network. Available at: www.who.int/social_determinants/knowledge_networks/final_reports/sekn_final%20report_042008.pdf?ua=1

Popkin, B.M., Adair, L.S., Ng, S.W. (2012) 'Global nutrition transition and the pandemic of obesity in developing countries', *Nutrition Reviews*, vol 70, no 1, pp 3–21. doi: 10.1111/j.1753-4887.2011.00456.x.

Poux, X., Aubert, P.-M. (2018) *Une Europe agroécologique en 2050: une agriculture multifonctionnelle pour une alimentation saine, Enseignements d'une modélisation du système alimentaire européen*, Iddri-AScA, Study N°09/18, Paris, France.

Powell, B., Thilsted, S. H., Ickowitz, A., Termote, C., Sunderland, T., Herforth, A. (2015) 'Improving diets with wild and cultivated biodiversity from across the landscape', *Food Security*, vol 7, no 3, pp 535–554. doi: 10.1007/s12571-015-0466-5.

Pretty, J. (2006) 'Background paper for the world development report 2008 Agroecological Approaches to Agricultural Development', Rimisp-Latin American Center for Rural Development.

Pretty, J., Benton, T. G., Bharucha, Z. P., Dicks, L. V., Flora, C. B., Godfray, H. C. J., Goulson, D., Hartley, S., Lampkin, N., Morris, C., Pierzynski, G., Vara Prasad, P.V., Reganold, J., Rockström, J., Smith, P., Thorne, P., Wratten, S. (2018) 'Global assessment of agricultural system redesign for sustainable intensification', *Nature Sustainability*, vol 1, no 8, pp 441–446. doi: 10.1038/s41893-018-0114-0.

Pretty, J., Toulmin, C., Williams, S. (2011) 'Sustainable intensification in African agriculture', *International Journal of Agricultural Sustainability*, vol 9, no 1, pp 5–24. doi: 10.3763/ijas.2010.0583.

Prieto, I., Violle, C., Barre, P., Durand, J.-L., Ghesquiere, M., Litrico, I. (2015) 'Complementary effects of species and genetic diversity on productivity and stability of sown grasslands', *Nature Plants*, vol 1, no 4, article 15033, doi: 10.1038/nplants.2015.33.

Ray, D. K., Ramankutty, N., Mueller, N. D., West, P. C., Foley, J. A. (2012) 'Recent patterns of crop yield growth and stagnation', *Nature communications*, vol 3, article 1293, doi: 10.1038/ncomms2296.

Reardon, T., Timmer, C.P., Barrett, C.B., Berdegué, J. (2003) 'The rise of supermarkets in Africa, Asia, and Latin America', *American Journal of Agricultural Economic*, vol 85, no 5, pp 1140–1146. doi: 10.1111/j.0092-5853.2003.00520.x.

Reganold, J.P., Wachter, J.M. (2016) 'Organic agriculture in the twenty-first century', *Nature Plants*, vol 2, article 15221, doi: 10.1038/nplants.2015.221.

Remans, R., Flynn, D.F.B., DeClerck, F., Diru, W., Fanzo, J., Gaynor, K., Lambrecht, I., Mudiope, J., Mutuo, P.K., Nkhoma, P., Siriri, D., Sullivan, C., Palm, C.A. (2011) 'Assessing nutritional diversity of cropping systems in African villages', *PLoS ONE*, vol 6, no 6, e21235. doi: 10.1371/journal.pone.0021235.

Sautereau, N., Benoit, M. (2016) 'Quantification et chiffrage des externalités de l'agriculture biologique?', Institut Technique de l'Agriculture Biologique. Available at: www.itab.asso.fr/downloads/amenites/amenites-ab-rapport-nov2016.pdf

Sen, A. (1981) *Poverty and famines: an essay on entitlement and deprivation*, Oxford University Press, New York.

Shively, G., Sununtnasuk, C. (2015) 'Agricultural diversity and child stunting in Nepal', *Journal of Development Studies*, vol 51, no 8, pp 1078–1096.

Styger, E., Traoré, G. (2018) '50,000 farmers in 13 countries results from Scaling up the System of Rice Intensification in West Africa: Achievements and Regional Perspectives for SRI SRIWAAPP Project Summary Report, 2014–2016', The West and Central Africa Council for Agricultural Research and Development (CORAF/WECARD), Dakar, Senegal. Available at: https://sriwestafrica.files.wordpress.com/2018/04/sri-waapp-book-single-p-8mb.pdf

TEEB. (2018) *TEEB for Agriculture & Food: Scientific and Economic Foundations Report*, UN Environment, Geneva. Available at: http://teebweb.org/agrifood/wp-content/uploads/2018/11/Foundations_Report_Final_October.pdf

Tengö, M., Belfrage, K. (2004) 'Local management practices for dealing with change and uncertainty: A cross-scale comparison of cases in Sweden and Tanzania', *Ecology and Society*, vol 9, no 3, doi: 10.5751/ES-00672-090304.

Tirado, R., Cotter, J. (2010) *Ecological Farming: Drought-Resistant Agriculture*, Greenpeace Research Laboratories, University of Exeter, Exeter, UK.

UN Water. (2013) *The Post-2015 Water Thematic Consultation Report*, The World We Want. Available at: https://reliefweb.int/sites/reliefweb.int/files/resources/The%20post%202015%20water%20thematic%20consultation%20report.pdf

UNCTAD. (2008) *Economic Development in Africa 2008: Export Performance Following Trade Liberalization: Some Patterns and Policy Perspectives*, United Nations Conference on Trade and Development, New York, Geneva.

Union of Concerned Scientists. (2015) 'Scientist and expert statement of support for public investment in agroecological research'. Available at: www.ucsusa.org/sites/default/files/attach/2014/08/scientist-statement-agroecology-7-2-2014.pdf

USDA. (2014) 'New data reflects the continued demand for farmers markets', US Department of Agriculture. Available at: www.usda.gov/media/press-releases/2014/08/04/new-data-reflects-continued-demand-farmers-markets

USDA. (2016) 'Retail Trends', US Department of Agriculture Economic Research Service. Available at: www.ers.usda.gov/topics/food-markets-prices/retailing-wholesaling/retail-trends/

USDA. (2019) 'Highlights from the 2019 farm income forecast', US Department of Agriculture. Available at: www.ers.usda.gov/topics/farm-economy/farm-sector-income-finances/highlights-from-the-farm-income-forecast/

Van Toan, P., Sebesvari, Z., Bläsing, M., Rosendahl, I., Renaud, F.G. (2013) 'Pesticide management and their residues in sediments and surface and drinking water in the Mekong Delta, Vietnam', *Science of the Total Environment*, vol 452, pp 28–39. doi: 10.1016/j.scitotenv.2013.02.026.

Vermeulen, S.J., Campbell, B.M., Ingram, J.S.I. (2012) 'Climate change and food systems', *Annual Review of Environment and Resources*, vol 37, no 1, pp 195–222. doi: 10.1146/annurev-environ-020411-130608.

Vigouroux, Y., Barnaud, A., Scarcelli, N., Thuillet, A.C. (2011). 'Biodiversity, evolution and adaptation of cultivated crops', *Comptes rendus biologies*, vol 334, no 5–6, pp 450–457.

Virginia, A., Zamora, M., Barbera, A., Castro-Franco, M., Domenech, M., De Gerónimo, E., Costa, J.L. (2018) 'Industrial agriculture and agroecological transition systems: A comparative analysis of productivity results, organic matter and glyphosate in soil', *Agricultural Systems*, vol 167, pp 103–112. doi: 10.1016/j.agsy.2018.09.005.

Watts, N., Adger, W.N., Agnolucci, P., Blackstock, J., Byass, P., Cai, W., Chaytor, S., Colbourn, T., Collins, M., Cooper, A., Cox, P.M., Depledge, J., Drummond, P., Ekins, P., Galaz, V., Grace, D., Graham, H., Grubb, M., Haines, A., Hamilton, I., Hunter, A., Jiang, X., Li, M., Kelman, I., Liang, L., Lott, M., Lowe, R., Luo, Y., Mace, G., Maslin, M., Nilsson, M., Oreszczyn, T., Pye, S., Quinn, T., Svensdotter, M., Venevsky, S., Warner, K., Xu, B., Yang, J., Yin, Y., Yu, C., Zhang, Q., Gong, P., Montgomery, H., Costello, A. (2015) 'Health and climate change: Policy responses to protect public health', *The Lancet*, vol 386, no 10006, pp 1861–1914. doi: 10.1016/S0140-6736(15)60854-6.

WHO. (2015) 'Obesity and overweight', World Health Organization. Available at: www.who.int/mediacentre/factsheets/fs311/en/

Ziska, L., Crimmins, A., Auclair, A., DeGrasse, S., Garofalo, J.F., Khan, A.S., Loladze, I., Pérez de León, A.A., Showler, A., Thurston, J., Walls, I. (2016) 'Chapter 7: Food Safety, Nutrition, and Distribution', in Basu, R., English, P. et al. (eds) *The Impacts of Climate Change on Human Health in the United States: A Scientific Assessment*, pp 189–216. U.S. Global Change Research Program, Washington, DC.

PART II

The Biodiversity for Food and Nutrition (BFN) Project

In this section:

- An overview of the global BFN Project, a multi-country, cross-sectoral initiative that promoted indigenous food biodiversity by: providing evidence, influencing policies and markets, and raising awareness.

  A) The conceptual and historical context in which the project developed

  B) The necessary steps to plan the project: situational analysis, development of partnerships and identification of entry points

  C) The actions taken to implement the project with country-specific highlights, lessons learned and best practices.

5

THE ABC OF MAINSTREAMING BIODIVERSITY FOR FOOD AND NUTRITION

Concepts, theory and practice

Eliot Gee, Teresa Borelli, Daniela Moura de Oliveira Beltrame, Camila Neves Soares Oliveira, Lidio Coradin, Victor Wasike, Aurillia Manjella Ndiwa, Gamini Samarasinghe, Birgül Güner, Ayfer Tan, Kürşad Özbek, Saadet Tuğrul Ay, Sevinç Karabak, Nurcan Aysar Güzelsoy and Danny Hunter

Overview

So far, Part I of this book has illustrated the potential of food biodiversity in contributing to much-needed food system change. It is clear that strategic promotion and use of food biodiversity is critical in uniting attempts to address conservation, nutrition and livelihood concerns. However, the knowledge base on much of this food biodiversity remains fragmented and unrecognized by decision-makers and the wider public, with few examples of initiatives that have successfully implemented approaches to assess, deliver or mobilize biodiversity.

In Part II, we turn to the Biodiversity for Food and Nutrition (BFN) Project, a cross-sectoral initiative spearheaded by Brazil, Kenya, Sri Lanka and Turkey to research and promote local food biodiversity. As introduced in Chapter 1, the BFN Project represents a unique approach to link biodiversity, diets and nutrition through three main components: providing evidence, influencing policy and markets, and raising awareness (illustrated in Figure 5.1: the BFN approach). This approach, adapted to the context of each country, offers numerous insights into how to best mainstream biodiversity into wider food systems (see Box 5.1).

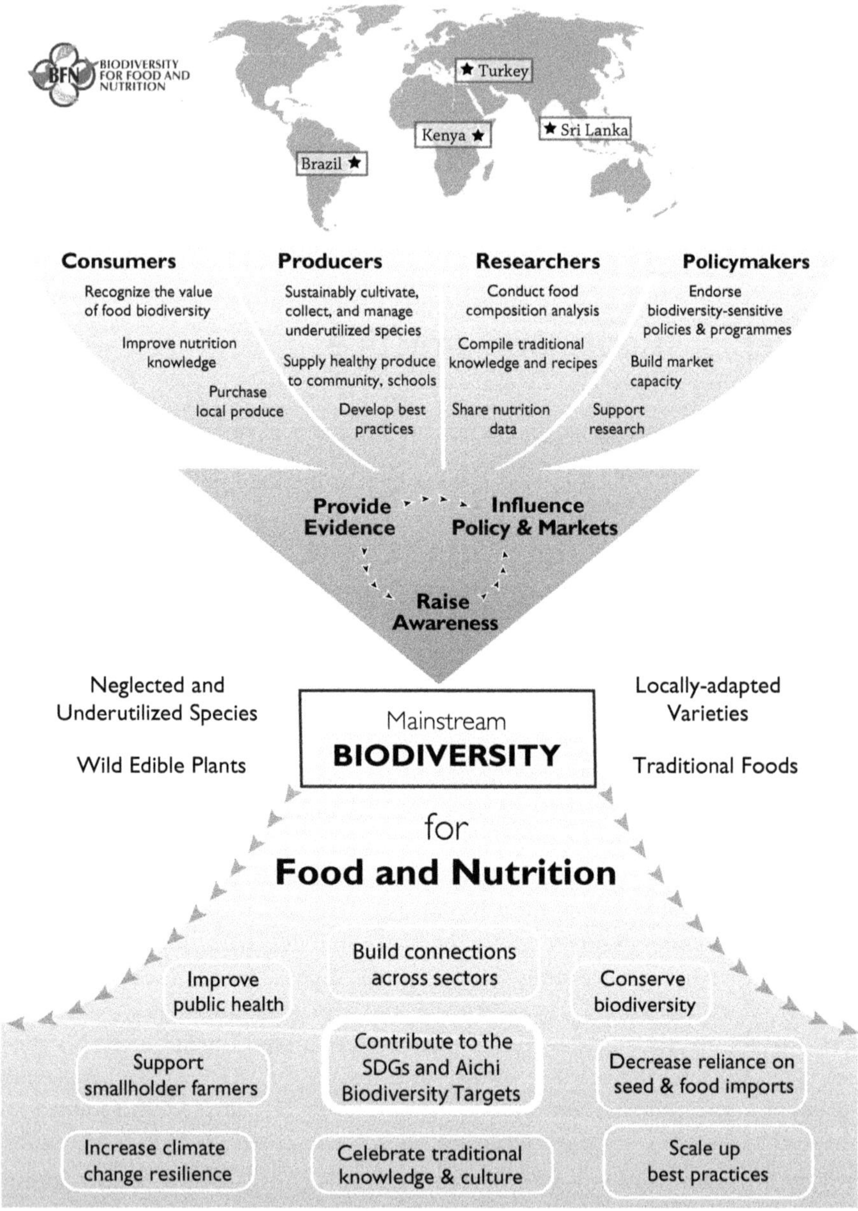

FIGURE 5.1 An overview of the BFN Project approach

Source: N. Ching, E. Gee, N. Lauridsen

BOX 5.1 WHAT WE MEAN BY: 'MAINSTREAMING BIODIVERSITY'

'Mainstreaming' biodiversity means integrating actions that conserve and sustainably use biodiversity into strategies for agriculture, fisheries, forestry and other related sectors.

Including biodiversity awareness in plans, policies and programmes increases recognition of the importance of biodiversity for human well-being, while ensuring the sustainability of the production sectors and the people who depend on them for their livelihoods.

Mainstreaming also refers to the inclusion of biodiversity considerations into plans for poverty reduction, national sustainable development and climate change adaptation.

In the context of food biodiversity – by which we mean 'the diversity of plants, animals and other organisms used as food, covering the genetic resources within species, between species and provided by ecosystems' (FAO, 2010) – mainstreaming refers to the appropriate integrated use of a wide range of edible plant varieties (including wild, neglected and underutilized species) and animal breeds with the main goal of reducing malnutrition.

Source: Adapted from The BFN Mainstreaming Toolkit and Decision XIII/3, Conference of the Parties to the Convention on Biological Diversity

Part II is a single chapter divided into ABC of the BFN Project:

A. Elaborates on the **context** that guided the formation of the BFN Project, followed by the situation in the four target countries, and the three components that guided the project approach.
B. Outlines the **planning** phase of the BFN Project, with particular emphasis on developing partnerships to support the successful implementation of the project at all levels.
C. Details how the BFN Project approach was put into **action**. The general practices associated with each component are illustrated with country-specific case studies and highlights. These promote dialogue between the countries and provide an understanding of how these activities can be up-scaled and adapted to country and regional contexts.

A Biodiversity for food and nutrition in context

1 The rising concept of biodiversity for food and nutrition

> **BOX 5.A1 PLAYERS AND DECISIONS DISCUSSED IN THIS SECTION**
>
> Key players: Convention on Biodiversity (CBD), the Global Environment Facility (GEF), Food and Agriculture Organization of the United Nations (FAO), United Nations Environment Programme (UNEP), Bioversity International, Brazil, Kenya, Sri Lanka and Turkey.
>
> Decisions of relevance to the project context: CBD Cross-cutting initiative on biodiversity for food and nutrition, Commission on Genetic Resources for Food and Agriculture (CGRFA), FAO Voluntary Guidelines (2015), The International Conference on Nutrition (ICN2), UN Decades on Biodiversity and Nutrition, Sustainable Development Goals (SDGs) and Aichi Biodiversity Targets.

The Biodiversity for Food and Nutrition (BFN) Project spanned seven years, from April 2012 to 2019, and stemmed from a build-up of fresh thinking in the late 1990s and early 2000s to redress years of ineffective agricultural and health sector interventions aimed at solving global malnutrition (Beltrame et al., 2019). The concepts emerging called for synergistic approaches towards integrating agriculture and food production, ecology and economics with nutrition and human health, environmental health and economic development, to promote sustainable agricultural and food production with minimal impacts on the environment. Concurrently, there was growing recognition of the important links between biodiversity, nutrition and health, and realization that their combined integration into public health, conservation and sustainability strategies would contribute not only to improved health and biodiversity outcomes, but also to poverty alleviation, disaster-risk reduction and, more broadly, to sustainable development, in keeping with global conversations at the time around the post-2015 UN development agenda, and the subsequent formulation of the UN Sustainable Development Goals (SDGs).

The origins of the BFN Project were indirectly linked to the launch in 1996 by the Conference of the Parties (COP) to the Convention on Biological Diversity (CBD) – the chief multilateral treaty targeting sustainable development and aimed at the conservation of biological diversity – of a Programme of Work (PoW) on Agricultural Biodiversity[1]. The PoW recognized agricultural biodiversity as a contributor to farmers' livelihoods and recognized the central role farmers and indigenous communities played in the conservation and sustainable use of this diversity. Within the PoW, three international cross-cutting initiatives were adopted focusing on declining or overexploited agricultural ecosystem services.

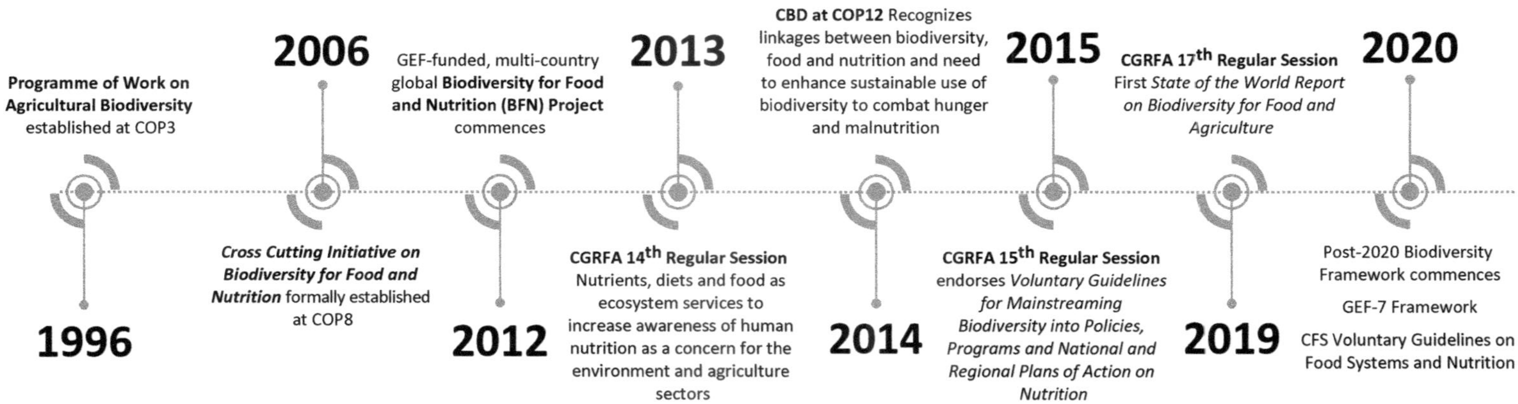

FIGURE 5.A1 Timeline: key steps in the development of biodiversity for food and nutrition

Source: D. Hunter

The one most relevant to BFN, and which led to the formulation of the themes that underpin the project, is the *Cross-cutting initiative on biodiversity for food and nutrition* formally established in 2006 by decision VIII/23 A of the Conference of the Parties.[2] Led by the Food and Agriculture Organization of the United Nations (FAO) and Bioversity International, the initiative aims to:

- Develop and document knowledge on the composition and consumption of food genetic resources, as well as the relationship between biodiversity and nutrition
- Integrate biodiversity, food and nutrition issues into research and policy instruments
- Conserve and promote wider use of biodiversity for food and nutrition
- Increase public awareness of the importance of biodiversity.

BOX 5.A2 WHAT IS GEF AND HOW DO YOU APPLY FOR GEF FUNDING?

Funding for the BFN Project was channelled via the Global Environment Facility (GEF), the institutional financing mechanism of the CBD that provides financial support to the 196 countries that are parties to the CBD for the implementation of the Convention and its Protocols and for the achievement of the globally-agreed objectives, goals and targets, and associated global benefits. By the time it was approved, the BFN Project fell under the umbrella of the GEF-6 four-year funding cycle (2014–2018), which had a strong focus on tackling the drivers of environmental degradation and particularly biodiversity conservation and climate change mitigation. The application process for GEF funding can be lengthy and cumbersome, and generally requires a negotiation process between countries and GEF that can take between 12 to 18 months, sometimes longer. The process is summarized (Figure 5.A2) for those interested in accessing future GEF funding. It should be noted, however, that from the time of concept development to final project approval, GEF country allocations may have been redirected to competing national projects and limited funds remain for project implementation. This was the case for one of the BFN countries, Kenya, which requested an initial allocation of approximately US$1M at the project proposal phase, but ended up receiving US$250K to implement a five-year project. While representing a barrier at first and forcing the country to downsize activities in order to meet a more limited number of project objectives and outcomes, the shortfall eventually turned into an opportunity, leading to greater creativity, to resource optimization and the quest for alternative sources of funding. For more information, see: https://wwfgeftracks.com.

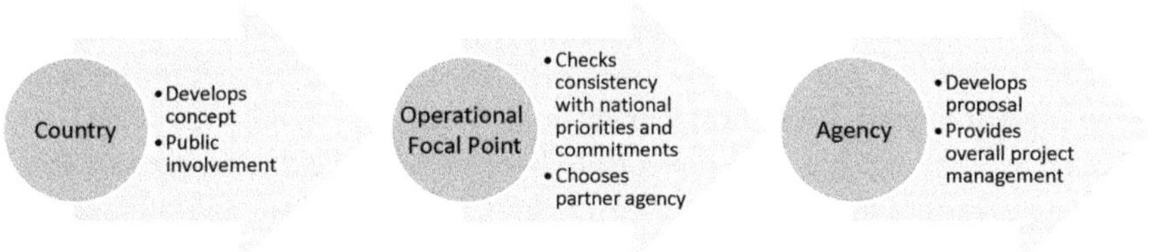

gef

FIGURE 5.A2 The Global Environment Facility (GEF) Process
Adapted from https://wwfgeftracks.com/project-cycle

Increasing recognition of food biodiversity

Biodiversity was also the main topic of the First International Scientific Symposium on Biodiversity and Sustainable Diets held in 2010 at FAO, in Rome, which served the purpose of positioning sustainable diets, nutrition and biodiversity as central to sustainable development. It also promoted 'the use of food biodiversity, including traditional and local foods, with their many nutritionally rich species and varieties' (FAO, 2012). During the Symposium, an outline for a potential code of conduct was also developed that recognized the key role of *food* – as opposed to single nutrients – as central to human nutrition but, more importantly, 'recognizing that the conservation and sustainable use of food biodiversity is an important part of human well-being'. Another important outcome of the Symposium was the establishment of a scientific, multi-stakeholder platform aimed at improving the evidence base for biodiversity and nutrition, and to develop methods and indicators for the characterization of different agroecological zones for sustainable diets (Dernini et al., 2013). Research partners of the platform include FAO, the Centre International de Hautes Études Agronomiques Méditerranéennes (CIHEAM), Bioversity International and the Institut National de la Recherche Agronomique (INRA), among others, who are developing new thinking and actions to better understand the links between human nutrition, biodiversity, ecosystems and the environment. Examples include developing new metrics for nutritional diversity of cropping systems (Remans et al., 2011), inter- and intra-species diversity within major food crops (Burlingame et al., 2009; Hunter et al., 2019), food chain sustainability, traditional food system and nutrition security, underutilized fruit for human nutrition and sustainable diets (Kehlenbeck et al., 2013), and conservation systems for plant biodiversity for sustainable diets (FAO, 2012a).

So, when the project started in 2012, global conventions unanimously agreed that agriculture, health and the environment are inextricably linked. They also accepted the sustainable diets rationale put forward by FAO and partners, and the role food systems play in safeguarding ecosystems and biodiversity, and promoting climate justice (UN, 2012). Global consensus was also that broader, multi-sectoral approaches were to be adopted to effectively address the underlying causes of food and nutrition insecurity, but cooperation efforts remained largely uncoordinated both horizontally and vertically. While the CBD and WHO bilaterally engaged on issues of biodiversity and health, WHO and FAO discussed agriculture and health, and FAO and CBD agriculture and biodiversity. In addition, few preceding examples existed on multi-sectoral cooperation using a cross-cutting, international perspective.

Strengthening the nutrition link

Project activities had only just begun in the four countries when, in 2013, the Commission on Genetic Resources for Food and Agriculture (CGRFA), at its 14th Regular Session, recognized the importance of 'linking food biodiversity and the environment sector to human nutrition and healthy diets', as well as recognizing nutrients, diets and food as essential ecosystem services (see Chapter 2). In 2014, the Second International Conference on Nutrition (ICN2), jointly convened by FAO and WHO, focused on policies aimed at eradicating malnutrition by transforming food systems to make nutritious diets available to all. While not openly mentioning the potential use of biodiversity or genetic resources for food and agriculture to address malnutrition, it endorsed the Rome Declaration on Nutrition and the Framework for Action, which include several recommendations that support the better utilization of neglected and underutilized species. Recommendation 10 urges countries to 'promote the diversification of crops including underutilized traditional crops, more production of fruits and vegetables, and appropriate production of animal source products as needed, applying sustainable food production and natural resource management practices', while recommendation 42 calls for the 'improved intake of micronutrients through consumption of nutrient-dense foods, especially foods rich in iron and the promotion of healthy and diversified diets'. Recommendation 8 clearly invites countries to adopt a multi-sectoral approach to tackle malnutrition by 'reviewing national policies and investments and integrating nutrition objectives into food and agriculture policy, programme design and implementation, to enhance nutrition sensitive agriculture, ensure food security and enable healthy diets'. Less than a year later, the CGRFA, at its 15th Regular Session, endorsed the *Voluntary Guidelines for Mainstreaming Biodiversity into Policies, Programmes and National and Regional Plans of Action on Nutrition*[3], supporting the step-wise approach that BFN was already adopting and urging countries to integrate biodiversity into relevant policies and programmes.

Working towards international goals

As the project progressed, UNEP and FAO, acting as the global implementing agencies, and Bioversity International, as global executing agency of the BFN

Project, were well embedded in relevant global processes and mechanisms and were able to ensure that project results contributed to the tracking of global indicators in the areas of biodiversity, health, agriculture and food security. Outcomes were used to monitor progress towards implementation of the CBD Strategic Plan for Biodiversity 2011–2020 and its Aichi Biodiversity Targets, as well as tracking the agricultural biodiversity indicators of the Global Strategy for Plant Conservation (GSPC), of the Global Plan of Action (GPA), of the International Treaty on Plant Genetic Resources for Food and Agriculture (ITPGRFA), and tracking the relevant core indicators of the Sustainable Development Goals (SDGs) and the Committee on World Food Security (CFS). Since the BFN Project's inception, support for biodiversity for food and nutrition has rapidly increased, alongside the realization that significant amounts of local biodiversity for food and nutrition, including wild species, remain untapped.

According to the recent State of the World's Report on Biodiversity for Food and Agriculture (FAO, 2019), most countries have put in place legal, policy and institutional frameworks for the sustainable use and conservation of biodiversity, both *in situ* – in protected areas and on farm – and *ex situ*, in gene-banks and botanic gardens. However, the report notes that these efforts are often insufficient or ineffectively implemented and that improved knowledge of biodiversity for food and agriculture is needed as many information gaps remain.

In this respect, with food composition provided for 188 plant species, the BFN Project is probably the largest single contributor of nutrition information on neglected and underutilized species and wild foods. Many more species remain unidentified and undescribed, particularly invertebrates and micro-organisms, and their impact on food and agriculture remains unknown. The BFN Project also leads the way in creating enabling frameworks, promoting pro-biodiversity initiatives and addressing the main drivers of biodiversity loss. As we will see in Section C, it has improved collaboration among policymakers, producer organizations, consumers, the private sector and civil society organizations across biodiversity, food and agriculture, and environment, as well as many other relevant sectors.

Moving ahead: CBD and COP15

The timeframe in Figure 5.A1 includes another milestone that has not yet been reached, but which represents a unique opportunity to better harness biodiversity for food and nutrition to ensure greater resilience of ecosystems and the people living within them (see Chapter 11).

In 2020, the United Nations Decade on Biodiversity 2011–2020 draws to a close, and global biodiversity experts will meet in Kunming, China, for the 15th meeting of the Conference of the Parties to the CBD, otherwise known as the 2020 UN Biodiversity Conference. It is foreseen that by that point many of the 196 signatory nations to the Convention will have failed to meet many of the Aichi Biodiversity Targets – the 20 global targets aimed at ending biodiversity loss and restoring ecosystems by 2020. The meeting will therefore be a critical moment for environmental decision-making and for the future of our planet. Slow or uneven progress towards meeting the Aichi targets will also put the attainment of the Sustainable Development Goals (SDGs) at risk as well as the planet's ability to sustain itself in the long term (IPBES, 2019).

Biodiversity and healthy ecosystems underpin several of the SDGs because they ensure that the natural resources we depend on are maintained and restored. Not only does the achievement of the Aichi targets impact SDG 12 (Consumption and Production), SDG 14 (Life in water) and SDG 15 (Life on land); critical biodiversity links exist with SDG 2 (Zero hunger). Every year, biodiversity provides the global economy with trillions of US dollars' worth of ecosystem services, such as carbon sequestration and water quality regulation, as well as material goods (e.g. food, medicines and raw materials) and intangible contributions such as spiritual, cultural and recreational experiences that support human well-being. For example, it is estimated that pollinator services by insects and other animals sustained by biodiversity contribute between US$235 billion and US$577 billion worth of annual global food production (IPBES, 2016).

Yet, the cross-cutting importance of biodiversity for healthy agriculture, food systems and diets is only weakly reflected in the wording of the Sustainable Development Goals (SDGs), with the exception, possibly, of Target 2.5 of SDG 2[4]. Although concerned mainly with the *ex situ* conservation of biodiversity for food and nutrition, the target recognizes the need to:

> maintain the genetic diversity of seeds, cultivated plants and farmed and domesticated animals and their related wild species, including through soundly managed and diversified seed and plant banks at the national, regional and international levels, and promote access to and fair and equitable sharing of benefits arising from the utilization of genetic resources and associated traditional knowledge, as internationally agreed.

With increasing knowledge of the importance of biodiversity for food and nutrition, it is felt that indicators should rise above the conventional metrics of agricultural production and yield to better measure nutritional quality (Hunter et al., 2016), nutritional diversity of food systems (Remans et al., 2014; Bogard et al., 2018) and dietary diversity (Lartey, 2016). The UN Biodiversity Conference in 2020 offers the chance to rectify this shortcoming by ensuring that biodiversity metrics are incorporated into relevant biodiversity indicators and targets for the next decades, so that the multiple ecosystem goods and services biodiversity delivers for human well-being are better tracked. Around the same time, representatives of the Committee on World Food Security will be called upon to endorse the *Voluntary Guidelines on Food Systems and Nutrition* (CFS 2019/46/2)[5] to help governments design and enact effective policies, investments and institutional arrangements to address malnutrition in all its forms using 'a systemic, holistic and evidence-based approach that considers food systems in their totality'. Projects such as BFN can help countries move beyond business-as-usual and offer tried and tested examples of how to influence the three constituent elements of food systems food supply chains, food environments and consumer behaviour for a food system transformation that nourishes people and sustains the planet.

2 Choosing the countries

- Rationale for a multi-country project, and the selection of the countries Brazil, Kenya, Sri Lanka and Turkey.

Country selection under the GEF funding mechanism, or any other mechanism for that matter, is not as straightforward as it may seem and may have to undergo several iterations before the official country list is established. In the BFN Project's initial stage in 2006, Brazil, Cameroon, India, Indonesia, Kenya, Peru and South Africa had been chosen based on the extraordinary diversity of ecosystems and food species these countries supported – the same criteria that subsequently led to the selection of Sri Lanka and Turkey. Six years later, the project was approved in its current form with the governments of Brazil, Kenya, Sri Lanka and Turkey spearheading activities.

As mentioned, the four countries were selected for reasons including the uniqueness and global significance of their biodiversity and ecosystems, and the potential for mobilizing agricultural biodiversity as a resource for food security and livelihoods. An enormous array of biological diversity exists collectively between the four countries including many endemic plant species of potentially high value from a nutritional and livelihoods perspective. The agricultural biodiversity within agroecosystems and surrounding landscapes – including diversity in species and crop varieties, neglected or orphan crops, home gardens and niche crops, including aquatic animal resources and forest margins or patches – is vitally important as a global source of micronutrients and dietary diversity. Yet, the use of these indigenous, largely plant genetic resources is still scarcely explored, appreciated or conserved, despite their potential contributions to food security, nutrition and the reduction of malnutrition (issues outlined in Figure 5.A3). This biological diversity is also increasingly exposed to threats and barriers in all four countries, including unsustainable harvesting, land degradation, urbanization, changes in land use, droughts and floods including neglect as a result of the marketing and promotion of simplified food systems (see Chapter 4).

Although Brazil, Kenya, Sri Lanka and Turkey had projects and activities underway that addressed biodiversity with high nutritional potential, this was carried out in a limited, uncoordinated and often fragmentary way, leading

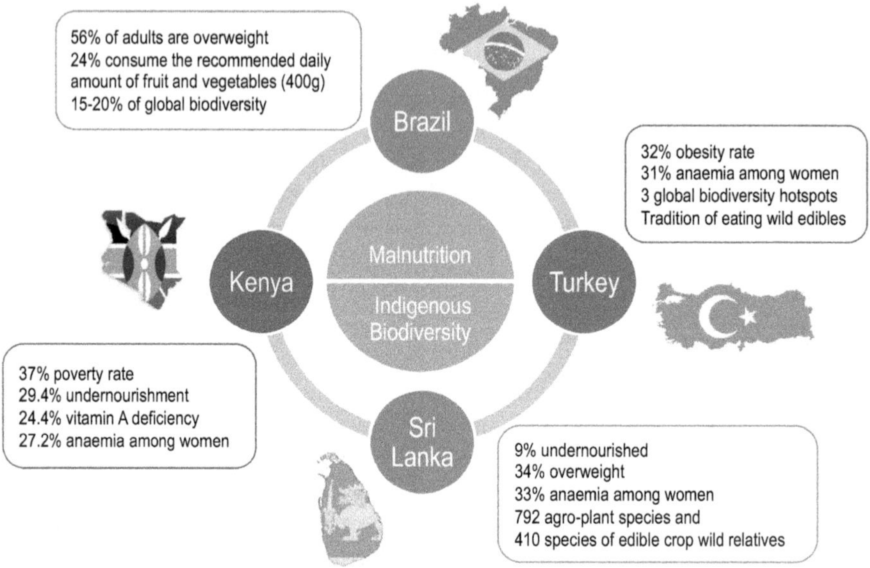

FIGURE 5.A3 Overview of the country contexts

(*Sources*: MoH Brazil, 2008–2017; Katulanda et al., 2010; MoH Kenya, 2011; Şekercioğlu et al., 2011; World Bank, 2015; MMA, 2017; Tan et al., 2017; FAO, 2019; Statista, 2019).

to missed opportunities for linkages and synergies with relevant national, regional and global initiatives. When the project commenced in 2012, the countries had all undertaken limited activities in areas closely related to the project. Largely unheard of outside of the country, Brazil had set up a national *Plants for the Future* Initiative (detailed in Box 5.B4) specifically to focus on neglected local biodiversity, and had carried out initial prioritization, research and documentation work that provided an excellent institutional template for dealing with the types of biodiversity under consideration. In Kenya, the Kenya Agricultural and Livestock Research Organization (KALRO), the Kenyan Resource Centre for Indigenous Knowledge (KENRIK) and the National Museums of Kenya (NMK) had all undertaken work on local biodiversity for food and nutrition, carrying out some research and development work, as well as promotion and awareness especially on indigenous African leafy vegetables. In Sri Lanka, the Department of Agriculture had spearheaded most research and development on biodiversity with high nutritional potential.

The countries had also been vocal in national, regional and global fora that drew attention to the importance of biodiversity for food and nutrition and, in 2008, all four countries had actively supported the establishment of the CBD's *Cross-cutting initiative on biodiversity for food and nutrition* framework[6] (previously described in Chapter 2). However, at the institutional, sectoral and policy level, Brazil was the only country that had established a globally recognized cross-sectoral institutional platform for dealing with the complexity of food security that had promising results (Box 5.B6). The remaining countries had rudimentary policy frameworks covering biodiversity, agriculture, nutrition and food security, and had explored in a limited fashion value chains for agricultural biodiversity and nutrition products. Most of these initiatives, which did offer something to build on, tended to operate in a sectoral vacuum without the multi-disciplinary and cross-sectoral efforts necessary to effectively mobilize biodiversity for food and nutrition. Furthermore, the lack of sectoral integration in the key areas where biodiversity can make a positive impact (e.g. agriculture, health and environment) meant limited financial and political support within countries, leading to limited resources, capacity and skills to better harness this diversity. Hence, the conservation and sustainable utilization of biodiversity for food and nutrition continued to meet many of the barriers detailed in Section B (Figure 5.B2).

The global nature of the project facilitated cross-country sharing of knowledge and experiences, ultimately bringing project outcomes and lessons learned to a much wider international arena for greater impact. Considerable South-South cooperation enabled countries to learn from each other, replicating approaches and fostering collaboration and healthy competition as a catalyst for innovation and improvement, both nationally and globally.

3 Country backgrounds

- Project implementation will vary significantly depending on geography and political context.
- The following sections provide background information on the four countries outlining the initial status of biodiversity, agriculture, diets and health in each country.

Brazil

Brazil is home to 15–20% of the world's varieties of flora and the largest remnants of tropical ecosystems (Myers et al., 2000; Ulloa et al., 2017; MMA, 2017). Out of the vast quantity of plant species in Brazil (46,713 by a recent count) (Flora do Brasil, 2019), an estimated 700 can be used in food production (FAO, 2017). Native fruits such as camu-camu (*Myrciaria dubia*), which contains at least 40 times more vitamin C than the common orange (*Citrus sinensis*), are available in the wild or on family-farmed plots (Hunter et al., 2019).

Brazil is also made up of 305 diverse traditional peoples and communities, such as indigenous, Quilombolas, riverine communities, artisanal fisherfolk, caiçaras, sertanejos and pantaneiros, who act as guardians of traditional knowledge about the use and conservation of native plants (MMA, 2018). Indigenous and family farmers are responsible for producing 70% of the food consumed domestically, however, these small farms only represent 25% of the total agricultural land (IBGE, 2006). The remainder is dominated by large farms focused on export agriculture, growing only a small selection of commercial crops (notably sugar cane, coffee, beans, rice, soy, orange, corn and wheat). Since the colonial period, Brazilian agriculture has focused on these largely non-native crops and, with the Green Revolution and modernization, productivity has taken precedence over sustainable land use (Redin and Fialho, 2010; Pereira et al., 2012; Bacha and de Carvalho, 2014). In the wealthier areas of Brazil's South, Southeast and Centre-West, large monoculture and commercial farms have flourished at the expense of smallholders and polyculture (Agra and Santos, 2001). Elsewhere, such as in north-eastern Bahia, large farms focus on commodities such as soy, cotton and maize.

Production of a smaller number of species has simplified diets and reduced intake of food and nutrients needed to promote, develop and maintain health (Padulosi et al., 2002). Only 25% of Brazilians reach the World Health Organization (WHO) recommended minimum targets of consuming 400g of fruits and vegetables per day for the intake of essential vitamins, minerals and fibre (WHO, 2003; IBGE, 2011). Over the last 30 years, natural food items have been replaced by processed, high-sodium, high-protein diets, both in urban and rural areas (Levy-Costa et al., 2005; Souza et al., 2013). Measures of nutritional status gathered over the last 35 years from anthropometric survey data (IBGE, 2011) indicate that, while overall the alarming undernutrition rates observed in the poorest regions of the country have decreased, women and children continue to suffer from micronutrient deficiencies. For example, despite the high rates of red meat consumption – well above the global average – Brazilians are still affected by iron deficiency: according to the 2017 Global Nutrition Report, anaemia affects one third of Brazilian women, with the deficiencies persisting, while overweight and obesity are on the rise; in the last two decades, the overweight prevalence for children and adolescents has tripled, while over half of the adult population is overweight (Hawkes and Fanzo, 2017).

It is against this backdrop that the re-introduction of native Brazilian food species (see Figure 5.A4) into sustainable production systems and diets is able to offer a multitude of benefits such as: (i) the reduction of malnutrition (Frison et al., 2011; Beltrame et al., 2016); (ii) a convincing justification for the protection of important natural habitats such as the biodiversity-rich Amazon (Valencia et al., 2019); (iii) income generation and livelihood improvement with the rising market demand for new food products; and (iv) renewed interest and appreciation of the varied Brazilian cultural heritage.

	DESCRIPTION AND USE	NUTRITION
	Juicy, low-calorie fruit that can be used in juices, jams, cakes, puddings, sorbets, liqueurs and wine. In traditional medicine, the peel and leaves are boiled in water and used to treat diarrhoea.	Rich in fibre, antioxidants, vitamin C, vitamin A and carotenoids
	The seeds – in the past the main staple for many indigenous tribes in Southern Brazil – are either cooked in water or grilled, and ground to produce flour for the making of bread, biscuits and cakes. Heavily exploited for its wood, less than 4% of the endangered tree's original forest cover remains. Growing demand for the seeds in restaurants has encouraged re-planting efforts.	Starchy, rich in fibre, B vitamins, magnesium and potassium
	A sweet fruit from the Cerrado region that can be used in juices, and processed into jams and ice cream.	Rich in vitamin C, vitamin E, potassium, and carotenoids
	The fruit pulp is consumed cooked in savoury dishes, and it can also be used for edible oil or as a traditional remedy. Its nuts are also starting to be used by chefs and cooks.	The pulp is rich in carotenoids, fibre, B vitamins and magnesium. The nuts are rich in protein and minerals
	The leaves are commonly used as vegetables in rural parts of Brazil. The fruit can be consumed fresh, however it is generally prepared as jam.	Leaves are high in protein, fibre, iron, manganese and carotenoids

FIGURE 5.A4 Highlight: selected BFN Brazil species

Kenya

Kenya's diverse biomes – including grasslands, savannahs, forests, wetlands and temperate deserts – are home to an estimated 35,000 known species of animals, plants and microbes. Of the 7,500 plant species growing naturally across Kenya, 475 are endemic and 258 are classified as threatened (Mwamodenyi and Omondi, 2012). Native fruits, vegetables, grains, roots and tubers include underutilized species such as Bambara groundnut, millet and numerous African leafy vegetables, which are high in nutrients and well-adapted to local ecosystems. However, cultivation and consumption of these species are in decline for numerous reasons such as habitat destruction, limited awareness and an emphasis on increasing agricultural production of a few energy-rich staples, cash crops (e.g. tea, coffee, cut flowers, sugar cane and cotton) and bio-energy crops (Republic of Kenya, 2015).

Currently, most Kenyans rely on only a handful of food crops for sustenance: maize, beans, banana, rice and potatoes (Hajjar et al., 2008), consuming very small amounts of fruits and vegetables, much lower than the WHO recommended intake (Ruel et al., 2005). Low quality, homogenous diets are having serious repercussions on health and life expectancy: one in four (26%) children suffer from chronic undernutrition, resulting in impaired development and growth. Overweight is also a problem for Kenyan adults, especially for women, with over one third of the female adult population exceeding healthy weight limits (IFPRI, 2016). Prevalent deficiencies are of iodine, iron (anaemia), vitamin A and zinc (MoH Kenya, 2011).

This malnutrition is accompanied by high rates of poverty, particularly for rural smallholder farmers. According to a welfare survey conducted in 2015/2016, around 40% of the rural population and 28% of the urban population is considered poor (Kenya National Bureau of Statistics, 2018). Busia County, where the BFN Project has focused its project implementation, is among the poorest in Kenya with almost 70% of the population considered poor and 27% considered extremely poor (ibid.).

Fortunately, native crops such as protein-rich Bambara groundnut or iron-rich African nightshades (see Figure 5.A5) hold untapped potential to improve nutrition, especially for the 1.5 million schoolchildren who could be reached through school meals and agriculture/nutrition education initiatives. Many of these species are climate adaptive, drought and pest-resistant, and require few inputs (Abukutsa-Onyango, 2007; Muthoni et al., 2010; Kehlenbeck et al., 2013). Additional research indicates that farm production systems with high agrobiodiversity contribute to food security among smallholder farmers in Kenya (Mburu et al., 2016).

Sri Lanka

As one of the world's biodiversity hotspots, Sri Lanka is home to around 4,500 edible species, of which 900 are endemic (MoMD&E, 2016). The country is divided into 46 different agroecological regions – including distinct dry and wet zones – where an array of grains, fruits, vegetables, tubers and pulses can be found. Many of these species are traditionally associated with health benefits and feature in Sri Lanka's tradition of Ayurvedic medicine. For example, *Centella asiatica*, known locally as 'gotukola', is a leafy green that has received scientific attention for its potential to improve cognitive function and resolve skin and digestive afflictions (Prakash et al., 2017).

Sri Lankan farmers have maintained this rich agricultural diversity in gardens for thousands of years, but urbanization, deforestation, colonial influence and generational changes in food preferences and occupations have shifted food production and consumption habits, resulting in a population that carries the double burden of malnutrition (Rajapaksha, 1998). While in the last decade the country has managed to halve its prevalence of undernourishment from 18% to 9%,

	DESCRIPTION AND USE	NUTRITION
	Hardy, nitrogen-fixing pulse that grows well in poor soil, and is pest- and drought-resistant. It can be boiled or roasted as a snack, and is often ground into a flour for cakes and porridge. It is traditionally used to treat digestive problems.	A complete source of proteins, carbohydrates and fats; also rich in iron, calcium and potassium
	A valuable and versatile indigenous leafy vegetable used as a folk remedy. Leaves and tender stems are eaten boiled, stewed, stir-fried, or in soup.	High levels of beta-carotene, vitamin C, folic acid, calcium and iron
	This drought-tolerant cereal is commonly used in porridge ideal for feeding to infants and the elderly, as well as many sweet and savoury dishes.	High in calcium, rich in iron and contains the amino acid methionine, important for joint health
	Though the plant is typically grown for its beans (which, like Bambara groundnut, are a "poor man's meat"), the leaves are also a source of food that is available year-round in Kenya.	A good source of protein and calcium
	A resilient plant that can grow wild in many parts of the world, in Kenya amaranth seeds are ground into a powder to bake biscuits and cakes, while the leaves are usually cooked with bitter leaf vegetables. Although it is widely consumed, producers have little access to the market.	Provides proteins, fibers, calcium, iron, riboflavin, niacin and vitamin C and is an excellent source of lysine (an essential amino acid)

FIGURE 5.A5 Highlight: selected BFN Kenya species

overweight and obesity are on the rise, with 34% of the population recorded as overweight (Katulanda et al., 2010; FAO, IFAD, UNICEF, WFP, WHO, 2019). Preschool children suffer from high rates of iron (34%) and vitamin A (29–38%) deficiency, while one in four young women are anaemic (Abeywickrama et al., 2018).

Amidst environmental concerns related to climate change and deforestation, many species are being neglected and pushed towards extinction (Williams and Haq, 2002). Nature reserves, botanical gardens and seed banks attempt to conserve Sri Lanka's biodiversity; however, it is recognized that the nutrition potential of the many local species needs further exploration (Dahanayake, 2015) (see Figure 5.A6).

Currently the Sri Lankan economy is growing, positively affecting the demand for high-quality food. This presents an opportunity to expand the production of

	DESCRIPTION AND USE	NUTRITION
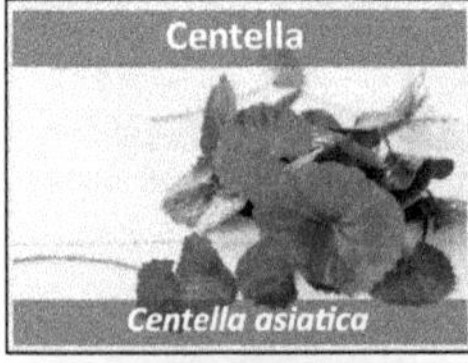	A flowering plant with leaves and stems eaten in soups, salads, juices, porridges and chutneys. Centella has a long history of Ayurvedic use, and is the ongoing subject of scientific studies for healing minor wounds, lowering blood pressure, and improving cognitive function. Research has also suggested that its phytoextraction abilities can remove heavy metals from soil.	Besides its many phytonutrients, it is a good source of carotenes, potassium, calcium, Vitamin B and C.
	One of seven rice varieties studied, this traditional white rice has a pleasant aroma and milky taste that makes it part of festive occasions. It contains more vitamins than many common rice varieties.	High in carbohydrates, protein and glutamic acid (important for muscle and mental health)
	Widely cultivated for use in local dishes, the distinct bitter tasting pod has vermifuge (anti-parasitic) uses. Other plant parts (leaves, flowers) can be consumed as well, or made into a tea or beverage. It is also known for having anti-diabetic properties.	Good source of iron, calcium, phosphorus and vitamin B
	One of five banana varieties studied, this local banana has reddish-purplish skin and light pink flesh with a slight raspberry flavour. The fruit is often eaten raw, added to fruit salads and desserts. It can also be baked, toasted or dried.	Highest potassium content out of the varieties studied, also contains considerable amounts of calcium and magnesium

FIGURE 5.A6 Highlight: selected BFN Sri Lanka species

local fruits and vegetables to both domestic and international markets, thereby contributing to increase national income, employment opportunities, and the population's nutrition and health status (Dahanayake, 2015).

Turkey

Turkey is a 'biodiversity superpower' containing 12,000 species of plants, with 10% estimated as being edible; so far 300 have been recorded as foods (Şekercioğlu et al., 2011; Güner et al., 2012; Republic of Turkey, 2014; Ertug, 2014). Particularly in rural areas, wild edible plants are collected and used for home consumption or sold in local markets, complementing people's diets and representing a source of additional income for foraging households (Dogan, 2012). Although the tradition is still widespread among older generations, the use of wild edibles is declining as young people migrate to urban areas and food collection from the wild is no longer possible or convenient (Tan et al., 2017). This shift is considered to be detrimental both for dietary and cultural diversity, as well as a missed economic opportunity for many families (Erem, 2015; Adanacioğlu et al., 2017; Karabak, 2017). In addition, increasing urbanization and overexploitation of natural resources are threatening the habitats where many edible species grow such as Mediterranean maquis, grasslands, coastal areas, wetlands, rivers, old growth forests, steppes and rangelands (Tan, 2009; Şekercioğlu et al., 2011).

Not only do wild edible species hold economic potential, but they can be used as a healthy alternative to imported and over-processed foods; this is particularly relevant as Turkish people increasingly suffer from dietary-related diseases (Erem, 2015; Candari et al., 2017). In 2016, obesity was recorded in 20% of the population above 15 years of age while more than 30% was overweight (Turkstat, 2017). A further concern is iron deficiency, with 31% of women of reproductive age recorded as iron-deficient (FAO, IFAD, UNICEF, WFP, WHO, 2019). Many wild edible plants are rich in macro and micronutrients and can help meet daily requirements of dietary fibre, vitamin C, iron, potassium and phosphorus (Sánchez-Mata and Tardío, 2016) (see Figure 5.A7).

4 The BFN Project approach

In a nutshell, the BFN Project approach demonstrates the contribution of nutritious biodiversity to diet diversification and to livelihood improvement (particularly in

	DESCRIPTION AND USE	NUTRITION
	A flowering plant in the mustard family with leaves that can be used in salads or cooked as vegetables and seed pods that can be used as a substitute for mustard seeds. The herb is also dried and used to make a tea that can be used as a natural remedy for diarrhea, and both internal and external bleeding.	Contains significant quantities of iron, beta-carotene, and potassium
	This colourful perennial's tender shoots, buds and leaves are cooked and used to prepare pastries, soups, stews and roasted dishes.	Source of beta-carotene, Vitamin A, and potassium
	A weedy annual plant with edible leaves and young shoots, and seeds that can be used to make porridge, cakes, and a fermented alcoholic beverage.	The seeds are high in calcium, vitamin A, potassium, and phosphorous.
	This small dioecious evergreen takes 2-3 years to form hard, fleshy fruits that, once ripe, are collected and turned into *pekmez*, a high-energy molasses	Contains potassium, calcium, phosphorous, magnesium, sodium, and zinc

FIGURE 5.A7 Highlight: selected BFN Turkey species

rural areas by linking farmers to markets), encouraging people to grow, use and safeguard these nutritious yet neglected species. These actions contribute to global biodiversity conservation as well as food system change.

The diet-related public health concerns in the four countries, accompanied by widespread environmental degradation and poverty described above, provided a rationale for the better integration of traditional food sources, which had largely untapped potential.

Nutritious local biodiversity offered a key to unlock benefits ranging from increased diversity in diets, improved livelihoods, healthier ecosystems and market development, calling for a dynamic approach that could take these interrelated issues into account. Elaborated on in Section C, the approach consists of three components:

Component 1: providing evidence

Research was key to the project, with teams engaging people across Brazil, Kenya, Sri Lanka and Turkey to implement community-based participatory approaches and community-based mapping of their agricultural landscape to document local biodiversity with nutritional value and its associated traditional knowledge. The extent to which that biodiversity is contributing to household diets, as well as barriers to current use and opportunities for greater awareness, promotion and utilization of biodiversity foods was also recorded, as were opportunities to better link farmers to markets for this diversity. Knowledge gaps relating to country-specific nutrition issues, food composition data, traditional knowledge and the benefits of using plant and animal species to tackle malnutrition were documented. With 188 species analysed across the four countries, the project has added a significant body of knowledge to existing information on the nutritional value of food biodiversity in food systems for human health and nutrition.

Component 2: influencing policy and markets

Information generated under component 1 was used to support the development of an enabling policy environment for the promotion of biodiversity for food and nutrition. In this component, it was essential to work with key national stakeholder groups and partners (e.g. consumers, producers, researchers and policymakers, Table 5.B1) who could positively influence the sectors relevant to biodiversity. These key actors developed planning, policy and regulatory instruments that encouraged the use of this diversity as part of national nutrition and food security programmes and strategies, as well as policies and actions aimed at enhancing the conservation and sustainable use of this diversity. The various cross-sectoral working groups and platforms set up within the project framework also used the information generated under component 1 to explore favourable marketing and trade of biodiversity for food and related products.

TABLE 5.B1 Stakeholders, their roles and involvement in the BFN Project

Stakeholders	*Type of involvement*
Decision-makers: • Project International Steering Committee • Project National Steering Committees	Make appropriate policy decisions and provide necessary guidance and advice to the project.
Ministries: • Environment and Forestry • Agriculture • Health • Education • Rural Development • Social Development • Other relevant ministries e.g. Finance and Planning	Representatives participate in project management and execution, and are invited to take part in project consultations, seminars, meetings and workshops for relevant training and ongoing awareness-raising and policy dialogues.
Scientific community (including academic and national institutions): • Relevant national research and development agencies/institutions (e.g. EMBRAPA in Brazil) • Botanical gardens • Universities in all partner countries • Relevant international universities • Centres of Excellence in biodiversity • National museums e.g. National Museum of Kenya (NMK) • National genebanks • Scientific and technology research councils • National federations (e.g. nutritionists) • Relevant global scientific and technical networks	Support the project by providing scientific and technical inputs, and collaboration in research and the development of methods and approaches. Invited to participate in BFN Project consultations, seminars, conferences and workshops, and to assist with development and delivery of training. Collaborate in the development of outreach materials and the dissemination and up-scaling of project outputs through peer-reviewed scientific publications.
Non-governmental organizations (NGOs) and civil society: • NGOs such as the Rural Outreach Programme in Kenya, The Green Movement and the Community Development Centre in Sri Lanka	Help facilitate consultations and collaborations with communities at pilot sites and assist in mobilizing participatory action research. Invited to take part in project consultations and meetings, and used extensively in the dissemination of outreach materials at the grassroots level.

(*Continued*)

TABLE 5.B1 (Cont.)

Stakeholders	*Type of involvement*
Local communities: • Community-based organizations (CBOs) • Women's groups • Youth Groups • Farmer Organizations and Groups	Involved in participatory appraisals and community-based activities to map nutritious agricultural biodiversity and to mobilize relevant biodiversity-based interventions. Have access to training and capacity building and other benefits arising through the project. Assist in the documentation of information. Involvement in activities pertaining to conservation and sustainable management of biodiversity for food and nutrition.

Component 3: awareness

Information generated as part of the project was used to improve the promotion, awareness and marketing of biodiversity for food and nutrition, and to ensure the replicability of results beyond the scope and boundaries of the project itself.

As sections B and C below illustrate, the implementation of this approach varied depending on country context. However, the holistic focus inherent in the approach ensured that a system-wide perspective guided all project activities, as partners engaged with research, policy and awareness-raising simultaneously and in accord with one another. For example, traditional knowledge regarding the health benefits of particular species were verified using food composition data, motivating decision-makers to support procurement policies for local crops. These policies could then influence activities such as national school meals programmes and pre-existing strategy and action plans, and would in turn be accompanied by training sessions, education initiatives and cultural events to improve producer and market capacity and facilitate effective implementation.

B Planning BFN

- This section covers the work that preceded project implementation: situational analysis, identification of stakeholders, development of partnerships and viable entry points.

Establishing an inclusive approach

Long-running, global multi-sectoral initiatives like the BFN Project, which ran from 2012 to 2019, encompass many sites, partners, disciplines and sectors. Therefore, success depends on careful and inclusive preparation, expectation management

FIGURE 5.B1 Photo of 'BFN Family' members during an International Steering Committee meeting, Turkey, 2018
Source: BFN Turkey

and well-established roles and responsibilities from the very outset. The early involvement of as many national stakeholders as possible in the planning process is important in ensuring heightened country ownership and buy-in from the many partners involved. For example, over the 18-month design phase (2009–2011) of the BFN Project, two global meetings enabled country partners to fully participate in the development of the logical framework and results framework – the main tools used to monitor and evaluate project progress. Partners proposed, negotiated and collectively agreed upon the project's objectives, outcomes, activities, budget allocations and reporting requirements, as well as roles and responsibilities. Participants included mainly senior policy and decision-makers from the national agencies responsible for implementing the project, the designated national project coordinators and potential international partners (see Figure 5.B1).

Pre-existing factors to consider

Contextual factors should be considered in each target country, from pre-existing scientific knowledge and capacity, to political will, governance, cooperation and mutual supportiveness between biodiversity and development priorities. Attention must be given to establishing what has already been accomplished in the target countries that might be built upon, such as which biodiversity species have been well researched, which have not, who is carrying out the research, and who to partner with, especially to better plan future interventions and optimize resources.

An example of an overlapping project comes from Sri Lanka, where the BFN Project partnered with another GEF-supported project ['Mainstreaming agrobiodiversity conservation and use in Sri Lankan agro-ecosystems for livelihoods and adaptation to climate change' (BACC)] that was being implemented in the country around the same period and at the same sites. Collaboration between the two projects ensured sharing of baseline data, thus saving time, resources and duplication of efforts; in addition to providing enhanced climate-resilience focus to the BFN Project.

Projects looking to promote the use of agricultural biodiversity though school gardens and school meal programmes might consider looking into and partnering with the health and education sectors that are carrying out activities such as gathering of child anthropometric data or setting up and maintaining school gardens in the study areas selected. BFN Brazil's initial strategy, which later grew to encompass the broader, federal picture, was to consider sites where schools were already included in the National School Meals Programme (PNAE) and in the Food Procurement Programme (PAA). Preference was given to schools located in rural areas where community-led cooperatives and/or producers' associations were active and registered with the National Plan for the Promotion of Socio-biodiversity Products (PNBSB). These community-led initiatives were ideally located near a federal or state university, to foster greater research collaboration with universities and contribute to graduate and post-graduate training.

KEY STEPS TO BIODIVERSITY MAINSTREAMING:

- Select sites and species on a basis of potential, using a pre-determined set of criteria
- Understand the context of each site through surveys, literature reviews and focus group discussions
- Supply factual evidence that shows the tangible benefits of biodiversity, especially for improved health and nutrition
- Identify nutrition and biodiversity outcomes (benefits attainable at various levels)
- Gain the backing of stakeholders and partners who can collaborate and support implementation
- Target promising sectors and policies
- Integrate biodiversity products and awareness into school feeding and procurement programmes
- Select activities best suited for effective implementation in the respective contexts
- Develop a wide-ranging awareness-raising strategy
- Identify methods for building capacity of markets, research institutes, etc.
- Evaluate project implementation and impact.

1 Situational analysis

- Background studies evaluated the status of relevant biodiversity and nutrition work, plus barriers that needed to be overcome.

Early in its preparatory phase, the BFN Project undertook a series of background studies, which indicated that some work in the area of biodiversity and nutrition had been undertaken in the participating countries with varying degrees of success. The number of potential species studied was limited in some countries, or the piecemeal and fragmented nature of these pre-existing initiatives translated into minimal impact and were largely disregarded by the wider development community. For example, Brazil's mammoth effort to catalogue and describe 674 native plants of potential nutritional and economic value under the *Plants for the Future Initiative* (Box 5.B4) was mostly limited to Brazilian research circles. What these initiatives did tend to suggest, however, was that there was considerable potential for both the promotion of local food biodiversity as a significant component for diversifying diets, as well as considerable potential for improved marketing of such biodiversity with a focus on improving nutrition.

The following are examples of background studies/activities carried out by the project to gain an initial baseline understanding of country context:

- Analysis of the state of the targeted food biodiversity in the selected ecosystems, including threats
- Literature survey on the nutritional content of cultivated and wild food biodiversity and identification of knowledge gaps in nutritional content information
- Assessment of the state of documentation of indigenous knowledge of food biodiversity
- Analysis of existing ethnobotanical methodologies
- Strengths, weaknesses, opportunities and threats (SWOT) analysis of current national, regional and global policy and legislative frameworks regulating the mainstreaming of biodiversity conservation into policies and strategies
- SWOT analysis of the current national, regional and global policy and legislative frameworks regarding the integration of nutrition and health incentives into sustainable conservation policies
- Analysis of market opportunities and obstacles for biodiversity-sourced food products
- A review of national nutritional status and diet quality, as well as existing poverty levels that an approach similar to the BFN Project could help address.

Where information might already exist

Much of this background information is readily accessible in national agricultural and biodiversity surveys, which are used for country reporting to major global conventions, such as the national reports to monitor progress on the implementation of the Aichi Biodiversity Targets of the Convention on Biological Diversity (CBD)[7] or country reporting to the Commission on Genetic Resources for Food and Agriculture (CGRFA) in preparation for its global assessments on the status of biodiversity[8]. To assess poverty levels and nutrition status, demographic and nutrition/health surveys are also useful to gather preliminary information, although the information can be outdated. For nutrition composition information, the first port of call would be national food composition tables (FCTs), databases that record the chemical composition, energy and nutrient yield of foods, and that provide consumers with nutritional information about the food they eat. However, national FCTs tend to contain data for mainstream crops and foods, and are more likely to reveal greater data gaps than provide information on potentially useful biodiversity for food and nutrition. Partnering with national universities and research agencies (such as in Brazil, Box 5.C3) can therefore be a useful strategy to gain access to grey literature such as MSc and PhD theses, and other information that may not be publicly available.

Gaps and barriers for improved use of biodiversity for food and nutrition

Background studies can reveal limited understanding of the potential benefits of using biodiversity for food and nutrition, and are useful to identify the gaps and barriers that hinder wider appreciation of such biodiversity in health, nutrition, agriculture and food security programmes and strategies. Figure 5.B2 illustrates

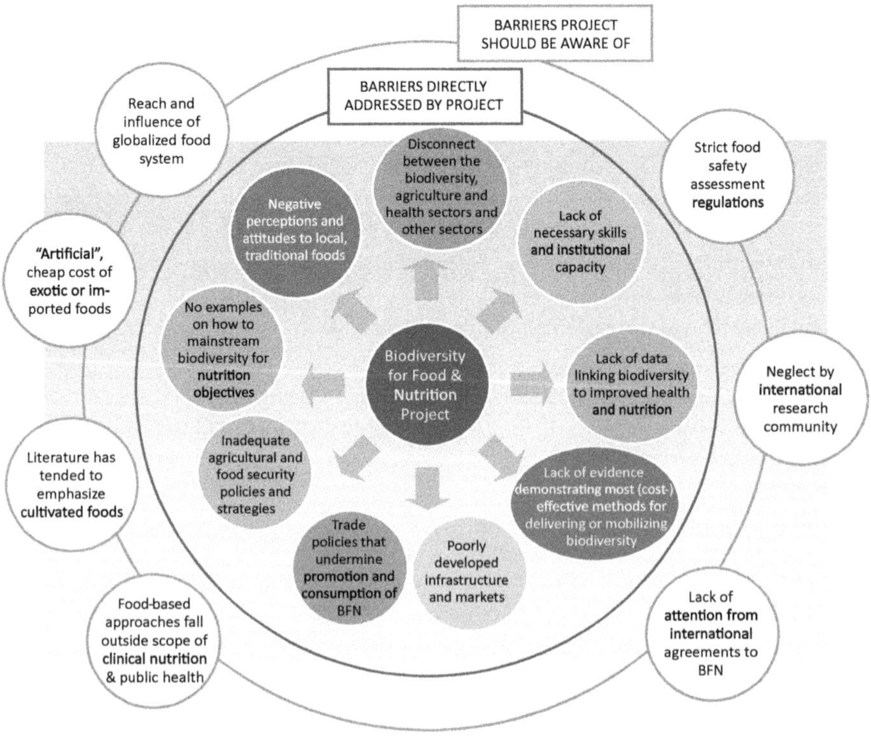

FIGURE 5.B2 Barriers to mainstreaming biodiversity for food and nutrition identified by the BFN Project

the gaps and barriers identified by the BFN Project to be considered during the project's planning and implementation phases.

- *Weak evidence of the benefits of biodiversity for food and nutrition, and absence of an integrated and accessible information management system.* Previous studies were unable to fully measure the impacts and outcomes of biodiversity in terms of its contribution to dietary diversity and nutrition status. Existing food composition data on biodiversity, along with other relevant information, was either fragmented or inaccessible to specific target groups, such as policy and decision-makers.
- *Poor partnerships and limited examples of effective cross-sectoral and multidisciplinary platforms to make the case for biodiversity for food and nutrition.* As noted earlier, a critical gap was the poor collaboration and coordination between relevant sectors (e.g. agriculture, health, environment, etc.). Few or no examples of cross-sectoral platforms existed in the participating countries, except Brazil, or at the global level for that matter.
- *General lack of enabling policy and regulatory frameworks for the promotion and sustainable use of biodiversity for food and nutrition.* Greater inclusion of biodiversity was not viewed as an effective intervention strategy for the improvement of dietary diversity, food and nutritional security strategies in all four countries.
- *Insufficient market infrastructure and development, and poor market information.* Although there were some good examples in participating countries, e.g. marketing of African leafy vegetables in Kenya, there were many gaps and barriers that needed to be overcome in all four countries to achieve broader marketing of biodiversity for food and nutrition.
- *Limited understanding of the most effective approaches to promote awareness and education on biodiversity for food and nutrition.* The background studies indicated that, despite the efforts to promote this diversity, limited attempts had been made to monitor and evaluate the impact and effectiveness of these approaches on behavioural change. This was seen as an essential component of implementation in order to understand what would and would not work in particular situations.
- *Lack of data on the effectiveness of possible delivery mechanisms for mobilizing biodiversity for food and nutrition.* While many initiatives at the country level referred to the use of home gardens and demonstration plots for mobilizing biodiversity, often in combination with other approaches such as marketing and education, few studies had compared the effectiveness of these approaches (e.g. in which situations they might work best; their limitations; or how they might be conveyed using mobile technology). There was also limited understanding of the feasibility of using school feeding programmes and other school-based activities such as gardens to promote this diversity.

- *Little understanding or knowledge of how to scale-up delivery mechanisms or to mobilize biodiversity on a much wider scale.* As a result of many of the above gaps and challenges, there was little understanding of the options for and effectiveness of promoting biodiversity for food and nutrition as a viable alternative to other 'silver bullet' options such as bio-fortification or vitamin supplements.

The gaps highlighted by the situational analysis were then used to better formulate and plan the project interventions.

BOX 5.B2 ADDRESSING A GAP: NUTRITIONAL INDICATORS

The situational analysis revealed an evident gap in the countries' capacity to compile national *Nutritional Indicators for Biodiversity in Food Composition and Consumption.* These indicators are a tool developed by FAO in collaboration with Bioversity International and other partners[9], to assess progress on the use of biodiversity and its role in nutrition and food security strategies. In simple terms, the exercise consists of counting and compiling information on biodiversity for food and nutrition available at the species, subspecies and variety/cultivar/breed level extracted from well-documented scientific literature, including national and regional food composition and consumption databases, and grey literature such as MSc and PhD theses. Reporting is then carried out via the FAO/INFOODS Regional Data Centre Coordinators[10] and uploaded into an open-source, global database. Often, relevant data is published in non-English reports and papers in national scholarly journals, never reaching the international scientific community and achieving its intended impact. Thus, as part of the project, global and national workshops were organized by FAO to strengthen national capacity to publish this national data and broaden global knowledge on available biodiversity for food and nutrition that might otherwise have been overlooked. The project's data on 188 species is possibly the single largest contribution to the FAO/INFOODS database.

2 Identification of stakeholders and partnership development

Interdisciplinary projects such as BFN depend on the inclusive involvement of stakeholders from a broad range of sectors, from grassroots to senior policy and decision-makers, from institutional to individual as well as NGOs, farmer and community groups, public institutions such as schools and others. Partnerships are crucial to the effective implementation and sustainability of any project and the strength of the partnerships significantly increases the likelihood of the project's impact outliving its implementation (Box 5.B3). To determine the best possible match for the project, a stakeholder mapping exercise was carried out during the planning phase.

BOX 5.B3 IN THE WORDS OF: CAMILA OLIVEIRA, BFN BRAZIL NATIONAL PROJECT MANAGER

When I first got in touch with the BFN Project, I was already working in the Ministry of the Environment for more than five years and many of the themes I was involved with were filled with conflicts with other agendas, sectors and ministries. I was absorbed by the BFN and it was a breath of fresh air, which created a big wave of optimism and a great working atmosphere. In Brazil, during the first meetings of the BFN, we were supposed to discuss and choose pilot sites for the implementation of the Project, but since the very beginning, some of the partners highlighted that a national scope would be of greater advantage, benefiting all the regions of the country and assisting a greater number of people. Also, many benefits could arise from the integration of efforts for mainstreaming biodiversity into the already existing federal initiatives linked to food and nutritional security in Brazil. The BFN Project was a perfect example of a win-win situation and it was always very easy to advocate for the theme and engage other institutions and biodiversity champions. We were promoting biodiversity conservation, better nutrition and health, and also helping to generate more income to poor communities. Who would complain? It is hard to explain the special atmosphere present during the partner meetings, especially with the universities and how they describe the BFN changed their lives. And all of this becomes even more special when we have the chance to taste the different and unique flavours of recipes developed by the BFN chefs based on our delicious native species. Being part of the BFN was a life changing, multi-level learning experience!

Stakeholder mapping

Guidelines were developed describing the types of stakeholders and actors the project should ideally engage. Once selected, the national executing agencies, together with Bioversity International, undertook extensive stakeholder consultations with promising partners and actors (at both the national and international levels) to explore roles, inputs and ways of creating added value and synergies. Some of the key stakeholders and partners identified came from those listed in Table 5.B1, where roles and responsibilities, level and type of involvement in project activities were defined.

The stakeholder mapping exercise resulted in the selection of relevant government ministries, research centres, non-governmental organizations (NGOs) and community-based organizations with a wide range of resources and skills to offer one another. A list of key project partners in Brazil, Kenya, Sri Lanka and Turkey can be found in Table 5.B2. For an in-depth review on developing and

TABLE 5.B2 The BFN Project's national partners in Brazil, Kenya, Sri Lanka and Turkey

Key partners	*Brazil*	*Kenya*	*Sri Lanka*	*Turkey*
Political partners	– Ministry of Environment (MMA) – The National Food and Nutrition Security Council (CONSEA) – Ministry for Agrarian Development (MDA) – Ministry of Social Development and Fight Against Hunger (MDS) – Ministry of Agriculture, Livestock and Food Supply (MAPA) – National Supply Company (CONAB) – Ministry of Education (MEC) – Ministry of Health (MS)	– Ministry of Health (MoH) – Division of Human Nutrition and Dietetics and National AIDS and STI Control Programme (NASCOP) – Ministry of Agriculture – Policy Department – National Environment Management Authority (NEMA)	– Ministry of Environment and Mahaweli Development (MOE) – Ministry of Agriculture (MOA) – Ministry of Health and Nutrition (MOHN) – Dept. of Health (Nutrition Coordination Division) and Medical Research Institute (MRI) – The Forest Department (FD) – Department of Animal Production and Health (DAPH) – Ministry of Rural Industries and Self Employment – Ministry of agricultural Development and Agrarian Services – Dept. of Agriculture (DOA) and Dept. of Export Agriculture (DEA)	– Ministry of Agriculture and Rural Affairs (MARA) – General Directorate of Agricultural Research (TAGEM) and of Protection and Control (KKGM) – Ministry of Environment and Forestry (MEF) – General Directorate of Nature Protection and National Parks – Ministry of Health General Directorate of Primary Health Care Services – Ministry of National Education (MEB) – Undersecretary of State Planning Organization

Research partners	– Brazilian Agricultural Research Corporation (EMBRAPA) – National Federation of Nutritionists (FNN) – São Paulo University (USP) – Campinas State University (UNICAMP) – Federal University of Pará (UFPA) – Federal University of Ceará (UFC) – Ceará State University (UECE) – Federal University of Goiás (UFG) – Federal University of São Paulo (UNIFESP) – Federal University of Rio Grande do Sul (UFRGS) – Mackenzie Presbyterian University	– KALRO Socio-Economics Division – KALRO-Headquarters Marketing Unit – National Genebank of Kenya (NGBK) – KALRO-Kakamega – Kenyatta University (KU)	– Bandaranayake Memorial Ayurvedic Research Institute (BMARI) – Dept. of National Botanic Gardens (DNBG) – University of Peradeniya, Faculty of Agriculture (FOAUP) – University of Ruhuna (Faculty of Agriculture) – Wayamba University – Dept. of Nutrition and Community Resources Management (NCRM) – Institute of Fundamental Studies (IFS)	– Aegean Agricultural Research Institute (ETAE) – West Mediterranean Agricultural Research Institute (BATEM) – Central Research Institute for Field Crops (TARM) – Bursa Food Control and Central Research Institute – Black Sea Agricultural Research Institute (KATAE) – The Scientific and Technological Research Council of Turkey (TÜBİTAK) – Erciyes University – Selçuk University – Gazi University – Ege University – Akdeniz University – Ankara University – Hacettepe University

(*Continued*)

TABLE 5.B2 (Cont.)

Key partners	*Brazil*	*Kenya*	*Sri Lanka*	*Turkey*
NGOs, other partners	– Brazilian Biodiversity Fund (FUNBIO	– National Museums of Kenya (NMK) – Rural Outreach Programme	– Biodiversity Secretariat (BDS) – Green Movement Sri Lanka (GMSL) – Community Development Centre (CDC) – Saaraketha Lanka Pvt. Ltd. – Sewalanka Foundation Agriculture project	– Turkish Association for the Conservation of Nature and Natural Resources (TTKD) – Union of Turkish Chambers of Agriculture – Association of Turkish Dieticians – İstanbul Commodity Exchange (İSTİB)

nurturing partnerships, we refer readers to Chapter 4 of a previous publication in this series, *Crop Wild Relatives: A Manual of In Situ Conservation* (Hunter and Heywood, 2011).

Building and sustaining long-term partnerships

The breadth and scope of these partnerships required constant attention, not only to build and maintain relationships, but to ensure that involvement was mutually beneficial. National champions (Boxes 5.B4, 5.B5) who can contribute local knowledge and expertise through their strategic positioning in each country are key for nurturing and sustaining existing partnerships and for reaching out to potential new partners.

BOX 5.B4 PARTNERING WITH PLANTS FOR THE FUTURE (BRAZIL)

Similarly to the BFN Project, *Plants for the Future* (PFF) seeks to increase knowledge and promote the use of native Brazilian flora, with the end goal of enabling commercial production. Coordinated by the Ministry of the Environment, PFF has involved over 500 research partners nationally since 2004. Focusing on species with actual or potential economic value, partners have gathered information on species' uses, management, production chains, regional importance, ecology and geographic distribution. Prioritization was based on assessments of traditional use, markets, extractive use/cropping, existing studies, endemism and multi-functionality (see following table).

Over PFF's lifetime, several of its selected species have grown in terms of use; for example, the açaí berry has become a popular 'superfood' beyond the Brazilian market (Coradin et al., 2011). The case of açaí in particular demonstrates the role of health claims in increasing products' value; therefore, it is important to develop reliable nutrition data while promoting these species. It is also vital to ensure that the pricing of the targeted species does not compromise accessibility. The 77 food species that were selected for BFN food composition research had been previously surveyed by PFF and shown to have high nutrition and economic potential.

Plants for the Future: stages of the Prioritization Process

Goals: Identify native plants of current or potential economic value, increase knowledge, promote more widespread use of Brazilian flora, diversify options available to different sectors in creating new opportunities for use and development of new products

Implementation Stage	*Level of involvement*	*Activities*
1. Administration	Ministry of Environment	Write baselines & Terms of Reference, obtain funding, call for and approve proposals
2. Mobilization	Regional Working Groups and Partners	Mobilize technical teams, establish partnerships with regional institutions, organize multi-disciplinary regional working groups, structure logistics for and systemize data and bibliographic collection, collect data in the field (fairs, farmer markets and distribution centres), organize technical meetings, prepare final discussion workshops
3. Compilation	Regional Working Groups	Systematize information on the species prioritized during the workshops, prepare preliminary portfolios and technical reports to send to Ministry of the Environment
4. Evaluation, review, sharing	National Coordination (Brasilia)	Conduct in-depth analysis of technical reports, following guidelines agreed on by national coordinator and regional coordinators, reassess prioritized species, approve final list of over 800 species (over 100 native food species) and organize authors to publish regional books

Plants for the Future's work represents a leap forwards for sharing information on regional biodiversity and its related traditional knowledge. One of the benefits for BFN Brazil, was the opportunity to have the support of PFF coordinator Lidio Coradin who as BFN Project Director in Brazil, played a key role in facilitating BFN Brazil's inclusive approach that focused attention to species' value chains, awareness, conservation, and the need for greater integration into diets (see Figure 5.B3).

FIGURE 5.B3 Lidio Coradin presenting a banner on healthy biodiversity, with Camila Oliveira in 2014

Source: BFN Brazil

In the words of: Lidio Coradin, Project Director of BFN Brazil/*Plants for the Future*:

> I grew up in the interior of the South Region of Brazil, in the middle of the subtropical forest, which was dominated by *Araucaria angustifolia*, today known as the famous 'pinhão'. The forest was, in fact, our garden, and I loved to eat 'pinhão'. All this made me learn, at a very early age, to value the forest, biodiversity, its products, flavors and aromas. I learned to love the forest and the importance of its sustainable use. This was what put me on the path to a career studying plants and biodiversity.
>
> Later, as a researcher from Embrapa, I had the opportunity to develop my first research project on the conservation and use of genetic resources, which allowed me to explore the country and its regions, and to cross the various Brazilian biomes and ecosystems. I was engaged in the gathering of plant germplasm of different groups of use, from tiny plants to large trees.
>
> The participation and coordination of dozens of national and international expeditions, in Brazil and abroad, coupled with my childhood experience, provided enormous knowledge that were the stimulus and key ingredients for the creation of the Plants for the Future Initiative. I always thought that the Brazilian population should get to know the Brazilian species, the possibilities and opportunities of their use, and that they should also have the right to rescue the regional traditions and the plants from the past. This experience and the engagement with Plants for the Future were fundamental for my involvement as the National Project Director of the BFN Project in Brazil.

BOX 5.B5 WHAT DOES A NATIONAL CHAMPION LOOK LIKE?

Dr. Victor Wasike, Kenya's national project coordinator from the Kenya Agricultural and Livestock Research Organization (KALRO) (Figure 5.B4), was instrumental in advocating the use of biodiversity in food-based interventions to tackle malnutrition within the country's Nutrition Inter-Agency Coordinating Committee (NICC), operating under the Scaling Up Nutrition (SUN) umbrella, to bring together stakeholders from government ministries, donors, UN agencies, civil society and business organizations, to link nutrition to agriculture. According to Dr. Wasike:

FIGURE 5.B4 Dr. Victor Wasike speaking at the BFN Symposium in Brasilia, 2017

> Participating in committees such as the NICC provided an excellent opportunity to showcase the work that the BFN Project was undertaking at the county and grassroots level to promote nutrient-rich biodiversity. It can also help significantly in efforts to mainstream biodiversity into national plans and action and the scaling up of local activities.

Dr. Wasike's coordinating abilities and familiarity with local government and the community were a significant asset in developing Busia County's Biodiversity Conservation Policy (covered in Section 5.C10), a major outcome of the project's second component.

As the project progressed and gained visibility, thanks to greater evidence of the benefits of biodiversity for food and nutrition, opportunities arose to broaden the partnership to novel, previously unthought-of stakeholder groups, such as like-minded initiatives and programmes, as well as public and private sector partners.

Other non-conventional partners in an initiative like the BFN Project might include schools (Box 5.C19) or vocational training colleges (referred to as 'green' education in Turkey, see Box 5.C22). In Kenya, the work undertaken to provide farmer groups with reliable and stable markets for African leafy vegetables opened up collaborative opportunities with local schools, which then became champions in their own right for the biodiversity for food and nutrition cause.

The ripple effect can include progressively greater engagement with the food industry (Chapter 8), chefs and the gastronomy sector in all countries (Boxes 5. C14; 5.C18; 5.C22) following rising interest in ethnic foods and new sources of nutrients, as well as the role of biodiversity in the global food systems' transformation, as discussed in Section A. In Brazil, project collaboration with chefs, federal and state universities and private sector actors has resulted in the publication of a new recipe book entitled *Biodiversidade Brasileira. Sabores e Aromas* (Brazilian Biodiversity. Tastes and flavours) containing 335 recipes using 64 nutrient-rich underutilized plant species from the six Brazilian regions. This encyclopaedic feat represents a huge contribution to the greater promotion of local species and the sustainable use of native biodiversity in general. Kenya, Sri Lanka and Turkey have been equally productive and have put together a massive body of knowledge in the area of sustainable gastronomy, developing new recipes and revitalizing old preparations rich in biodiversity and local food cultures, and making these underutilized species desirable again.

Finding space to nurture partnerships

Once partnerships have been established, clear mechanisms for participation, partnership building and effective communication are essential to ensure full inclusion of all relevant stakeholders. Spaces and time must be set up for partners to work together effectively, making sure all stakeholders are kept fully informed of project planning, implementation and progress, as well as ensuring they understand the importance of the project outputs. In the experience of the BFN Project, working closely with national partners during the implementation of activities ensured they took ownership of the project and were able to run activities when political unrest and other economic and environmental vulnerability and shocks arose. As any project implementer will no doubt be aware, during a multi-year project such as BFN, a multitude of problems are likely to arise both at the political (e.g. country elections, political unrest) and at the institutional level (e.g. staff turnover), but also at the environmental level (e.g. extreme climatic events); issues which of course are beyond the project's control, but which may slow down project progress.

Organizing the project

To facilitate global coordination, Bioversity International established a Global Project Management Unit (GPMU), which undertook coordinating activities, technical back-stopping, reporting and facilitating the timely exchange of information internally and externally, and managing the project's overall financial aspects. The GPMU also established reporting guidelines for all partners to ensure submission of quality reports to the donors. The GPMU was instrumental in using the BFN Project's successes and achievements to inform relevant global processes and initiatives including the CBD, FAO CGRFA, CFS and United Nations System Standing Committee on Nutrition (UNSCN).

The BFN Project also established an International Steering Committee (ISC), which met once a year, hosted in turn by each project country, composed of representatives from UNEP and FAO as implementing agencies, Bioversity International as the global executing agency, and representatives from the national executing agencies. These formal convenings were important to keep track of project implementation and served as an excellent adaptive management tool to review the feasibility of activities and outputs. The logical framework and work plan were used as monitoring and evaluation tools, and were discussed during each ISC meeting. Occasionally the wording of the global work plan and logframe were revised to ensure milestones and targets were measurable and realistically achievable within the given timeframe.

ISC meetings were also a convenient way for partners to share innovative approaches to reach project targets that could be adopted in another country if deemed appropriate. Another chance offered by these meetings, and one which is often underestimated, was the opportunity to establish long-term, meaningful and friendly relationships with project partners. Since the BFN Project's inception, the ISC meeting offered the host country a chance to showcase its work in detail and for country and global teams to bond with one another, particularly during the informal meetings, meals and field trips that were organized outside the more formal ISC setting. As mentioned, the ISC meetings were hosted in a different country each year, providing the opportunity for national partners to familiarize themselves with different food, culture and traditions. The very nature of the project was appealing in itself and opened up considerable opportunities for cross-cultural learning and exchange; furthermore, leisure time outside of the formal environment fostered enjoyment and positive collaboration, to the extent that the BFN Project soon became known as the 'BFN Family'. Opportunities were taken to celebrate wins and special occasions, such as a team's success in implementing a specific activity or global achievements, but also anniversaries and more personal events. Despite inevitable changes in national project staff throughout the project's lifetime, a feeling of camaraderie was established so that all newcomers immediately felt at ease and ready to give their utmost to achieve the project's outcomes and objectives.

Between one ISC meeting and the next, the Global Coordinator would often travel to meet with national steering committees to ensure individual country progress was on track, but also to foster trust and maintain the high commitment levels. When face-to-face interactions where not possible, clear and effective communication was maintained via online calls, so that all project partners understood what was expected of one another. In essence, a virtual open door policy was established, fostering trust and facilitating the flow of important feedback and information.

The GPMU was also very active in identifying national and regional opportunities where the national teams could present the project and its results to a broader audience, promoting the agriculture-nutrition narrative and innovative cross-sectoral perspectives. With a very small global team, this was doubly important in terms of multiplying the project's public engagement capacity in national, regional and international fora.

At the country level, project implementation was led and coordinated by a National Project Management Unit (NPMU). Of the four countries, only Brazil's NPMU was hosted by the Ministry of Environment. The remaining three were hosted by the Agricultural sector. The lead implementing ministries and organizations for the four countries were:

- Biodiversity Conservation Department, Biodiversity and Forestry Secretariat, Ministry of Environment, Brazil
- Kenya Agricultural and Livestock Research Organization (KALRO), Nairobi, Kenya
- Ministry of Environment through the Department of Agriculture, Sri Lanka
- General Directorate of Agricultural Research, Ministry of Agricultural and Rural Affairs, Ankara, Turkey.

The NPMUs consisted of the National Project Coordinator (NPC), a Project Assistant (PA) and thematic consultants (on a needs basis). The NPMU served as the critical link between the project pilot sites and district and national committees and the GPMU to ensure that lessons learned were shared among sites and within national committees and between countries, and to provide visibility of the project at the national and international level.

Each country, guided by the NPMU, established a National Steering Committee (NSC) and, where relevant, thematic and other committees. Each NSC consisted of representatives of major partners actively involved in project activities such as representatives from the National Executing Agency, Government agencies (Agriculture, Environment, Health and Natural Resources), private institutions, local institutions, NGOs, civil society organizations, and academia depending on the country's focus and needs. These formal implementation arrangements ensured a constant flow of information among the national partners and countries, as well as with the executing and implementing agencies at international level (see Figure 5.B5).

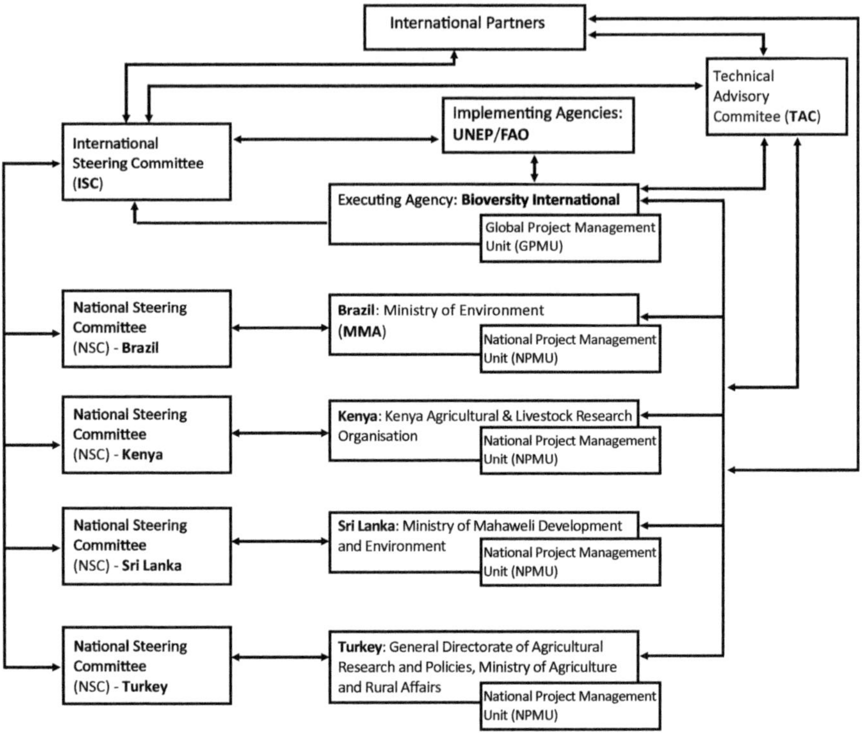

FIGURE 5.B5 Elaborated structure of the BFN Project

3 Identification of entry points

- Different strategic approaches were taken to find the most viable entry points for project interventions per country.
- The example of Brazil shows how the different components were adapted to the country's context.

Following the establishment of the formal structures of the ISC, NSC and general partnerships at the national and international level, the project was able to identify key opportunities for intervention across the key components of the project (establishing the evidence base; influencing policy; identifying markets; and creating awareness). Besides the partners identified in the previous section, the project continued to look across these thematic areas to identify useful experts who could contribute necessary skills and knowledge, as well as pre-existing policies (such as health interventions, National Biodiversity Strategy and Action Plans (NBSAPs), dietary guidelines) and programmes (such as school meals, marketing schemes, education initiatives) that could guarantee successful implementation. Identifying enabling policies can engender positive downstream effects that help to broaden the partnership and reach other useful and unplanned stakeholders and target groups. Naturally, starting points may differ in each country and two examples are offered (see Boxes 5.B6 and 5.B7), one on each side of the spectrum.

BOX 5.B6 BUILDING UPON A PRE-EXISTING FRAMEWORK (BRAZIL)

Brazil is probably the best example of how the BFN approach and its three components can be adapted to country context when pre-exiting platforms are in place.

The Zero Hunger Strategy, established by Brazil's federal government in 2003 and recognized internationally over the years as an innovative and successful multi-sectoral institutional framework addressing poverty and malnutrition (Chmielewska and Souza, 2011; Maluf et al., 2015; Von Grebmer et al., 2016), launched and strengthened a number of initiatives that increase access to food for the poorest Brazilians, while supporting small-scale and family farmers (Fraundorfer, 2013). BFN Brazil took advantage of the horizontal and cross-sectoral governance mechanisms already in place under the Zero Hunger umbrella and began by strategically targeting relevant public policies and instruments with potential to mainstream agricultural biodiversity:

- The Food Acquisition Program (PAA): pays 30% more for organic and agroecological food.
- The National School Meals Programme (PNAE): decreed in 2009 that at least 30% of food purchased must be from family farmers.
- The National Food and Nutrition Policy (PNAN): first policy to place the human right to adequate food into a food and nutrition security context.
- The Minimum Price Guarantee Policy for Biodiversity Products (PGPM-Bio): uses subsidies to establish fair prices for products of plant origin regardless of rural credit.

The next step was to provide a stronger evidence and knowledge base for native biodiversity that could increase its value within these programmes. Pre-existing portfolios from the *Plants for the Future Initiative* (Box 5.B4) enabled the BFN team to focus on filling previously identified data gaps by partnering with federal universities and national research agencies to carry out nutritional composition analysis of 70 edible plant species. Collaboration with universities, which host Collaboration Centers on Food and Nutrition (CECANEs), again provided a useful entry point to the PNAE, which is coordinated by the National Fund for Education Development of the Ministry of Education. These CECANEs provide research and technical backstopping to those involved with delivering the national school meals programme, which feeds over 40 million children on a daily basis. As Section C (Actions) shows, these pre-existing entry points all provided fruitful pathways to navigate a complicated national landscape with positive outcomes, including an official list of biodiverse species that could be used in institutional procurement, and the publication of a species database as well as recipe books and sustainable management manuals.

BOX 5.B7 EXPLORING POLICIES WHEN NO ENABLING PLATFORMS ARE IN PLACE (KENYA)

Prior to the project, Kenya lacked a cross-sectoral policy platform and an enabling policy environment for the promotion and sustainable utilization of biodiversity for food and nutrition (see Table 5.B3). To identify suitable policy entry points, the BFN Project therefore set out to:

1. Review biodiversity conservation, food and nutrition policies at the (i) national, (ii) county and (iii) institutional level to assess levels of coordination
2. Review legislation and institutional arrangements for implementing policies that could influence biodiversity conservation and sustainable use
3. Identify policy and institutional barriers and opportunities for mainstreaming biodiversity conservation and sustainable use
4. Suggest options for integrating biodiversity conservation and sustainable use into national, county and institutional policies.

TABLE 5.B3 Relevant policies existing in Kenya at the project outset

Parent Ministry	*Implementing Institution*	*Legislation*	*Relevant International Conventions*
Ministry of Environment & Mineral Resources	National Environment Management Authority (NEMA)	Environmental & Coordination Act 1999	Convention on Biological Diversity (CBD) – Main
	Kenya Wildlife Service	Wildlife Act, 1989, Cap. 376	CBD Convention on International Trade in Endangered Species (CITES)
	Kenya Forest Service	Forest Act, 2005	CBD
Ministry of Higher Education, Science & Technology	National Council for Science& Technology	Science & Technology Act, 1977, Cap. 250	
Ministry of Agriculture, Livestock & Fisheries	Fisheries Department	Fisheries Protection Act, 1977, Cap. 379	CBD

(Continued)

TABLE 5.B3 (Cont.)

Parent Ministry	*Implementing Institution*	*Legislation*	*Relevant International Conventions*
	Kenya Agricultural and Livestock Research Organization (KALRO)	Kenya Agricultural and Livestock Research Act No 17 of 2013	International Treaty for Plant Genetic Resources for Food and Agriculture (ITPGRFA), CBD
	Kenya Plant Health Inspectorate Service (KEPHIS)	Seed & Plant Variety Act, Cap. 326	International Union for the Protection of New Varieties of Plants (UPOV) World Trade Organization Trade-Related Aspects of Intellectual Property Rights (WTO-TRIPs) Agreement
Ministry of National Heritage	National Museums of Kenya (NMK)	National Museums & Heritage Act, 2006	CBD
Ministry of Industrialization	Kenya Industrial Property Institute (KIPI)	Industrial Property Act, 2001	WTO-TRIPs Agreement

The review of the national policy framework identified four policies with a bearing on biodiversity conservation:

- The Vision 2030 blueprint that recognizes value addition of biodiversity as a means of increasing rural household incomes
- The Agricultural Sector Development Strategy that advocates the marketing of biodiversity to improve farmers' livelihoods
- The National Food and Nutrition Security Policy that promotes increased sustainable production of diversified, affordable food that helps meet nutritional requirements
- The National Land Policy that recognizes the importance of land in the context of agricultural productivity, equity, environmental sustainability and conservation of traditional cultures.

Results showed that despite the presence of a wide range of policies and laws that regulate activities in the agriculture, environment, livestock, wildlife, forestry, fishery, plant protection, plant breeding, and trade and industry sectors, overlaps existed with regard to biodiversity conservation at the national and international level with multiple agencies acting as focal points to the main biodiversity conservation conventions (something that can probably be said for a great many countries). Furthermore, the study highlighted the narrow focus, disconnect and limited coordination targeting biodiversity conservation across these systems. Being confronted with this situation made it all the more urgent to establish a sound evidence and knowledge base for the importance of biodiversity for food and nutrition.

BOX 5.B8 IN THE WORDS OF: BFN COUNTRY PARTNERS, ON LESSONS LEARNED

> Assess budgetary limitations from the beginning to agree on realistic possible actions. Communicate with public stakeholders in accessible and relevant language: encourage questions such as, "How is this in my interest as a doctor, or mother, or...?" For example, make it possible for a pregnant mother to identify exactly how African Leafy Vegetables could help her and her baby's health. Translate official English into local languages.
>
> *Kenya*

> The needs of target beneficiaries must be a top priority, with considerations given to traditional food systems, nutrition, livelihoods, and links to employment.
>
> *Turkey*

Part C: BFN actions

This section tracks how the BFN Project approach was implemented in Brazil, Kenya, Sri Lanka and Turkey around the project's three components: providing evidence, influencing policy and markets, and raising awareness (introduced in A.4). General descriptions of actions taken are accompanied by highlights from the countries. Although these examples represent only a small selection of the many activities undertaken by each country, they are intended to illustrate the breadth and depth of the approach and provide readers with useful insights that can be adapted to other contexts (see Figure 5.C1).

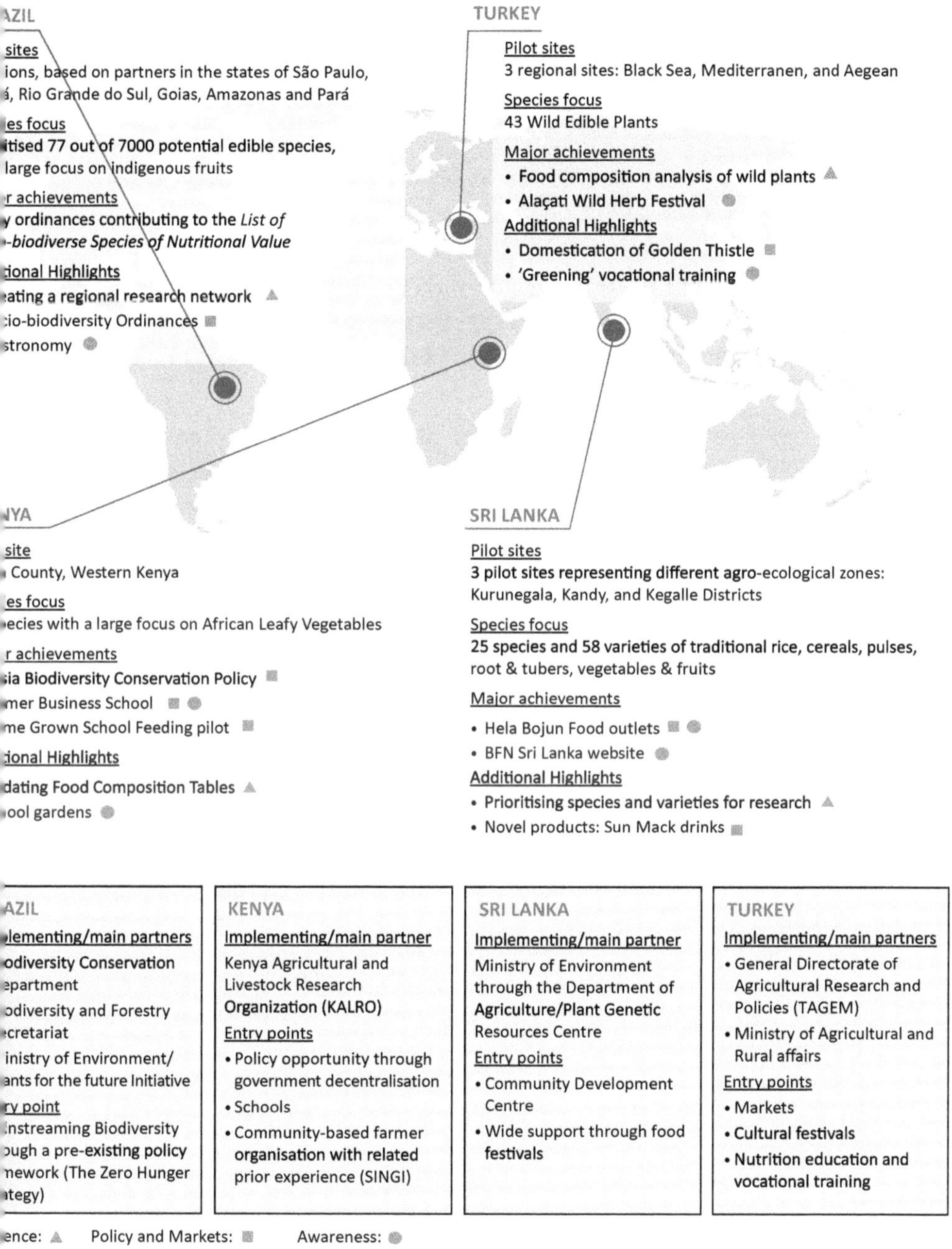

RE 5.C1 An overview of implementation in the four BFN countries
e: E. Gee and N. Lauridsen

1 Providing evidence

- Research partners brought together fragmented information and conducted new food composition analysis in order to establish a strong evidence base for neglected and underutilized species.
- Key elements in the research process were: surveys, species selection, sampling, analysis and sharing.

BOX 5.C1 IN THE WORDS OF: PROF. DR. ADRIANA CAMURÇA PONTES SIQUEIRA, FEDERAL UNIVERSITY OF CEARÁ

BFN had a great impact for all of those involved in several areas. Regarding scientific and popular knowledge, we got to know and study several species of great nutritional importance, which were being lost simply because their real nutritional value was unknown. Related to human resources, BFN helped with the development of several PhD, masters and graduate students. All of them have become multipliers of the value of our biodiversity. BFN was important for the local economy and social development of the region, since many products and recipes were created using the prioritized species and this enabled people from poor communities to prepare and commercialize these products, which present unparalleled flavor and great nutritional value.

As mentioned in Section B, the implementation of project activities was guided by a global work plan, which was developed in agreement with all major country stakeholders prior to the running of project activities. Central to all work plans is the definition of the direct, tangible and measurable products (i.e. outputs) that the activities will generate. For BFN, these are illustrated in Tables 5.C1, 5.C5 and 5.C7 for each component, helping the reader navigate the text and providing a list of potential knowledge products and/or activities to consider for better mainstreaming of biodiversity for food and nutrition.

Before we start describing the activities undertaken under the first component, linked to improving the evidence base for biodiversity for food and nutrition, it must be clarified that the species targeted by the project are often referred to as 'neglected and underutilized' because they are 'marginalized, if not entirely ignored, by researchers, breeders and policy makers' (Padulosi et al., 2013); mainly due to a lack of trustworthy information on many traditional and local food species. Many high-potential species remain virtually unknown, without clear consensus regarding their names, much less their health properties. For example, case studies conducted

TABLE 5.C1 Research component and outputs

Component 1
KNOWLEDGE BASE

Output 1.1:	**Output 1.2:**	**Output 1.3:**
Assessments of nutritional value of agrobiodiversity and associated traditional knowledge of prioritised species is carried out in 3 ecosystems in Turkey and Sri Lanka, 1 ecosystem in Kenya and at national level in Brazil	National portal on local foods, containing databases on nutritional properties of agrobiodiversity and associated traditional knowledge, developed in each country relying on pre-existing infrastructure and linked to relevant national and global nutritional databases	Information generated by the project contributes to global knowledge generation and is reflected in an increase of the Nutritional Indicators for Biodiversity on food composition and consumption

by FAO and collaborators of indigenous peoples in rural Asia, prior to the BFN Project, found that of 716 traditional food species reported by the communities, 93 still required original scientific identification, and for approximately 147 species there was not even the most basic nutrient data on record (Kuhnlein, 2003). By building on traditional knowledge with verifiable nutritional data, BFN Project research intended to address this gap. Beyond collecting and verifying pre-existing information, it was necessary to conduct food composition analysis for the first time on many selected species. Each country developed its own research approach around the following elements:

Surveys

- On-site assessments helped researchers understand species' availability and use.

Ethnobotanical approaches, which bring together social, nutrition, medical and botanical sciences, were used by the countries to understand the range of plants in a community or area, their uses and importance in a social, cultural and economic context. Local knowledge, land tenure and gender issues are important players in this type of survey, as are issues of seasonality when dealing with wild plants, edible fungi and even edible insects. Additionally, issues of labour, economics, health and nutrition, crop diversity and productivity are important in more intensively managed agricultural systems.

A review of ethnobotanical methods and approaches undertaken by Bioversity International was shared with partners at the beginning of the project. The review included information on plant specimen collection, rapid assessment tools of available plant resources – such as market visits and surveys – and participatory rural appraisal tools such as prioritizing and ranking, participatory mapping, and seasonal calendars. Many of these methods to assess agrobiodiversity have now been compiled by the Platform for Agrobiodiversity Research (PAR) in a recently-released compendium[11] and online resource[12] that draws on experiences from around the world. Created to support the documentation, co-creation and sharing of knowledge about diversity and its management, the compendium provides guidelines for the collection and analysis of data about the diversity of crops, livestock, pollinators and harvested wild plants. Another useful resource to draw from is the manual *Improving Nutrition with Agricultural Diversity*, published by Bioversity International in 2011.

Where possible, countries selected their target sites to capture the widest range of climatic, environmental and cultural conditions. For example, Sri Lanka selected three ecologically-diverse villages to represent the wet/dry zones of the island, while Turkey chose three distinct regions where different wild edible plants have traditionally been harvested and consumed. Additional considerations for site selection included:

- incidence of nutritional-related problems
- levels of poverty and food insecurity
- quality of diets
- available local biodiversity
- infrastructure and capacity at site
- willingness of communities at site to collaborate with the project
- level of ecosystem degradation/biodiversity loss; and cultural diversity.

Examples of surveys that were undertaken include:

- Eco-geographic: conducted by specialized agronomists at local markets, restaurants, supermarkets, villages, and through on-site natural flora visits
- Ethnobotanical: face-to-face interviews in villages and cities, with direct field observations in local restaurants, cross-checked with informant consensus, group discussions and questionnaires
- Socio-economic: monographic research questionnaires developed by regional coordinators, with input from university faculties' departments of agricultural economy and the Ministry of Health.

Results from these surveys helped fill gaps emerging from the background studies (see B.1 Situational analysis), which were usually conducted by students or research consultants. Together, these tools helped paint a picture of each site's status of agrobiodiversity, associated traditional knowledge, dietary diversity, impact of the food production system on food security at the household level, and socio-economic status.

Species selection

- Criteria were developed to assess the potential of the shortlisted species, which were then ranked and prioritized.

Information from the surveys was then used to narrow down a list of species and varieties with the highest mainstreaming potential. To achieve a balanced selection, numerous factors had to be taken into account such as economic and market value, practical use, cultural acceptance, ease of cultivation and dietary benefits. To avoid reinventing the wheel (and optimizing resources), it is worth spending time to explore whether one can build on pre-existing or parallel initiatives that have developed a list of species of potential nutritional and/or economic value, as done by Brazil, for example, when making use of the prior *Plants for the Future* inventory (see Box 5.B4).

With no such list in place in Turkey, researchers had to start from scratch and compiled a list of 43 more commonly consumed wild edible species based on the results of market surveys and interviews with 2,334 local wild plant gatherers, sellers and consumers across the three pilot sites. In addition, a custom-made sustainability index (Table 5.C2) was developed to select three species for value chain upgrading and market development. Each of the shortlisted 43 species was

TABLE 5.C2 Criteria used by BFN Turkey to assess potential species for value chain upgrading and market development (Sustainability Index)

Environmental	*Economic*	*Food & nutrition*
1. **Conservation** • *ex situ*, *in situ*/on farm 2. **Cultivation** • Ease of production • Growth rate • High adaptability • Vegetation period • Annual growth 3. **Disappearance/threat** • Habitat destruction and fragmentation • Pollution, exploitation • Destructive practices 4. **Widespread distribution** 5. **Habitat preferences**	1. **Collection/production continuity** • Collection/production constraints • Distance from collection/production site • Collection/production continuity 2. **Market characteristics** • Recognizable • Easily packed • Suitable for storage 3. **Processing industry available** 4. **Marketing opportunities of collected/produced species** 5. **Distance from market (km)**	1. **Iron content** 2. **Calcium content** 3. **Fibre ratio** 4. **Antioxidant ratio** 5. **Vitamin A**

numerically ranked according to the listed environmental, economic, and food and nutrition criteria, and einkorn wheat (*Triticum monococcum*), golden thistle (*Scolymus hispanicus*) and foxtail lily (*Eremurus spectabilis*) selected for the Black Sea, Aegean and Mediterranean regions respectively.

Sampling, transport and preparation for lab analysis

- Procedures were developed for consistent collection of species from the field and in the wild.

Once species had been selected, countries undertook an extensive literature review to compile all available information on the species' nutrient and anti-nutrient components. Food composition analysis, described in the next section, is a costly exercise. Standard analytical laboratory tests for food nutrition labelling (including calories, protein, carbohydrate, calories from total fat, dietary fibre, sugars, total fat, saturated fat, trans fat, sodium, cholesterol, vitamin A, vitamin C, calcium and iron) can cost up to US$1,000 for a single species and it stands to reason, especially when resources are limited, to try and obtain as much information as possible from existing sources, such as food composition tables and scientific literature, prior to embarking on expensive lab tests. However, when information is missing, unreliable or incomplete, countries will need to collect and prepare samples for laboratory analysis. For the data to be accurate, reliable and useful to others – particularly for inclusion in global databases such as FAO/INFOODS (Box 5B.2) – sampling and testing should be carried out using standard methodologies and protocols as well as certified laboratories.

BFN Project sampling was undertaken following FAO methodologies and guidelines (Greenfield and Southgate, 2003). Standardized protocols were developed in each country and staff trained in collecting (quantity, batches to capture variation, size and property of samples), storing (temperature regulation) and transferring (packaging, means of transport) plant samples to laboratories. Collectors in the field or wild sites completed a sampling form that recorded food names, sampling region, origin of sample, date and other relevant details.

In Brazil, samples were collected directly from smallholder farmers, informal and/or municipal markets or from the wild. For each fruit, three independent samples were collected (from different cities, markets region, depending on availability and seasonality). Each sample consisted of a minimum of ten units and was analysed in triplicates to obtain nine sub-samples. Following collection, the foods were transported to the lab, identified by a professional taxonomist, sorted, washed and processed to obtain an edible portion for analysis. Examples of data recorded at each stage of the protocol are presented in Tables 5.C3 and 5.C4.

TABLE 5.C3 Data recorded during sample collection/purchase

Common name	*Scientific name (with description of variety/below species level)*	*Collection date/Date of purchase*	*Place of collection/Purchase (e.g. municipality, shop, producer)*	*Quantity (kg)*	*Level of ripeness*	*Place of cultivation (if possible with GPS location)*	*Date of collection by producer/farmer*

TABLE 5.C4 Data recorded during transport and processing in preparation for lab analysis

Common name	*Transport*				*Processing*
Scientific name (with description of variety/below species level)	*Transport time (hrs)*	*Temperature (refrigerated or not)*	*Form of packaging*	*Processing description*[1]	*Yield (weight prior and after processing)*

1 Record the different processing stages for each species (e.g. washing, peeling, pitting/stoning, removing seed etc.). Analysis should then be carried out on the edible portion.

Analysis

- Food composition analysis took place in different labs depending on the capacity of each country.

Analysis represented both the BFN Project's most significant research contribution, and a momentous challenge to implement. Aside from the immediate cost associated with obtaining reliable data, food composition analysis requires lab capacity beyond what may be readily available in most countries. Some of these challenges are described by Greenfield and Southgate (2003):

> The use of some of the most advanced methods may require substantial capital investment to provide the necessary instrumentation. Considerable resources are also required in the form of trained staff to operate and maintain the instrumentation.

And again,

> Laboratories in developing countries may lack funds for capital outlay (especially as foreign currency) and lack the resources for the specialized maintenance and supplies necessary for high-technology instrumentation. On the other hand, local funds may be available for technical staff with the necessary background for carrying out non-instrumental methods that provide valid data.

Countries would therefore need to decide whether to:

1) Carry out in-country sampling and dispatch the samples to a central laboratory (outside the country) for testing
2) Build on and develop in-house capacity to undertake the testing and invest in appropriate instrumentation

The decision will very much depend on specific situations and context, but the BFN countries chose the second option, coming up with creative solutions to optimize limited resources. Brazil and Sri Lanka, for example, used the nutritional analysis work to create research networks, and expand the capacity-building element of the project (by work being implemented in more than one central location). Both countries partnered with universities and engaged students to analyse nutrition data and the outreach and capacity building achieved through these partnerships is impressive (for Brazil, see Box 5.C1, Box 5.C2 and Box 5.C3).

Creative solutions to deal with limited resources and capacity included making the most of South-South cooperation opportunities offered by the BFN platform. Kenya, which started the project on a much lower budget than other countries (see Box 5.A2 on GEF), decided to invest its limited resources on building capacity to collect target species, then shipped its samples to Turkey, where they were analysed at a subsidized cost in certified laboratories managed by BFN Turkey partners.

It is arguable that if the analyses had been carried out at a central laboratory for all countries, nutrition data would have been obtained more quickly and implementation delays might have been avoided. Sri Lanka, for example, took two years to produce nutrition data which was then used in the identification of target species. However, the adoption of a centralized approach might have generated logistical challenges with regards to transport, sample quality and integrity, while reducing the capacity building and partnership opportunities as well as negatively influencing country buy-in and ownership, which is likely to translate into or contribute to sustainability of project activities beyond project closure.

BOX 5.C2 IN THE WORDS OF: DANIELA MOURA OLIVEIRA BELTRAME, BFN BRAZIL NATIONAL PROJECT COORDINATOR

'I spent my childhood in a small city in the middle of the Amazon, and grew up surrounded by fruits with strange names and strong flavours, such as açaí, cupuaçu, pitanga, buriti, and cajá. Later on, I moved to the Southeast region of Brazil to study Nutrition, and got a Masters and a PhD degree in Food Sciences. During my studies, I only heard of or ate most of those fruits when I came back home to visit my parents. It wasn't until 2013, when I joined the BFN Project, that I realised how "absent" they were from nutrition books, scientific papers, food composition tables, university classrooms, and from the day-to-day work of young nutritionists who, like me, had no idea of how nutritious these foods can be.

The narrative of the BFN Project instantly made a lot of sense to me: it is necessary to generate knowledge about our native foods, publicize and use our biodiversity in order to promote its conservation and, in turn, these foods can improve our nutrition, the sustainability of our food systems and the livelihoods of our food producers, especially family farmers and traditional communities who still consume and safeguard the knowledge about these native species.'

Sharing

- Evidence was made widely available through databases and other knowledge-sharing platforms.

The four countries ultimately compiled and shared data on 188 species and varieties. National portals and databases were created and hosted by relevant government ministry websites for long-term dissemination of information.

Information sharing and data exchange is another delicate area, with many projects and programmes often experiencing challenges. Prior to implementation, efforts should be made to establish data sharing agreements with national and international project partners to avoid problems such as refusal or reluctance to share, or partners using information from other partners without permission or acknowledgement.

An important strategy of the BFN Project was to address these concerns by sensitising partners to these issues and ensuring that the project would not misuse, appropriate or distribute information which might be of a sensitive nature without the prior informed consent of key partners and full disclosure and acknowledgement of information sources. The process was guided by Bioversity, UNEP, FAO and collaborating partners, which have extensive experience in working with data sharing and exchange agreements. An equally important approach was to promote community sensitization and empowerment activities at the outset of project implementation, to help shape ownership of relevant activities and outputs and an understanding and appreciation that with the project came a decision to share and exchange information. Where necessary, prior informed consent and data sharing agreements were signed.

Once obtained, generating this information and making it available in appropriate ways and in different languages is fundamental to the task of increasing demand for biodiversity for food and nutrition, as a means of achieving the conservation, and mainstreaming of biodiversity into a wide range of sectors, programmes, strategies and activities covering health and food security. Providing an updated and reliable data source that includes local and traditional foods can provide government agencies (e.g. state and municipal authorities) with guidance for setting up and implementing locally relevant public policies and actions that benefit local communities while providing income for family farms. In Brazil, for example, special curriculum guidelines exist for Quilombola[13] schools. The guidelines provide schools in Quilombola areas with 50% more funds than other schools for the purchase of school meals that take in account the ethno-cultural background of their students. Regrettably, surveys undertaken by the project between 2014 and 2015 showed that activities focusing on Quilombola culture were few and poorly integrated within the school curriculum and that Quilombola food culture was rarely taken into account in school meal planning and preparation, partly due to the lack of information on these foods. The database and associated preparation methods documented by the BFN Project for some of these 'forgotten foods' are helping expert nutritionists to incorporate traditional Quilombola foods into school menus while awareness-raising activities are ongoing to revive interest in African and Quilombola history and culture (Beltrame et al., 2016, Box 5.C3).

While national databases are important for reaching out to country government agencies, research and development agencies, and the wider audience, it is important to ensure that collected data is also stored in a central database. Long-term costs associated with website and portal maintenance beyond project end are seldom factored in by countries in their budgets, and it can happen that a national portal containing crucial information is suddenly shut down due to lack of funds. To pre-empt this, countries also contributed to the FAO/INFOODs database, the international repository for reliable, referenceable food composition and nutrient data (see also Box 5.B2 and Box 5.C4). These databases validate the nutrition potential of BFN species for a wide audience as they are often referred to by government agencies, nutrition scientists and educators, health and agriculture professionals, policy makers and planners, food producers/processors/retailers and consumers. A tailor-made central database was also created on the BFN Global portal (www.b4fn.org/resources/species-database/) for the 188 species.

Numerous scientific publications and presentations also highlighted the evidence base for the wider research community (See Box 5.C7: Publishing research for development). By placing renewed emphasis on dietary diversity, the food composition results substantially strengthened the link between the fields of biodiversity and nutrition. Research papers shared this data with multiple disciplines including nutrition and environmental sciences.

Additional evidence on traditional knowledge, culinary uses and cultivation techniques was shared through online platforms and manuals developed by the project (see also part 3 of this section on raising awareness). For research case studies per country, see Box 5.C3 (Brazil); Box 5.C4 (Kenya); Box 5.C5 (Sri Lanka), and Box 5.C6 (Turkey).

BOX 5.C3 CREATING A RESEARCH NETWORK (BRAZIL)

In Brazil, food composition analysis was carried out in partnership with public universities and research institutes under the guidance of the Ministry of Education. The universities were selected based on their links to the National School Feeding Program (PNAE) as Collaborating Centres on Food and Nutrition (CECANES), meaning that they already provided research and technical assistance to managers, nutritionists and cooks responsible for providing school meals. The BFN Project organized an introductory workshop in Brasilia that engaged interested professors from across the country and encouraged them to submit proposals for formal partnerships. Partners entered into a contractual agreement to compile and generate composition data, develop recipes and insert target species, and the concept of food biodiversity for sustainability, into CECANES and school feeding activities. Subsequent regional partnerships were formed in order to: (i) build capacity by linking universities without adequate laboratory facilities to local research centres; and (ii) increase the reach of research activities, along with community extension and interactions with state food councils.

Within this decentralized approach (see Figure 5.C2), BFN involved over 100 professors, students and researchers. Masters and doctorate students conducted the initial data and literature review, using FAO/INFOODS methodology (FAO/INFOODS, 2012) to identify gaps, such as the lack of reliable data on dietary fibre, vitamins and minerals for most fruit species. BFN provided scholarships and training to students in each region to support them over the course of the 1–2-year compilation work. Following this, sample collection and lab analysis engaged a range of universities and centres in each region, establishing 'Regional Centres for Food Composition Data' that acted as awareness and capacity multipliers within education and research institutions. Universities brought in other departments and gastronomic experts to gather additional information about recipes, production and traditional knowledge.

Traditional knowledge

To gather traditional knowledge, researchers administered questionnaires to a range of indigenous and traditional communities. Many of these groups are valuable knowledge holders but remain marginalized and underrepresented. For example, the 'Quilombolas' are descendants of Africans, typically former slaves, who established settlements in isolated forest areas and did not receive legal land rights until 1988. The Federal University of Goiás administered questionnaires to representatives of 20 Quilombola communities in Brazil's Centre-West region. Results showed the production of 17 priority fruit species for personal consumption across the communities, including pequi, mangaba and barú at each site. Despite 76% of respondents reporting having

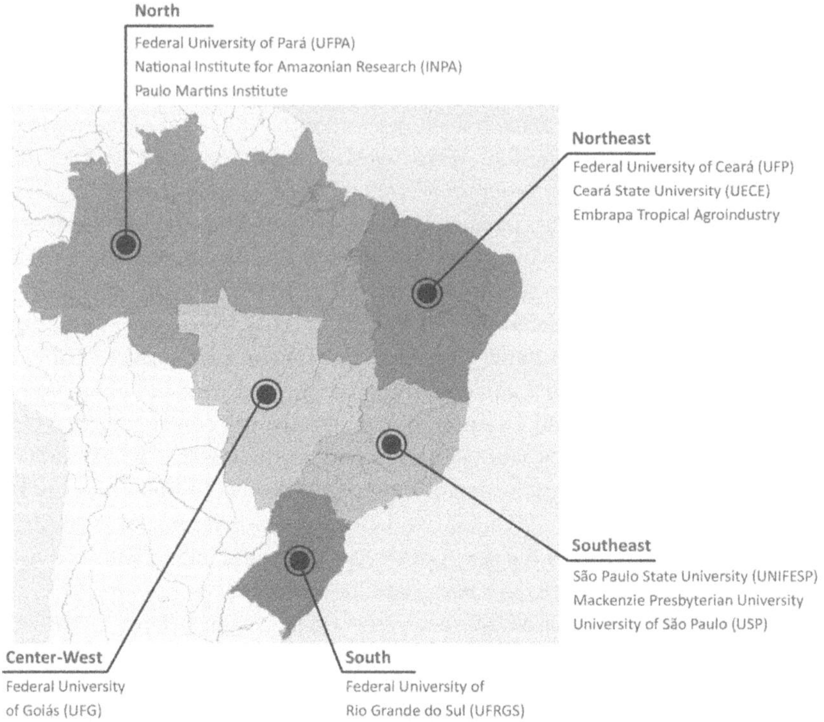

FIGURE 5.C2 A map of Brazil's regional research network of universities and research centres

eaten these fruits during their childhoods, all reported decreased consumption, especially amongst the younger generations. University researchers focused on one community (Pombal) to collect recipes, folkloric, and traditional knowledge on target species (Beltrame et al., 2016). This information was used to develop a recipe book on traditional Quilombola cuisine. Microbial food safety of fresh fruit and frozen pulps produced by the community was also assessed in order to develop guidelines and training workshops for food manufacturing processes.

The SiBBr database

When data analysis was complete, the universities and research centres shared their information on a food composition database hosted by the Information System on Brazilian Biodiversity (SiBBr), which is freely accessible online and hosted by the Ministry of Science, Technology and Innovation. There is now quantifiable data available for over 78 species and 180 local and regional foods (this includes different parts of the same species such as pulp, peel and seeds). The results have brought the nutrition and health potential of Brazil's native biodiversity to the attention of policy and decision-makers and offer irrefutable evidence that many native food species should be included in school feeding and procurement programmes, as well as other similar opportunities.

BOX 5.C4 UPDATING NATIONAL FOOD COMPOSITION TABLES (KENYA)

Before the BFN Project, the information that existed regarding the nutrition content of native Kenyan biodiversity was fragmented and out of date. Kenya's Food Composition Table (FCT), a central reference for nutrition-sensitive activities, had not been updated since its creation in 1993. Over 6 years, BFN partners directly contributed key information from nutrition data holders in Busia County, and composition analysis for four varieties each of Bambara groundnut (*Vigna subterranea*) and finger millet (*Eleusine coracana*). Additional funding made it possible to collect samples for and conduct analysis of sorghum (*Sorghum bicolor*), green gram (*Vigna mungo*), spider plant (*Cleome gynandra*), vine spinach (*Basella alba*), ethiopian kale (*Brassica carinata*), amaranth (*Amaranthus dubius*), jack fruit (*Artocarpus heterophyllus*), guava (*Psidium guajava*) and tilapia (*Tilapia* sp.). This research was overseen by a committee and working groups coordinated by KALRO and the BFN National Project Coordinator.

The result was an updated **Kenyan Food Composition Table**, published in 2018 by the Kenyan Ministries of Health and Agriculture and Irrigation with the support of FAO, which includes data on 522 foods including target species of the BFN Project. The preface states the importance of indigenous biodiversity: 'Understanding the quality of locally available foods is important in the development of strategies for combating food and nutrition insecurity' (FAO/Government of Kenya, 2018). Besides providing a unified national-level platform for species data that follows international standards, the FCT is intended to inform ongoing and future food and health-related programmes and policies.

The updated FCT was released online and accompanied by the **Kenya Recipe Book**, which contains 142 common recipes. Recipes were selected based primarily on frequency of use and ingredient availability corresponding to the FCT, then verified by focus groups consisting of 54 women from 20 different communities. The book includes details on preparation, cooking process and time, servings and nutrition per 100g of the dish, accompanied by a photograph for each dish. The book aims to facilitate inclusion of more nutritious ingredients in diversified everyday recipes.

The process for updating Kenya's Food Composition Table:

1. Established a participatory list of foods and nutrients for which compositional data was to be sought including raw foods, cooked foods, recipes, processed foods (including fortified foods) and beverages.
2. Compiled analytical data within the country (scientific literature, universities, labs etc.).

3. Updated knowledge through the FAO/INFOODS Food Composition Study Guide and/or a training course.
4. Compiled the data using the FAO/INFOODS Compilation Tool.
5. Completed the dataset.
6. Quality check overseen by FAO Rome.
7. Published the database/table including on INFOODS website as a PDF and Excel file.

See also:

FAO/Government of Kenya. 2018. Kenyan Food Recipes. A recipe book of common mixed dishes with nutrient value. Nairobi, FAO. 349 pp. www.fao.org/3/I8897EN/I8897en.pdf

FAO/Government of Kenya. 2018. Kenya Food Composition Tables. Nairobi, 254 pp. www.fao.org/3/I9120EN/i9120en.pdf

BOX 5.C5 PRIORITIZING SPECIES AND VARIETIES FOR RESEARCH (SRI LANKA)

Once Sri Lanka's pilot sites had been selected and surveyed, it was possible to prioritize species for further study. Members of the BFN Sri Lanka National Steering Committee contributed to a short list that considered the full range of cereals, pulses, vegetables, fruits and leafy vegetables, as well as their potential nutrient value, correlation with traditional knowledge and absence of prior research. Ultimately, 58 cultivars/varieties of 28 species were selected, enough to represent a complex and varied diet, though still only a fraction of native biodiversity.

Although during initial training FAO had suggested sending samples to the National Institute of Health in India, the BFN NPC suggested using the exercise as an opportunity to improve Sri Lankan research capacity. Food composition analysis was conducted through the joint collaboration of food scientists of the Department of Agriculture, University of Peradeniya, University of Wayamba, Industrial Technology Institute and Plant Genetic Resource Centre. This was a major achievement in building research capacity for the future, and was supplemented by evidence gathered through food festivals, fruit exhibitions and other outreach events which created a wider platform for both eliciting and disseminating crop information.

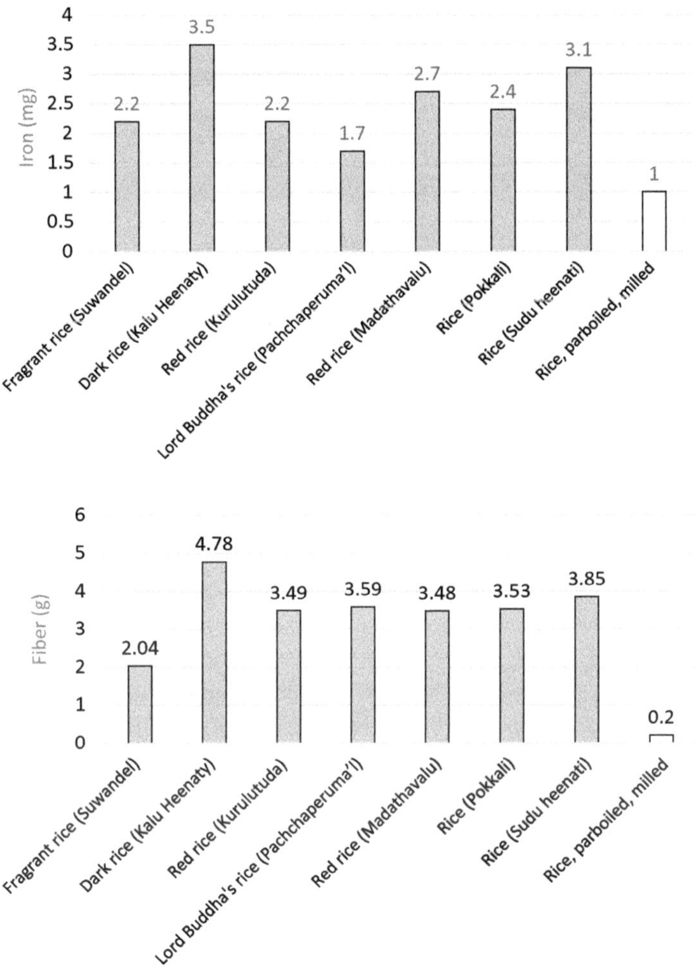

FIGURE 5.C3.1 Food composition analysis results comparing Sri Lankan varieties to commonly consumed white rice

Source: BFN Sri Lanka (Hunter et al., 2019)

FIGURE 5.C3.2 Samples of Sri Lankan rice varieties, 2014
Source: S. Landersz

Across the four countries, Sri Lanka had the strongest focus on intra-species diversity, with food composition data collected on multiple local varieties of fruit, tubers and grains. Most notably, seven traditional varieties of Sri Lankan rice were analysed. These results were graphically represented (Figures 5.C3.1 and 5.C3.2) and compared to milled white rice, which is a household staple in most Sri Lankan meals. The data revealed that consumption of these varieties could supply significantly higher quantities of iron (up to three times as much) and fibre (0.2g/serving from white rice vs. 2.04–4.78g/serving from native varieties). This data demonstrated the importance of diversified consumption of multiple varieties for higher micronutrient intake.

BOX 5.C6 RESEARCHING WILD EDIBLES (TURKEY)

As noted previously, Turkey is notable for its ongoing tradition of collecting and consuming wild edible plant species, especially in rural lower-income regions. Nutritious varieties such as foxtail lily and shepherd's purse are part of regional diets but there was little prior data on their nutrition (see Figure 5.A7). To make a selection of priority species, Turkish researchers used a sustainability index with multiple criteria (Table 5.C2). Standardized procedures were then used to collect and rapidly transport samples to research centres for laboratory analysis. Some results are shown below in Figure 5.C4. The final data was made publicly available in the Turkomp National Food Composition Database. In order to ensure sustainable collection, best practices for the management of wild and semi-wild species were developed and then disseminated to foragers and producers (this was also done in Brazil regarding wild-growing fruit trees). Domestication was also explored as an option, for example with golden thistle (covered in Box 5.C16).

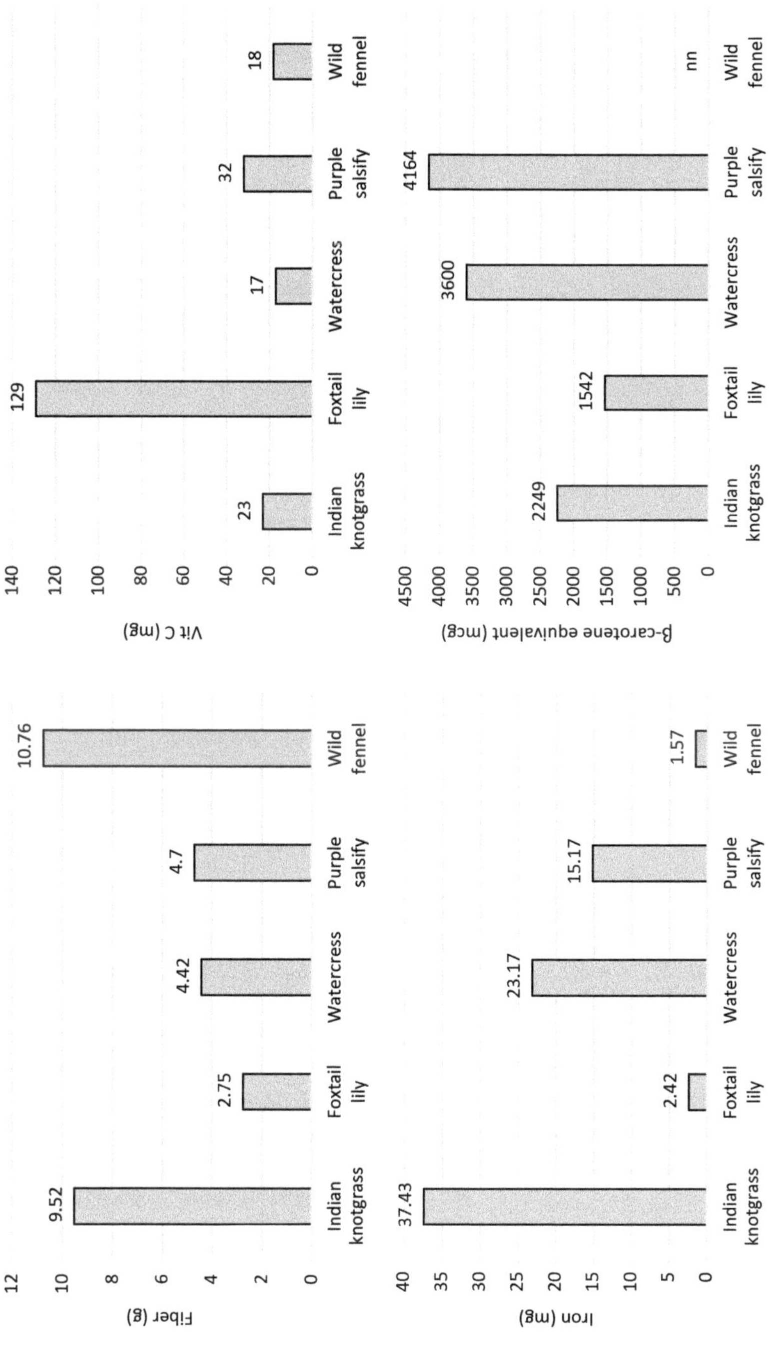

FIGURE 5.C4 An example of nutrition data produced on wild edible species

Source: BFN Turkey (Hunter et al., 2019)

BOX 5.C7 PUBLISHING RESEARCH FOR DEVELOPMENT (GLOBAL)

Each of the four countries developed studies and addressed research questions relevant to their particular context. In 2017, BFN Project researchers in Turkey filled nearly an entire scientific journal, the *ANADOLU Journal of the Aegean Agricultural Research Institute (AARI)*, with articles about wild edible species. The articles represented the economic and socio-cultural importance of wild edibles, the antioxidant capacities of a selection, and species-specific studies zooming in on the properties of sea beets (*Beta vulgaris* subsp. *maritima*), foxtail lily (*Eremurus spectabilis*) and saffron milk cap (*Lactarius deliciosus*). The Brazilian team shared their experiences on mainstreaming biodiversity into food procurement programmes in the Brazilian journal *Revista Raizes*, while the Kenyan team explored the genetic diversity of Bambara groundnut in the *African Journal of Biotechnology*. The Sri Lankan team published articles about the characterization of indigenous rice and evaluation of nutritional properties in yams and leafy vegetables in the Sri Lankan journals *Tropical Agriculturalist* and *Tropical Agricultural Research.* When assessed as beneficial for the local research community, the articles were also published in local languages.

Examples of research articles from the countries:

Brazil:

- Beltrame, D.M.O, Oliveira, C.N.S., Borelli, T., Santiago, R.A.C., Monego, E.S., Rosso, V.V., Coradin, L. Hunter, D. (2016) 'Diversifying institutional food procurement – Opportunities and barriers for integrating biodiversity for food and nutrition in Brazil'. *Revista Raizes*, vol 36, no 2, July–Dec 2016. Available at https://cgspace.cgiar.org/handle/10568/80207

Kenya:

- Odongo, F., Wasike, V., Owuoche, J.M., Laura, K., Tabu, I.M., Oyoo, M. (2015) 'Genetic diversity of bambara groundnut (*Vigna subterranea* (L.)) (Verdc) land races in Kenya using microsatellite markers'. *African Journal of Biotechnology*, vol 14, no 4, pp283–291. Available at: https://academicjournals.org/article/article1422524818_Odongo%20et%20al.pdf
- Hunter, D., Giyose, B., Pologalante, A., Tartanac, F., Bundy, D., Mitchell, A., Moleah, T., Friedrich, J., Al-derman, H., Drake, L., Kupka, R., Marshall, Q., Engesveen, K., Oenema, S. (2017) 'Schools as a system to improve nutrition: A new statement for school-based food and nutrition interventions'. United Nations System Standing Committee on Nutrition (UNSCN). Discussion Paper. Sept 2017. Available at: www.unscn.org/uploads/web/news/document/School-Paper-EN-WEB-nov2017.pdf

Sri Lanka:

- Samarasinghe, W.L.G., Dalshini, K., Sumanasinghe, V.A., Ubesekara, N. M.U., Hettiarrachchi, K. (2015) 'Characterization of suduru samba rice (*Oryza sativa* L.) accessions using genomic wide SSR poly-morphism and seed morphology'. *Annals of the Sri Lanka Department of Agriculture* 17, pp 36–39, *Tropical Agriculturalist*, vol 163. Available at: www.academia.edu/17020769/CHARACTERIZATION_OF_SUDURU_SAMBA_RICE_ORYZA_SATIVA_L._ACCESSIONS_USING_GENOME_WIDE_SSR_POLYMORPHISM_AND_SEED_MORPHOLOGY
- Nadeeshani, H., Wimalasiri, S., Samarasinghe, W.L.G., Silva, R., and Madhujith, T. (2018) 'Evaluation of the nutritional value of selected leafy vegetables grown in Sri Lanka'. *Tropical Agricultural Research*, vol 29, no 3, pp 255–267 www.pgia.ac.lk/files/Annual_congress/journel/v29/Journal-No%203/Papers/3-%2038.%20Ms.Harshani%20Nadeeshani-Final%20R.pdf

Turkey:

- ANADOLU. (2017) Volume: 27 – issue:2. Available at: https://dergipark.org.tr/en/pub/anadolu/issue/34063

Producing articles that emphasize how biodiversity and nutrition are linked has been essential to the project as these are scientifically often treated in two different categories – nutrition and environment – and thus the interlinking benefits are not acknowledged or exploited. Evidence of such linkages have been shared across a wide a range of books, articles and scientific conferences within the project:

- Fanzo, J., Hunter, D., Borelli, T., Mattei, F. (eds) (2013) *Diversifying food and diets: Using agricultural biodiversity to improve nutrition and health*, Earthscan, Routledge. Available at: www.bioversityinternational.org/fileadmin/_migrated/uploads/tx_news/Diversifying_food_and_diets_1688_02.pdf
- Kennedy, G., Hunter, D., Garrett, J., Padulosi, S. (2017) 'Leveraging agrobiodiversity to create sustainable food systems for healthier diets'. *UNSCN News,* vol 42, pp 23–31. Available at: https://cgspace.cgiar.org/handle/10568/89400
- Hunter, D., Borelli, T., Beltrame, D.M.O., Oliveira, C.N.S., Coradin, L., Wasike, V.W., Wasilwa, L., Mwai, J., Manjella, A., Samarasinghe, G.W.L., Madhujith, T., Nadeeshani, H.V.H., Tan, A., Tuğrul Ay, S., Güzelsoy N., Lauridsen, N., Gee, E., Tartanac, F. (2019) 'The potential of neglected and underutilized species for improving diets and nutrition'. Special issue on orphan crops: Contributions to current and future agriculture.

Planta. Available at: https://link.springer.com/article/10.1007/s00425-019-03169-4?wt_mc=Internal.Event.1.SEM.ArticleAuthorOnlineFirst&utm_source=ArticleAuthorOnlineFirst&utm_medium=email&utm_content=AA_en_06082018&ArticleAuthorOnlineFirst_20190428

There are many operational learning outcomes connected to a multi-country project like this that can benefit other initiatives (see Boxes 5.B8 and 5.C8). Considerations on methodology and experiences gained from the project across the four countries have been shared in a number of articles. With these the purpose is to aid and inspire other similar projects by for instance displaying how to take advantage of already existing political contexts in the countries or how to identify project entry points and opportunities and barriers. See the following examples:

- Hunter, D., Ozkan, I., Beltrame, D.M.O., Oliveira, C.N.S, Samarasinghe, W.L.G., Wasike, V.W., Charrondiere, U.R., Borelli, T., Sokolow, J. (2016) 'Enabled or disabled: Is the environment right for using biodi-versity to improve nutrition?' *Frontiers in Nutrition*, 06 Jun 2016. doi:10.3389/fnut.2016.00014
- Beltrame, D.M.O., Gee, E., Güner, B., Lauridsen, N., Samarasinghe, W.L.G., Wasike, V.W., Hunter, D., Borelli, T. (2019) 'Mainstreaming biodiversity for food and nutrition into policies and practices: Methodologies and lessons learned from four countries'. *ANADOLU, Journal of AARI*, vol 29, no 1, pp 25–38. Available at: https://dergipark.org.tr/anadolu/issue/45404/568795

BOX 5.C8 IN THE WORDS OF: BFN COUNTRY PARTNERS, ON LESSONS LEARNED

The volume of analysis was challenging; it required a high level of institutional and human input, and was costly in terms of time and resources (one should anticipate consistency and accountability difficulties, and delays in the purchase of materials). Hypothetically, a single specialized lab would have produced cheaper and more uniform results; however, the establishment of a new research network was very positive. BFN Project research has opened the door for further research on a large selection of underutilized species. Researchers and students now have harvesting and foraging experience (both in markets and nature) and have accumulated valuable material which is stored in labs. The samples and resources enable continued work with these neglected species (for example, they have already started examining other aspects such as different ways the crops can contribute to health). Other labs are designing new products for school feeding programmes.

Brazil

'Amongst National Steering Committee members, there were varying opinions about the most suitable target sites and species. The National Project Coordinator had to mediate diverse interests and establish clear guidelines that could comprehensively support the early identification of priority species, keeping in mind relevant nutrition problems and available traditional knowledge. In the end, BFN Project research has been an eye-opener, revealing the value of many traditional species. It has also underscored the need for further well-organized research.

Sri Lanka

C2 Influencing policy and markets

- Successful policies depend on supportive policy champions and/or pre-existing policy frameworks.
- School feeding and procurement was a major opportunity for expanding the market for biodiversity.
- Greater integration of food biodiversity into markets can provide economic empowerment for local smallholders and businesses, while benefitting consumer health.

TABLE 5.C5 Policy component and outputs including marketing options

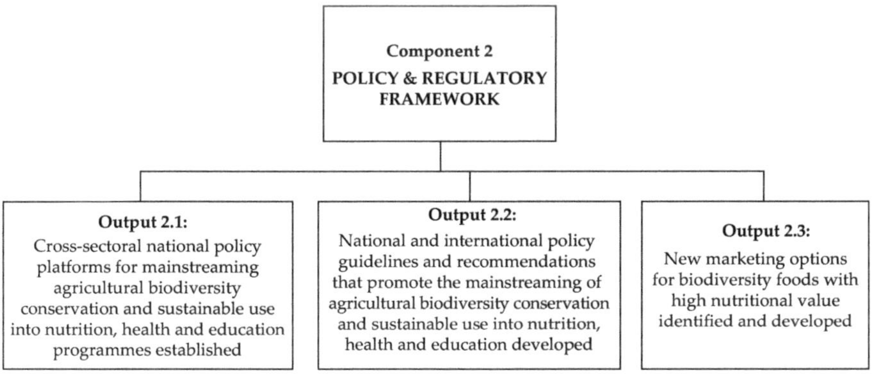

Policies

- Increased political interest in integrating nutrition objectives into food and agriculture meant numerous opportunities for policy change.

A stronger evidence base made it possible to approach policy and decision-makers who could advocate for the recognition of food biodiversity. As each BFN country established a comprehensive set of information, this was used to develop a convincing business case advocating for biodiversity mainstreaming. Previously identified entry points such as favourable procurement policies in Brazil (Box 5. C9), government decentralization in Kenya (Box 5.C10) and National Biodiversity Strategy and Action Plans (NBSAPs, Box 5.C11) were used to better mainstream biodiversity with the help of policy champions and other case-by-case connections. Champions were drawn from various spheres, but often included government representatives who were sensitive to project themes or whose ministerial and departmental goals strongly aligned with those of BFN. Sharing common objectives can be mutually advantageous, particularly when the champion is held personally accountable for the achievement of institutional targets that align with project outcomes. By the same token, government posts can be very volatile, particularly following elections and ensuing reorganizations. Champions can be removed at short notice and the time spent building personal relationships is made worthless, with the only option remaining to start again from square one.

Despite the challenges (see Box 5.C17), mainstreaming efforts are highly rewarding and the following two policy examples, from Brazil and Kenya, highlight policy achievements at the national and the county level, respectively. The examples demonstrate how, despite differing contexts, experience sharing between the two countries enabled the partners to identify opportunities, such as creating demand for biodiversity via school feeding programmes that source from local farmers (see section below).

BOX 5.C9 LISTING SPECIES WITH SOCIO-BIODIVERSITY ORDINANCES (BRAZIL)

In Brazil, the BFN Project's work directly contributed to a national policy ordinance which integrates underutilized biodiversity into public procurement. Up to this point, Ministry of Social Development employees were struggling to identify biodiversity species. Financial incentives reserved for regional foods were not being claimed due to uncertainty about which species qualified. To meet this need, the Secretary of Biodiversity, Plants for the Future and BFN Project staff in Brazil brought together their expertise to add 101 native food species to the *Official list of native Brazilian socio-biodiversity species of nutritional value*, an inventory that is collaboratively published by the Ministry of the Environment and the Ministry of Social Development (MMA, 2018) and guides public food procurement, school feeding and other public policies. The implementing policy, inter-ministerial ordinance 163/2016, marked the first official recognition of native species and their nutritional value in Brazil (it was followed by 284/2018 with additional species). Ministries now refer to this list to monitor PNAE and PAA institutional purchases that support biodiversity production and marketing. Specifically, the market consequence of the ordinance is that it guarantees purchases of BFN species at a fair price, thereby encouraging smallholders to further diversify their production with these viable fruits and vegetables.

Additional benefits of policies like the ordinance include local food product labelling, which can support the conservation and use of local species and crops; in this case with the creation of a 'socio-biodiversity label' indicating diversified production. This voluntary label increases product visibility for consumers and procurers and is linked to the pre-existing smallholder farmer label. PNAE also organized a biennial national recipe contest with school cooks in which the judges consider 'valuing local habits: use of regional foods'.

The ordinance is complemented by a food quality index, IQ COSAN (see below), which assigns point-based ratings to school meals based on criteria including dietary diversity and presence of listed species. This acts as both a monitoring mechanism and an incentive for school nutritionists and management to further integrate biodiversity.

Finally, having an official list of socio-biodiversity species facilitates the monitoring and tracking of biodiversity within the PAA, PNAE and PGPM-Bio programmes, a challenge encountered by BFN Brazil before such an ordinance was put in place. The lack of an adequate monitoring system to track and assess the purchase of socio-biodiversity food products by national food procurement programmes was another result of the poor understanding of the definition of 'biodiversity for food and nutrition', which the ordinance list is now helping to clarify (Beltrame et al., 2016).

Assessing healthy school meals with IQ COSAN: quality index of the Coordination of Food and Nutritional Security

One of the objectives of PNAE, Brazil's oldest food and nutrition policy, is to develop healthy eating habits in schoolchildren through education and the provision of meals meeting the children's nutritional needs throughout the school year. The menus prepared for PNAE are an important strategy for the improvement of the children's eating habits. The main objective of the IQ COSAN is the standardization of the school menu analysis, not only for the coordinating technical staff of the National Fund for the Development of Education (FNDE), but also for nutritionists and other actors involved with PNAE, to supply consistently healthy food.

IQ COSAN is an easy-to-use and accessible instrument that analyses school feeding menus, with points awarded for the presence of different food groups, regional and socio-biodiverse foods, weekly meal diversity and absence of restricted or prohibited foods like sweets. The tool refers to Ordinance N. 284/2018 to add an extra 2.5 points if listed species appear on school menus. At the end of the evaluation of these parameters, the instrument sums the points and calculates the weekly averages. The IQ COSAN score ranges from 0 to 95 points and classifies the menus as

Inadequate (0 to 45.9 points), Needs improvement (46 to 75.9 points) and Adequate (76 to 95 points). Beyond functioning as a control mechanism, this is an incentive for socio-biodiversity and support for the nutritionists and other actors from PNAE, providing guidelines when preparing the menus (FNDE, 2018).

BOX 5.C10 CREATING KENYA'S FIRST BIODIVERSITY CONSERVATION POLICY (KENYA)

Contrary to Brazil, Kenya did not have a pre-existing national policy framework. Instead, BFN Kenya faced obstacles including fragmented government ministries and very limited funds. Like many countries in Asia and sub-Saharan Africa, Kenya is undergoing political, administrative and financial decentralization (Hunter et al., 2016). BFN Kenya chose to use this as an opportunity to push through local level policy that would not have easily gained traction at the national level.

The first step was to identify and sensitize potential 'champions'; in this case, government ministry members who were receptive to indigenous crop integration or traditional dietary health practices and could convene other influential policymakers. For example, BFN representatives approached a minister who had personal experience cultivating Bambara groundnut, and invited local government leaders to a county seed-sharing festival. The interested parties formed the core of a cross-sectoral task force, which drafted a policy proposal designed to fill in the gaps of prior conservation, health, agriculture and education plans with a biodiversity focus. Over repeated workshops, the draft was revised to include the inputs of public stakeholders including community leaders, traditional healers, women and youth. By the time it reached the legislative assembly, the policy proposal encompassed three rounds of revisions and represented a shared vision and statement of intent by proponents of local biodiversity.

The process culminated in the March 2018 endorsement of *Busia County's Biodiversity Conservation Policy*, the first of its kind across Kenya's 47 counties. The policy has four focus points:

- Enhancing biodiversity conservation.
- Improving access to and ensuring equitable sharing of benefits accrued from access and utilization of county biodiversity.
- Promoting biodiversity utilization.
- Promoting biodiversity research and development.

The policy allocated resources to conserve regional food biodiversity and made specific provisions for designated conservation areas and further incorporation of native species into school meals, such as with direct farm-to-school procurement. This official recognition of the role of biodiversity is a significant achievement for Busia County, but it also has greater significance as it is meant to inspire additional counties to follow suit, in Kenya as well as in other countries where decentralization provides an opportunity for change.

The policy's success was only possible because of the concerted effort of bringing together multiple parties, with the engagement of public stakeholder groups as well as high-level officials. The core task force was a team of experts from the County Ministries of Agriculture, Health, Education, Environment, Public Health and Forestry, the Kenya Agricultural and Livestock Research Organization (KALRO) and members of the local community-based and farmer training organization Sustainable Income Generating Investment (SINGI).

TABLE 5.C6 The Busia Policy Process and Actors

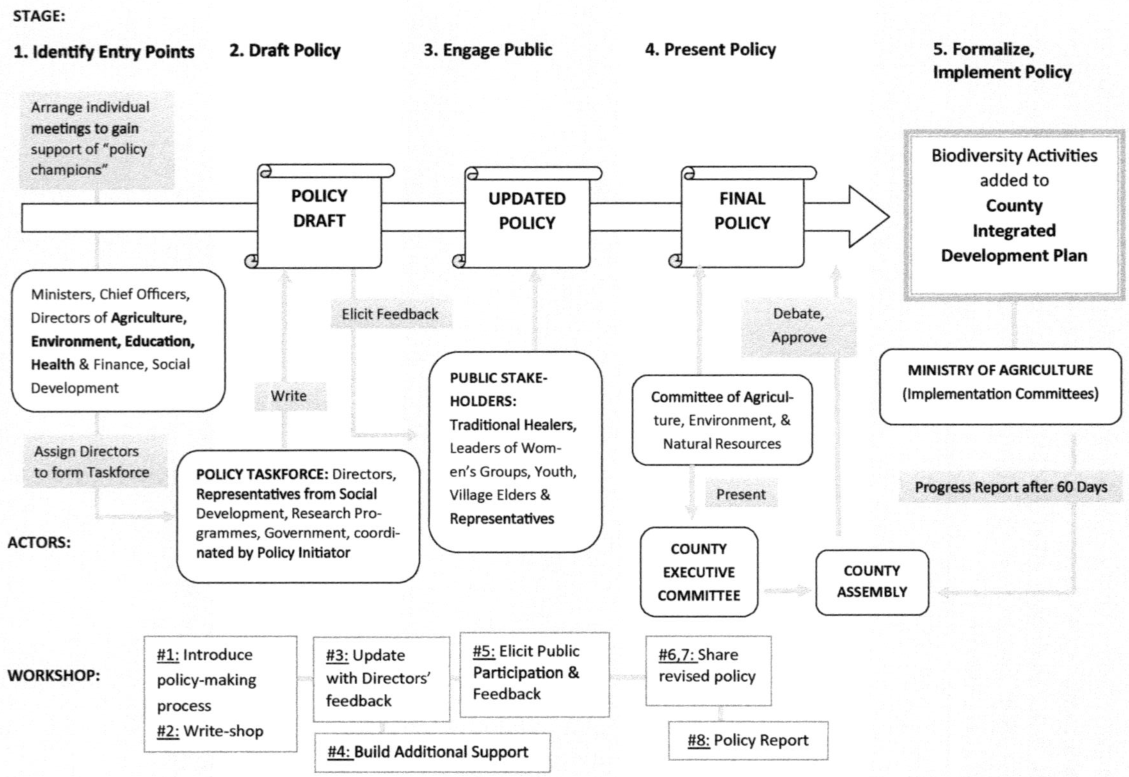

Source: E. Gee and V. Wasike

BOX 5.C11 NATIONAL BIODIVERSITY STRATEGY AND ACTION PLANS (GLOBAL, BRAZIL, SRI LANKA)

The NBSAP is a policy instrument that each country within the Convention of Biological Diversity (CBD) is obliged to develop in order to mobilize resources and activities to achieve the respective country's commitments to the CBD and associated Aichi Biodiversity Targets. An NBSAP gives an overview of the current state of biodiversity within individual countries and describes the threats leading to degradation of biodiversity and the sectoral actions within plans, programmes and policies that may ensure its conservation and sustainable use (CBD, 2019). The high recognition and potential impact of NBSAPs represented a strong policy entry point for the BFN Project, which joined discussions and development of the NBSAPs. An excellent example of this is the Brazilian NBSAP revision process where biodiversity for food and nutrition was integrated into the National Targets.

In Brazil, a broad policy consultation, *Dialogues on Biodiversity*, engaged 400 participants from institutions across five sectors: business, environmental NGOs, academia, federal and state government as well as indigenous peoples and traditional communities. During the process, 'limited appreciation of the use of biodiversity for food and nutrition' was included as one of the main causes for biodiversity loss in the country, with participants identifying 23 priority actions to reverse biodiversity loss and help achieve National Targets. Both the revision of the National Targets (2011–2020) and the establishment of a Governmental Action Plan and guidelines for the Multi-year Budget Planning for 2016–2019 provided opportunities to incorporate the theme of biodiversity for food and nutrition, and to ensure the resources and budgets to implement relevant actions (MMA, 2017). For example:

> National Target 13: By 2020, the genetic diversity of microorganisms, cultivated plants, farmed and domesticated animals and of wild relatives, including socio-economically as well as culturally valuable species, is maintained, and strategies have been developed and implemented for minimizing the loss of genetic diversity.

The proposed indicator for National Target 13 is the number of native species of Brazilian biodiversity included in the Biodiversity Nutritional Composition Database in the Brazilian Biodiversity Information System (SiBBr), the database containing nutrition information of a substantial selection of native Brazilian species including 78 species that the BFN Project has developed information for.

Additionally, the BFN Sri Lanka team integrated health and nutrition aspects into their country's NBSAP, concretized in National Target 9. This target states the need to *promote conservation of neglected, lesser known and underused food crops, livestock and their wild relatives such as vegetables, seeds, fruits, poultry, livestock and food fish* (MMDE, 2016). The target should be achieved by raising awareness through booklets, media articles, short video clips and building capacity through training.

School feeding

- Schools can be a major supplier of biodiverse food and food-based knowledge, with proper policy support.
- School feeding programmes also link to raising awareness and supporting farmer livelihoods.

Institutional food procurement has emerged as a massive opportunity for mainstreaming biodiversity, with schools acting as one of the most promising entry points. School meals are a vital source of daily nutrients for millions of children, and can also act as a vehicle for education on nutrition, health and food choices. It is estimated that school meals in Brazil impact 45 million children (Sidaner et al., 2012) and supply up to 70% of children's nutritional needs (Kitaoka, 2018), giving some sense of reach to the achievements listed previously in Box 5.C9. Likewise, school meals can be a determining factor in student attendance, and physical and mental development (see also Chapter 7 by Swensson and Tartanac).

Procurement programmes can also benefit local smallholder farmers while building market capacity for food biodiversity. Farmers are more likely to diversify their production if they are certain their crops will later be purchased at a fair price by schools, hospitals, prisons and other public institutions. Brazil's experiences demonstrated how this could be accomplished at the national level. In Kenya, BFN community partner SINGI led a farmer business school (see Box 5.C12) that empowered local farmers, while also trialling a direct procurement model. This form of home-grown school feeding, graphically explained in Figure 5.C5, has already shown to be profitable for both farmers and the schools that they sell to, cutting out market intermediaries and streamlining the delivery process. Though it requires training and standardization to improve consistency and safety at a larger scale, home-grown school feeding has many positive implications including for community livelihoods, ecosystem resilience and student health. It also holds strong potential to be adopted by other regions and supported by further ad hoc policies.

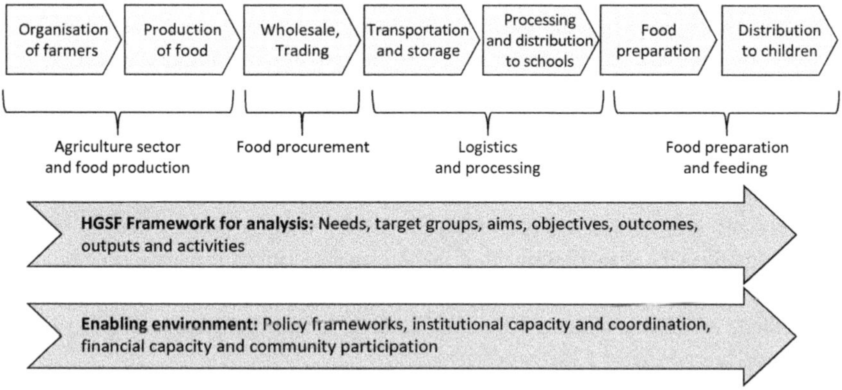

FIGURE 5.C5 Conceptual model for home-grown school feeding
Source: Gelli et al., 2010

BOX 5.C12 PIONEERING A FARMER BUSINESS SCHOOL AND DIRECT PROCUREMENT MODEL (KENYA)

Farmer Business School

To equip farmers with relevant knowledge and skills, Kenyan partner and community organization SINGI led classes for 4,000 farmers on sustainable agriculture methods, with a strong focus on maintaining local biodiversity. SINGI piloted a Farmer Business School (FBS) Model with funding from the Australian Centre for International Agricultural Research (ACIAR) and MacArthur Foundation. Twenty-five farmer groups received training on sustainable agriculture practices, with cultivation techniques such as mandala, keyhole and multi-story gardens incorporating traditional crops and varieties. Farmers learned about their nutritional value and how they could be incorporated into local markets, plus value addition and quality control. Perhaps most importantly, the FBS trained entrepreneurial farmers in Busia County on contract farming, business plans and the ability to respond to demands from institutional markets including local schools and clinics, meeting the ambitions of small-scale agricultural entrepreneurs, mostly women and youth, to grow their small enterprises by commercialising African leafy vegetables (ALVs) and other nutritious crops.

Home-grown school feeding

An important offshoot of the FBS was the trial of a Direct Food Procurement Model linking local farmer groups to school canteens that feed students and staff. In the schools where this was trialled, farmer groups visited local schools and presented their produce to the administration and catering staff. Subsequently, memorandums of understanding between the farmer and school regulated an arrangement which bypassed market operators for mutually beneficial and consistent pricing of indigenous produce (see Figure 5.C6).

To cut transport costs and avoid food losses, one farmer group arranged to grow vegetables directly on school land. As a result, the school has a reliable and constant source of quality ALVs, while farmers have reduced their costs and found a dependable buyer for their produce. Early projections for the dry season show that, when market prices for leafy greens are higher, the school can save up to 10 Ksh (USD$0.10) per kg of leafy greens supplied, which translates into weekly savings of approximately $9 per week and yearly savings of $360 a year.

Projected average profits for the smallholder farmer group, which supplies 91kg of ALVs to the school per week, amount to 15Ksh ($0.15) per kg of leafy green. This translates into a weekly profit of 1,365Ksh ($13) and yearly profits of approximately $540. While further monitoring of the economic benefits to the school and to the farmers is needed over time, it has already become apparent that farming of ALVs as a viable business is slowly gaining ground in Busia County, with farmers increasingly willing to invest resources in ALV production and marketing.

A facilitator's training manual was developed to extend the training on this approach to other farmers and community-based organizations. The potential of this approach matches current interest in home-grown school feeding (HGSF) and is being tested in other regions.

FIGURE 5.C6 A SINGI farmer group meeting
Source: BFN Kenya

BOX 5.C13 IN THE WORDS OF: MR. OBONYO, FORMER SCHOOL PRINCIPAL, ST. MARY'S MUNDIKA, KENYA:

'There is nothing more important than seeing my boys fed on a nutritious and balanced diet. This will not only improve their health but also increase their academic performance and reduce absenteeism due to sickness and diet-related diseases.'

St. Mary's Mundika secondary school in Busia County has become a best practice hub for sustainable school procurement in the area. In 2018, an academic and agricultural day was organized on the school premises attracting more than 800 participants from NGOs, universities, the agriculture and nutrition sectors, hospitals, neighbouring county government and schools, as well as farmers and parents to witness first-hand the benefits of its direct procurement approach for local biodiversity: the mutual economic benefits for farmers and the school – more than 12% savings reported per year for vegetable purchases alone – the improvement in the students' academic performance thanks to improved diets, as well as the more intangible gains such as the school's heightened corporate responsibility and the farmers' increased social standing within the community.

Markets

- Market capacity needs to increase for biodiverse species to be worth producing.
- Consumers, farmers and seed suppliers need practical and affordable ways to access biodiverse species.

FIGURE 5.C6T Conducting a market survey in Turkey, 2014
Source: Ayas et al. (2017), BFN Turkey

Policies such as those described in Boxes 5.C9, 5.C10 and 5.C11 play an important role in building market capacity, stimulating both production and consumption of targeted species. However, this market capacity can also be strengthened beyond the realm of policies, for example, where there is no enabling policy environment, or implementation is too slow or bureaucratic. In these instances, it is possible to leverage connections within government ministries, NGOs, or the private sector to explore other spaces for local food production and business. This is an important step in incentivizing production of and ensuring sustained support for biodiversity. BFN efforts within this component therefore also included establishing food outlets (Box 5.C14), developing novel and value-added products (Box 5.C15), and domesticating wild species for wider market sales (Box 5.C16).

BOX 5.C14 EMPOWERING WOMEN-RUN FOOD BUSINESSES WITH HELA BOJUN (SRI LANKA)

As in many other sites worldwide, fast food chains have increasingly degraded Sri Lankan dietary health, with obesity and diabetes rates rising while intake of fresh fruits and vegetables declines. In an effort to provide an alternative food option, Sri Lanka's Ministry of Agriculture has trained over 750 women to turn their home cooking skills into profitable businesses. Providing courses on food preparation, food hygiene, customer care and business management, The Women's Agriculture Extension Programme facilitated the creation of a network of local food outlets called 'Hela Bojun' or 'True Sri Lankan Taste'. Hela Bojun translates traditional home cooking into food stalls – largely near urban spots like government offices and school campuses – that offer people access to clean, nutritious and safe local food at reasonable prices. Designed to simultaneously promote local food, create a healthier young generation and enhance the economy of village women by spreading their food across the country, the stalls have expanded to a total of 22 locations, with some including adjoining shops that sell local food products. Some women operating the food outlets have since become the main breadwinners in their families, earning a living wage of USD$600–800 per month.

Since opening in Colombo and Peradeniya, the Hela Bojun outlets and shops have become immensely popular, earning highly positive reviews from locals and tourists on TripAdvisor. The shops promote underutilized crops identified by the BFN Project, including medicinal plants such as water lily (*Nymphaea pubescens*), used in Ayurvedic medicine, and pulses such as cowpea (*Vigna unguiculata*), black gram (*Vigna mungo*) and green gram (*Vigna radiata*), as well as traditional rice varieties (*Oryza* spp.). Furthermore, this has created market demand for traditional foods such as honey, jaggery, treacle, bananas and organic products, boosting production for smallholders.

Hela Bojun centres have also drawn on BFN research in order to publish nutrition facts for 91 traditional and local dishes. Efforts to make Hela Bojun snacks and offerings healthier include further uptake and adaptation of these dishes with a focus on diverse local ingredients (see Figure 5.C7). In an offshoot of Hela Bojun, Wayamba University trialled cooking demonstrations with locally available greens to raise awareness of nutrition, hygiene and food safety, and encourage the cultivation of diverse home gardens. Further work is being done to produce more shareable recipes and discuss potentially integrating the Hela Bojun model into school canteens with the support of the Ministry of Education.

FIGURE 5.C7 Cooking with the highly nutritious gotu kola
Source: S. Landersz

Recipe: Gotu kola sambol

Ingredients

Gotu kola (*Centella asiatica*) cleaned – 3 cups
Fresh grated coconut* – 3 tablespoons
Green chili – 1
Red pearl onions (ratu lunu)** – 3–4
Powdered Maldivian fish (optional) – 1 teaspoon
Lime – 1/2
Salt to taste.

Preparation

Shred the gotu kola finely. Finely chop the green chili and onions.
Mix all the ingredients and add lime juice and salt to taste.
Serve immediately.

* If fresh coconut is not available, you can use desiccated unsweetened coconut.

Add about 1 tbsp of water to it and briefly heat before using.

** If red pearl onions are not available, use about a quarter of a regular onion.

Source://bfnsrilanka.org/2016-07-18-07-20-55/gotukola-sambol

BOX 5.C15 DEVELOPING NOVEL PRODUCTS WITH SUN MACK DRINKS (SRI LANKA)

In Sri Lanka the potential for cultivating fruits for the domestic and export market is high. Most local fruits could be used in value-added products, but they remain underutilized without proper marketing strategies. An example of this is ready-to-serve fruit drinks, which are highly nutritious and high in fibre and antioxidants. However, these fruits are different from mainstream commodities in that they are limited to seasonal availability and time-consuming to harvest, increasing the products' final cost. With this in mind, the National Food Promotion Board selected three high-potential BFN species (Ceylon olive/weralu, mango and annoda/katuan-noda) to use for novel drink products (see Figure 5.C8). Over a period of five months, production protocols were developed for these three varieties; the juices were bottled, labelled and distributed at outlets of the National Food Promotion Board under the name 'Sun Mack' with accompanying brochures explaining their nutritional values. Further actions included sharing production protocols with small entrepreneurs, experimentation with different species and expanding into other market channels.

FIGURE 5.C8 Sun Mack printed label for a soursop (labelled as anoda) drink
Source: BFN Sri Lanka

BOX 5.C16 DOMESTICATING THE WILD GOLDEN THISTLE (TURKEY)

Golden thistle (*Scolymus hispanicus*) is a thorny wild edible plant that grows across Southern Europe in disturbed habitats ranging from fallow fields to roadsides. Its leaves are cooked in omelettes, its root is used as a coffee substitute, and its flowers as a replacement for saffron. It is highly nutritious and has traditional health uses, including as a licensed medicine for urinary issues through the early 20th century (see Figure 5.C9).

FIGURE 5.C9 Harvested golden thistle available at a Turkish market
Source: D. Hunter

Based on surveys and the sustainability index ranking from the planning phase (see Section B), golden thistle was one of three wild edible species selected in Turkey for market promotion. Its wide recognition and multiple uses made it a promising candidate; however, the most significant reason for golden thistle's selection is its potential for easy domestication (breeding for commercial cultivation). Therefore, it was possible to promote the species and increase demand without risking overexploitation (unsustainable harvesting).

Additional studies focused on the potential of golden thistle as a market product, such as post-harvest (cold storage capacity, shelf life and quality changes during storage) and biochemical analyses (organic acids, sugars, vitamin C content, phenolic compounds and antioxidant activities). Further research was conducted on determining the marketing channels, calculating the cost and potential income, identifying value chain actors and production opportunities. New and experienced golden thistle farmers trialled different seeds – half of which were supplied by the project – on 86 hectares of land which was monitored according to growing conditions (irrigation was found to moderately increase yield). The average production cost of golden thistle was estimated to be 1 euro/kg; in comparison with collecting from the wild (0.76 euro/kg), this is slightly but not significantly more expensive. Subsequently market channels were identified such as selling directly to traders and markets as well as to hotels and restaurants.

With these studies, there were clearer incentives for farmers to invest in golden thistle cultivation: high income, low production costs and high demand as the current market is eager to integrate new local products. In addition, it is a good crop to grow on small and marginal areas and it is highly compatible with sustainable and organic farming methods.

Read more about the species and its nutritional properties here: www.b4fn.org/resources/species-database/detail/scolymus-hispanicus/

Recipe: Golden thistle and lamb stew

Ingredients

1kg of golden thistle
400g of lamb meat (cubed)
1 onion, chopped
2 garlic cloves, chopped
3–4 tbsp of olive oil
Water
Salt (to taste)
1 tbsp of flour
Juice of 1 lemon
For the sauce:
Juice of 1 lemon
1 tbsp of flour

Preparation

Put the water, flour and lemon juice in a deep bowl. Strip the leaves of their thorns, peel the roots and remove their central woody part. Chop the cleaned golden thistle into 4–5cm pieces and put them in the mixture.

Heat the olive oil in a saucepan, add the cubed lamb meat and sauté until brown. Add the chopped onion, garlic and salt and cook for 4–5 minutes. Cover the meat with water and cook until tender.

Meanwhile, prepare the sauce: add the juice of 1 lemon and 1 tbsp of flour in a bowl and whisk until smooth.

Add the strained golden thistle to the meat. Add the sauce and cook a little longer if needed, then serve.

(*Source*: BFN Turkey team, Karabak, 2017; Güzelsoy et al., 2017; Ayas et al., 2017)

BOX 5.C17 IN THE WORDS OF: BFN COUNTRY PARTNERS, ON LESSONS LEARNED

Without an adequate enabling environment, recommendations and guidelines could not be integrated into existing policies such as the country National Biodiversity Strategy and Action Plan. A policy consultant could help identify promising policies and policymakers.

Sri Lanka

Reach out to receptive stakeholders and potential "champions" early on and arrange personal meetings to ensure support, but be prepared for positions and people to change.

Kenya

Assign a dedicated policy focal point person or group consistently responsible for tracking policy happenings and meeting with policymakers. While at first it was difficult to explain the concept of BFN to policymakers in Brazil, now it is understood, particularly at the federal level. This can be seen during the course of the National Steering Committee, which has now developed plans that include BFN-relevant species and actions. Now that baseline knowledge and recognition has been established, the concepts can be taken downstream to influence what is happening in the field and market. In other words, there can be local uptake in creating and building channels for BFN-analysed foods.

Species, and evidence about the value of biodiversity for food, conservation and livelihoods, are currently being integrated into markets with increased general consumption and awareness. The resultant growing demand is outpacing the capacity and consistency of production. Therefore, further work—technical assistance, quality and value chain investment—is needed to increase production and make it easier for consumers to find local foods in the market.

Brazil

3 Raising awareness

- Partners engaged with stakeholders across the producer-consumer spectrum, using context-specific approaches to raise awareness.
- Strategies included education initiatives, farmer training, gastronomy and cultural events.

BFN's most successful achievement was probably its final component: raising awareness. BFN partners used a range of strategies across the four countries to raise awareness around species diversity (see Box 5.C24). Working across the producer-consumer spectrum, partners developed mixed media (such as the recipe card in Figure 5.C11), events and activities that could increase understanding and practical use by farmers, teachers, parents, chefs, students and many others; these ranged from learning tools to cultural festivals. Figure 5.C10 below summarizes some of the approaches that were taken.

TABLE 5.C7 Awareness-raising component and outputs

Component 3
INCREASED AWARENESS & OUTSCALING

Output 3.1	**Output 3.2**	**Output 3.3**	**Output 3.4**	**Output 3.5**
Best practices for mobilizing nutritionally rich biodiversity to improve dietary diversity identified and promoted	Capacity of beneficiaries and stakeholders to deploy and benefit from nutritionally-rich biodiversity enhanced	Information events that foster greater appreciation of nutritionally-rich biodiversity as a resource for development and wellbeing conducted	Guidelines for improved use of nutritionally-rich foods from local biodiversity, including processing, food safety measures, and recipes adapted to modern lifestyles based on traditional food systems developed	Tools and methods for mainstreaming biodiversity into food and nutrition upscaled and disseminated

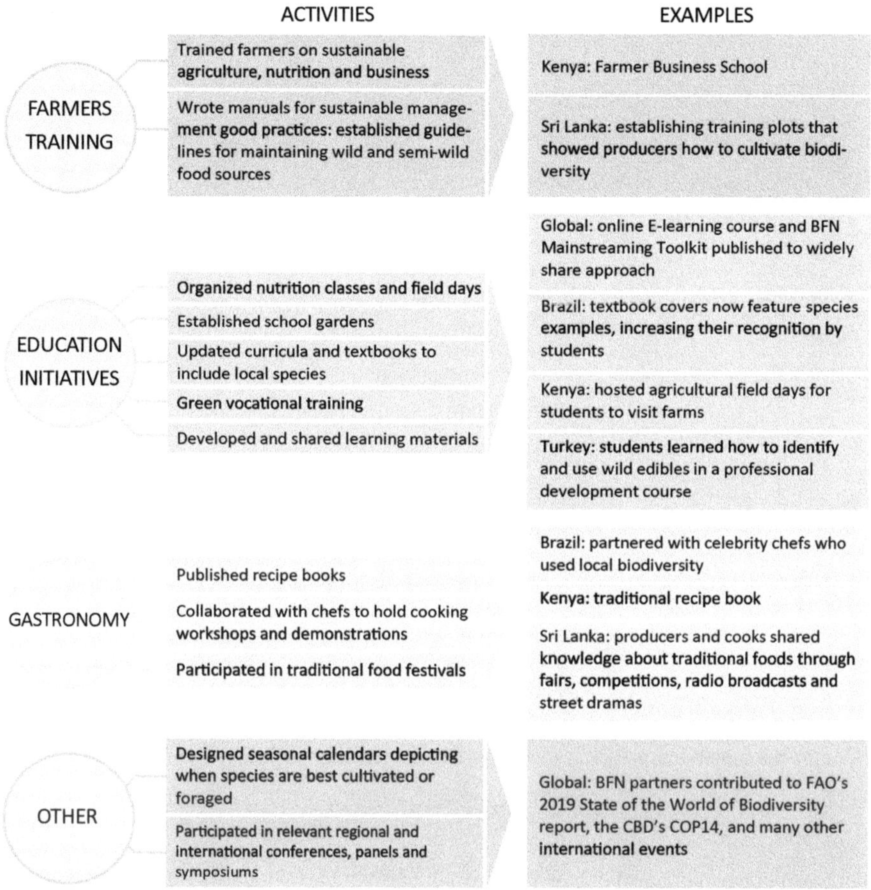

FIGURE 5.C10 Overview of awareness-raising activities

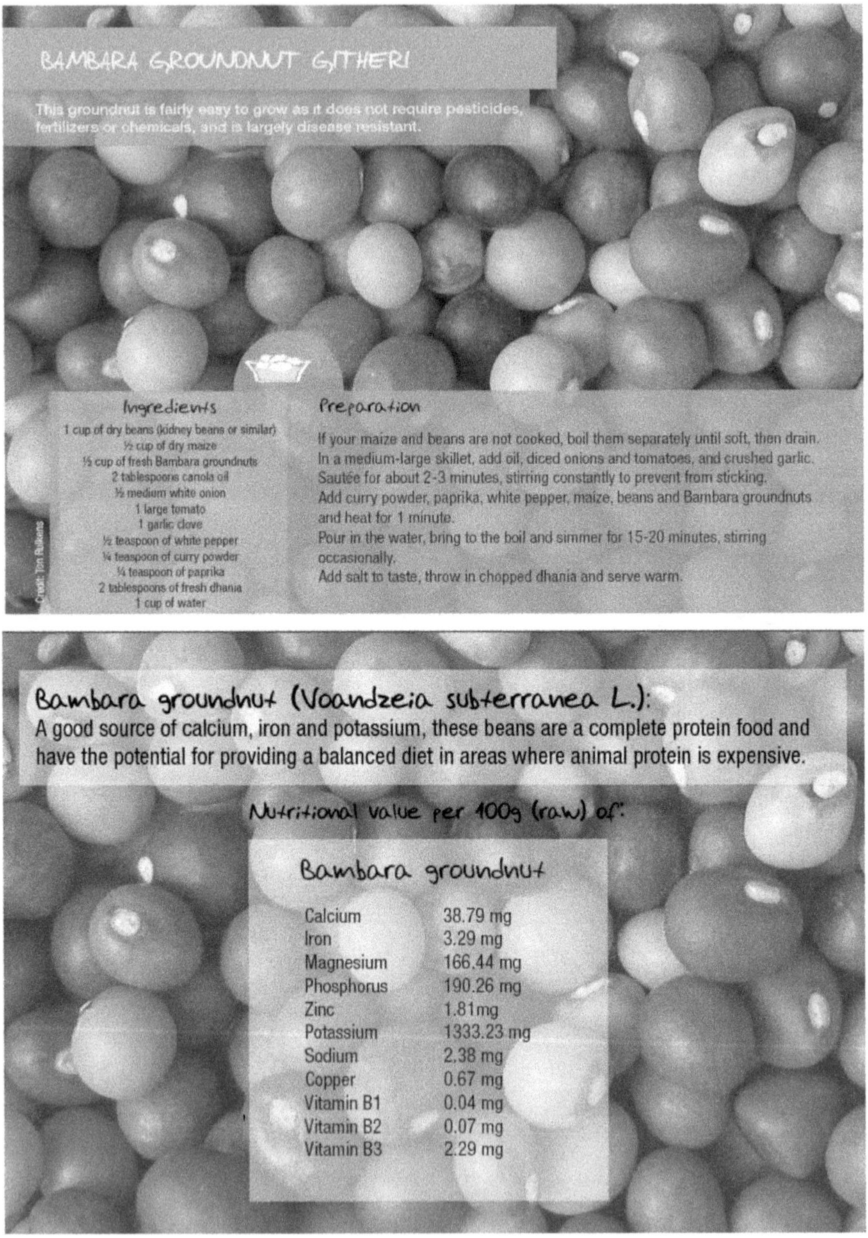

FIGURE 5.C11 Example of a recipe card designed by BFN Kenya

BOX 5.C18 PUTTING BIODIVERSITY ON THE TABLE: GASTRONOMY (BRAZIL)

The project's gastronomic elements represented an exciting and accessible way to bring biodiversity into people's everyday lives. While mobilizing Brazilian regional research teams for food composition analysis, the BFN coordinating professors reached out to gastronomy departments at their universities or nearby gastronomy institutes. They simultaneously recruited high-profile local chefs to develop a recipe book that showcased regional foods. As many of these foods were not well-known by the students or professors, the universities created groups that analysed their taste and flavour profiles, suggested different combinations of ingredients and, finally, created and tested the recipes. Much of this work was used by students for theses on gastronomy and nutrition, while the final book was published in 2019.

This gastronomy element was taken beyond academia with the production of informative media outreach materials including magazine reports, radio broadcasts, instructional cooking videos and appearances on TV cooking shows. Chefs used modern culinary influences to reinterpret traditional dishes, creating new and appealing adaptations that people would be interested in trying for themselves. In 2017, The Biodiversity for Food and Nutrition International Symposium in Brasilia (see Figure 5.C12) gathered participating chefs on stage to demonstrate these dishes, such as the Buriti flan recipe below). This tangible use of local ingredients with a modern twist translated the traditional dishes into something that people would be enthusiastic about eating at a restaurant or in their own home.

FIGURE 5.C12 Chef demonstration at the BFN Symposium in Brasilia, 2017

Source: S. Landersz

There were numerous other platforms to promote cooking and eating Brazilian biodiversity. In 2016, BFN had the unique opportunity to join the 'Rio Food Vision' (Rio Alimentação Sustentável) initiative, which added healthy and sustainable food awareness to the Rio Olympic and Paralympic Games. The initiative worked closely with the Olympic Committee and caterers to enable the inclusion of certified organic, fair trade and local biodiverse foods during the games. BFN participated alongside Olympic nutritionists and athletes to promote sustainable food value chains from 'dish to podium', as thousands of visitors flocked to street and indigenous markets to learn about Brazil's regional food and biodiversity first-hand. The global and media attention helped spread awareness beyond Brazil and contributed to a sustainable legacy for the Rio de Janeiro Olympics (Coradin and Beltrame, 2016).

Other BFN Brazil collaborations included the project 'Educating through School Gardens and Gastronomy', which helped schools establish native tree nurseries for native species and school gardens, as well as an online E-Learning course which provides interactive lessons on biodiversity.

Species highlight: Buriti

Merely ten species correspond to 91% of the total fruits consumed by Brazilians (IBGE, 2010). Of these, only pineapples and açaí are native to Brazil. Yet Brazil boasts a stunning variety of nutrient-rich fruits, which have many uses, even beyond contributing to a healthier diet. **Buriti** (*Mauritia flexuosa*) is a nutritious palm that produces fruit high in vitamin A and C, with five times more beta-carotene than the average carrot. Besides the edible pulp, the trunk can produce a flour called *ipurana*. The pulp is also used to feed farm animals and fish. Buriti oil is used for cosmetics including sunscreen and insect repellent, while the wood and straw makes furniture, construction materials, baskets, tablecloths, toys, kitchen utensils and other handicrafts. Buriti also provides important ecosystem services such as the maintenance of water springs and food for native parrots, and also has cultural and heritage significance to local communities. Due to these multiple uses, Buriti palm has earned the nickname 'Tree of Life'.

Recipe: Buriti flan (Tested by researchers at the Federal University of Goiás)

Ingredients

420g of milk (1½ large Tumbler glass)

85g of buriti paste (3½ level tbsp)

21g of corn starch (3½ level tbsp)

Preparation

Place all the ingredients in a saucepan on medium heat and stir until they are well combined and a smooth mixture is obtained. Pour the mixture in individual serving glasses. Chill in the refrigerator. Serve cold (Figure 5.C13).

FIGURE 5.C13 Buriti flan

Source: E. Soares & I. Sandoval, Federal University of Goiás

BOX 5.C19 COMBATTING STIGMAS WITH SCHOOL GARDENS (KENYA)

Mundika Special School for the Deaf (Busia County, Kenya) is a refuge for students with special needs. Many have been abandoned by their families due to social stigmas surrounding their disabilities as well as the perception that they are putting a strain on already limited resources. When BFN field consultant Aurillia Manjella visited the school by chance (see Figure 5.C14), the principal reported difficulty procuring enough food for school meals; funds for vegetables were insufficient, particularly during the dry season (when produce is more expensive). Community-based partner SINGI (for more on partners see Section B) identified this as an opportunity to connect the overlooked resources of the environment with the most marginalized youth in the community. Over several training sessions, SINGI empowered the staff and students with the skills to cultivate their own indigenous leafy vegetables. After learning about traditional plants and their health benefits, the students established their own garden in an empty plot of land, seeding it with several BFN species. Aurillia reported:

> Our sustainable agriculture training was smoothly implemented thanks to the hardworking and talented students. They were so quick to learn

that before we knew it, the garden was well prepared, and seeds were planted. Within a month the first vegetables were harvested and sent to the kitchen. Local nutritious vegetables such as cowpea leaves, amaranth, and African nightshade were cooked and served to the students at no extra cost. The agriculture students have continued with their gardening project and now can comfortably feed the entire school, day in and day out. During holidays they sell their vegetables and use the profits to buy seeds and other materials (such as school supplies and clothes). To them, rather than feeling idle at home, this useful skill helps them earn money for their basic needs and gain independence.

Beyond the immediate effect of supplying healthy meals for the 100 students (boosting their nutrition and development), the school garden demonstrates to students' families and neighbours that disability does not prevent contribution to the community.

FIGURE 5.C14 Aurillia with students at Mundika Special School

Source: A. Manjella

Source: Adapted from a testimonial by BFN/SINGI Consultant Aurillia Manjella

BOX 5.C20 MAXIMIZING OUTREACH, FROM A STREET DRAMA TO A WEBSITE (SRI LANKA)

Sri Lanka experimented with many different outreach approaches including food festivals, national and international symposiums to reach scientists and development workers, television and radio programmes to reach the general public, and a combination of interactive learning materials to reach schoolchildren.

Producer training and demonstration plots

Sri Lankan partners reached producers by conducting training sessions throughout the implementation process encompassing food composition analysis, school gardening, home gardening, integrated pest management, cultivation of root and tuber crops and food use and recipes. Three forms of training were employed:

- Urban model garden: invites community members to see strategic cultivation of local tubers, leafy greens and vegetables that combines traditional growing methods with new space-saving techniques such as aerial planting (with the National Food Promotion Board).
- School gardens: introduces students to traditional crops and trains them on cultivation methods, also encourages germplasm collection for unique species/varieties (with the Plant Genetic Resources Centre).
- Demonstration/Conservation plots: showcase biodiversity through the cultivation of 36 local roots and tubers, which are made into value-added products for income generation in 18 villages (with the Community Development Centre).

Consumer training and novel foods

Building on the success of the Hela Bojun food outlets (see Box 5.C14), awareness campaigns were developed to inform Sri Lankans of the concepts of calorie-based intake of food, nutrition labelling of processed food, formulation of therapeutic diets and highlighting the nutritional diversity amongst the traditional Sri Lankan dishes. Additional efforts popularized local recipes for the production of novel food products utilizing local root and tuber crops, including *kiri ala* chips, *mukkru, wal ala dodol*, toffee *and damdina ala* porridge.

Food festival

Coinciding with World Food Day 2017, BFN Sri Lanka and the Department of Agriculture hosted a massive festival that attracted thousands of participants and represented the diversity of 25 districts through cooking competitions (in which finalists contributed recipes for a published booklet), food stalls, and seminars focused on food and health (see Figures 5.C15–5.C16). Finally, a colourful street drama was scripted and performed (with traditional Sri Lankan characters alongside costumed chefs) on the negative effects of fast food and the health value of indigenous produce.

Free fruit and increased research capacity

BFN joined an annual fruit exhibition, sponsored by the Fruit Crop Research and Development Institute, where students sampled a display table of local fruit varieties, and learned about their medical and commercial uses. An associated seminar gathered experts from the Department of Agriculture, Department of Ayurveda and Peradeniya University to discuss fruit diversity as well as nutritional and medicinal values. Members of the Food Promotion board were present and were encouraged to sell ready-to-serve drinks made from local fruits (see Box 5.C15 for an example). The event simultaneously educated consumers and brought together specialists with knowledge of underutilized local fruits, many of whom were able to contribute information previously unknown to BFN researchers.

Bringing BFN online

Information gathered over the course of the project was compiled into the BFN Sri Lanka website. The site, created by the Department of Agriculture's National Information and Communication Centre, was designed to be reached both via computers and smartphones to encourage access for a wide audience. The easily navigable platform contains facts about traditional crops and varieties (identified in English, Sinhala and Tamil), food composition tables, traditional knowledge, nutrition, recipes, and resources from the BFN Project. As of September 2019, over 432,000 visitors had accessed the site: www.bfnsrilanka.org.

Additional media approaches used by the project included a radio quiz competition in which students answered questions about agriculture and biodiversity for food and nutrition. A multimedia CD was given to schools with information corresponding to the website, and an Android phone application was designed that shared information on nutritious and traditional foods available at food outlets.

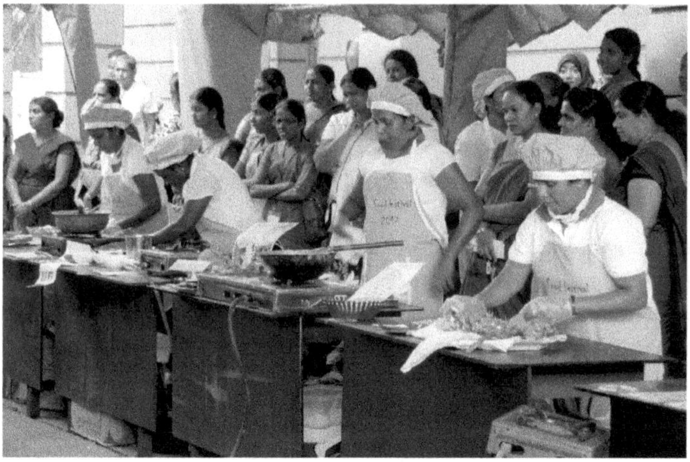

FIGURE 5.C15–16 Scenes from the Traditional Food Festival in Colombo, 2017
Source: D. Hunter

BOX 5.C21 REVIVING TRADITIONS WITH ALAÇATI WILD HERB FESTIVAL (TURKEY)

The Western region of Turkey is characterized by the lowest prevalence of undernutrition and diet-related illnesses. A significant factor for this is the regional diet of local vegetables, herbs and fruit, which includes a wide variety of wild edible plants. This biodiversity, and related food culture, is celebrated each year by collectors, farmers, chefs and thousands of visitors at the Alaçatı Herb Festival, a four-day event devoted entirely to wild edibles (see Figure 5.C17).

FIGURE 5.C17 A family-run food stall at Alaçatı Wild Herb Festival, 2015
Source: S. Landersz

The festival, at its 10th edition in 2019, acts as a platform for sharing knowledge about wild biodiversity and Turkish food culture. During the festival, a wild herb collecting competition is organized that rewards the gatherer of the highest number of edible plants; a separate prize is awarded to the best recipe using wild edible plants. The involvement of celebrity chefs during the festival has also helped popularize wild edible plants among younger generations. Furthermore, the festival offers seminars on nutrition and diets, exhibitions, selling of local products and plants, activities for children linked to the festival theme, food and cooking workshops and visits to the Wild Edible Plants Collection Garden.

The BFN Turkey team has successfully partnered with Alaçatı festival organizers to ensure that key messages and products are disseminated to the thousands of attendees. Awareness-raising material and recipe books produced by BFN Turkey for the occasion help visitors appreciate underutilized wild species and learn more about their nutritional value. The festival has opened up the market for wild edibles, an opportunity which has been mostly taken up by the enterprising women of Alaçatı who produce and sell local and traditional products/foods and have organized as a group.

BOX 5.C22 GREENING TOMORROW'S WORKFORCE (TURKEY)

If practical knowledge and use of wild edibles in Turkey is to survive, it must be passed from older to younger generations. 'Green' vocational training offers a good entry point for this transition, while providing unique skills to young people. In 2017, the Aegean Team of the BFN Project partnered with the Halim Foçali Anatolian Vocational and Technical College to deliver a three-day training programme to 16 student chefs. Lectures and handout materials taught students about the uses of aromatic and medicinal plants existing in Turkey, sustainable harvest techniques and production, the nutritional value of vegetables and wild edible plants and their importance in tackling malnutrition. Lectures were reinforced by practical activities and nature walks led by the BFN team, schoolteachers, local foragers and research staff from the Foça District National Education Directorate. Students were trained in the field to identify species, collect and photograph plants to start a school herbarium and use the plants in cooking demonstrations and practical sessions led by school staff for the preparation of traditional salads, roast dishes, pancakes and omelettes.

This newfound knowledge was shared during the Foça Science Education Festival where students involved in the project set up an exhibition stand, called 'Training on the international Biodiversity for Food and Nutrition (BFN) Project: Students in Nature and in the Kitchen' (see Figure 5.C18). Posters and other information on the different species, including recipes for their preparation, were distributed to visitors, teachers, students and parents. Neighbouring schools also took part in the Science Festival and thereby learned about the importance of wild edibles, their nutritional value and their importance in healthy diets. The longer-term implication of this training is that it helps students view 'green' work with biodiversity and traditional foods as a viable and applicable practical skill for future employment, with unique associated knowledge that makes students stand out.

FIGURE 5.C18 Turkish students sharing BFN species and recipe booklets at Halim Foçali science week
Source: BFN Turkey

The activities undertaken as part of this project have fostered interest from the National Education and Development Directorate of Foça to extend the programme to other schools and officially include traditional wild plants as part of the school curriculum.

> Our horizons have broadened as a result of this project. We have seen the plants and herbs we studied in their natural environment, collected them with our own hands and turned them into a meal. I believe the project will have lasting effects on our professional life.
>
> *Ahmet Sezer Şanlioğlu, Halim Foçali student*

BOX 5.C23 IN THE WORDS OF: AYFER TAN, BFN TURKEY AEGEAN REGIONAL COORDINATOR

'For a long time I coordinated the National Biodiversity and Genetic Resources Programme, conducting various projects for the conservation and sustainable utilization of biodiversity. During the implementation of the projects' ethnobotanical surveys, I visited villages and local markets and documented the tremendous number of local plants and underutilized species.

The BFN Project give us the opportunity to take a multidisciplinary approach collecting data on the traditional uses of these wild edible plants and mushrooms, and sharing knowledge on nutrition and dietary diversity. Due to their high nutritional value and inherent health benefits, edible wild species play a key role in traditional cookery and local diets, especially in the Aegean region. Traditional Aegean vegetable dishes – enriched by the culinary influence of the different civilizations and cultures that followed one another in the region – are no longer limited to the Aegean and have taken their place in world cuisine. This is an area that would continue to benefit from further study.' (see Figure 5.C19)

FIGURE 5.C19 Photo of Ayfer Tan at Alaçatı Wild Herb Festival, 2014
Source: D. Hunter

BOX 5.C24 IN THE WORDS OF: BFN COUNTRY PARTNERS, ON LESSONS LEARNED

There was no initial baseline awareness regarding the significance of 'biodiversity', therefore it was difficult to communicate the message. Terms such as 'local' might have evoked oranges and strawberries grown nearby rather than an indigenous species, while some target crops carried stigmas of being food for the "poor". From the beginning, recruit a dedicated communications specialist who can coordinate outreach activities and materials. Hire a press agency to get the word out: people (and not least, politicians and decision-makers) will only want to conserve something if they understand its value.

Brazil

Overcome low nutrition awareness and stigmas with clearly presented evidence. Prepare graphs that display the percentages of iron, zinc, vitamins that can be taken to schools, field days, agriculture shows and other events. "Seeing it with your own eyes" and experiential knowledge through hands-on activities can solidify knowledge in children and can subsequently reach their families.

Kenya

Develop a user-friendly website early on to organize and incorporate project information, making relevant information available to stakeholders in real time throughout the project.

Sri Lanka

Assessing the BFN project

BOX 5.C25 IN THE WORDS OF: DANNY HUNTER, BFN GLOBAL PROJECT COORDINATOR

'At the outset, the BFN Project asked, "Can nutrient-rich food biodiversity be mainstreamed in countries for the purpose of improving diets and nutrition?" This in itself was an unusual question for the Global Environment Facility to support. It was in fact the first time the GEF had funded a project with diet and nutrition objectives. At the end of the day, the BFN Project has clearly answered this question in the affirmative and provided a legacy of achievements and success.

The project has been unique and novel. It has brought together extensive research partnerships to establish a significant body of evidence on the nutritional value of biodiversity, and it has used this to persuasively inform policies and create new market opportunities. In doing so, it has generated important knowledge, policy and market innovations, many of which will no doubt positively influence the future actions of other countries. The project created innovative connections with institutional markets such as schools, public food procurement, linked to dietary guidelines and backed this up with great awareness-raising efforts by all countries. A key factor of success was having biodiversity mainstreaming champions with the vision and gravitas to mobilize people, partners and resources for a worthwhile vision.

There is no prescriptive approach to doing this, no one-size-fits-all. The BFN Project tailored its approach to local context and nuances, responding as we went along to many unanticipated opportunities and challenges. In many ways, this provides an exemplar roadmap for other countries to mainstream biodiversity for healthy diets and improved nutrition, which we hope will also inform the post-2020 CBD Biodiversity Strategy Framework. The BFN Project has set the bar high with its achievements and success, which will help realize food systems and landscapes that sustain the planet, drive prosperity, and nourish people.'

Notes

1 www.cbd.int/doc/bioday/2008/ibd-2008-factsheet-02-en.pdf
2 www.cbd.int/decision/cop/?id=11037
3 www.fao.org/3/a-i5248e.pdf
4 www.unstats.un.org/sdgs/metadata?Text=&Goal=2&Target=
5 www.fao.org/3/na698en/na698en.pdf
6 www.cbd.int/decision/cop/?id=11037
7 www.cbd.int/nbsap/
8 www.fao.org/cgrfa/assessments/global-assessments/sow/en/
9 www.fao.org/infoods/infoods/food-biodiversity/en/
10 www.fao.org/infoods/infoods/regional-data-centres/en/
11 www.agrobiodiversityplatform.org/files/2018/10/Assessing-Agrobiodiversity-A-Compendium-of-Methods-lowres.pdf
12 www.agrobiodiversityplatform.org/compendium/
13 Quilombolas are Afro-Brazilian residents of settlements first established by escaped slaves in Brazil.

References

Abeywickrama, H.M., Koyama, Y., Uchiyama, M., Shimizu, U., Iwasa, Y., Yamada, E., Ohashi, K., Mitobe, Y. (2018) 'Micronutrient status in Sri Lanka: A review', *Nutrients*, vol 10, no 11, article 1583. doi: 10.3390/nu10111583.

Abukutsa-Onyango, M.O. (2007) 'The diversity of cultivated African leafy vegetables in three communities in western Kenya', *African Journal of Food, Agriculture, Nutrition and Development*, vol 7, p 3.

Adanacioğlu, N., Tan, A., Karabak, S., Güzelsoy, N., Ayas, F., Aykas, L., Taş, N., Taylan, T. (2017) 'Economically important wild mushroom saffron milk cap (*Lactarius deliciosus* (L.) Gray) of the Aegean region, Turkey', *ANADOLU, Journal of AARI*, vol 27, no 2, pp 91–96.

Agra, N.G., Santos, R.F. (2001) 'Agricultura brasileira: situação atual e perspectivas de desenvolvimento', *Congresso da Sociedade brasileira de Economia e Sociologia Rural*, vol. 39, 2001, Recife, SOBER.

Ayas, F., Vuran, F.A., Yuksel, K., Çınar, O., Tugrul Ay, S., Karabak, S. (2017) 'The antioxidant capacities and per capita consumption of edible wild species and local varieties collected from Turkey within the GEF-funded Biodiversity for Food and Nutrition (BFN) project', *ANADOLU, Journal of AARI*, vol 27, no 2, pp 46–53.

Bacha, C.J.C., de Carvalho, L.V. (2014) *IRIBA Working Paper: 02. What explains the intensification and diversification of Brazil's agricultural production and exports from 1990 to 2012?* International Research Initiative on Brazil and Africa. Available at: www.brazil4africaorg/wp-content/uploads/publications/working_papers/IRIBA_WP02_What_Explains_The_Intensification_and_diversification_of_Brazil's_Agricultural_Production_and_Exports.pdf

Beltrame, D.M., Oliveira, C.N.S., Borelli, T., Santiago, A.C., Monego, E.T., de Rosso, V. V., Coradin, L., Hunter, D. (2016) 'Diversifying institutional food procurement—opportunities and barriers for integrating biodiversity for food and nutrition in Brazil', *Raízes*, vol 36, no 2, pp 55–69.

Beltrame, D.M.O., Gee, E., Güner, B., Lauridsen, N., Samarasinghe, W.L.G., Wasike, V. W., Hunter, D., Borelli, T. (2019) 'Mainstreaming biodiversity for food and nutrition into policies and practices: Methodologies and lessons learned from four countries', *ANADOLU, Journal of AARI*, vol 29, no 1, pp 25–38.

Bogard, J.R., Marks, G.C., Wood, S., Thilsted, S.H. (2018) 'Measuring nutritional quality of agricultural production systems: Application to fish production', *Global Food Security*, vol 16, pp 54–56. doi: 10.1016/j.gfs.2017.09.004.

Burlingame, B., Charrondière, R., Mouille, B. (2009) 'Food composition is fundamental to the cross-cutting initiative on biodiversity for food and nutrition', *Journal of Food Composition and Analysis*, vol 22, no 5, pp 361–365. doi: 10.1016/j.jfca.2009.05.003.

Candari, C., Cylus, J., Nolte, E. (2017) *Assessing the economic costs of unhealthy diets and low physical activity – An evidence review and proposed framework*, World Health Organization. Available at: www.euro.who.int/__data/assets/pdf_file/0004/342166/Unhealthy-Diets-ePDF-v1.pdf?ua=1

CBD. (2018) *Biodiversity at the heart of sustainable development input to the 2018 high-level political forum on sustainable development (HLPF)*, Secretariat of the Convention on Biological Diversity, Montreal, Canada. Available at: https://sustainabledevelopment.un.org/content/documents/18277CBD_input_to_2018_HLPF.pdf (Accessed 9 September 2019).

CBD. (2019) 'What is an NBSAP', Convention on Biological Biodiversity, Montreal, Canada. Available at: www.cbd.int/nbsap/introduction.shtml

Chmielewska, D., Souza, D. (2011) *The food security policy context in Brazil (No. 22) Country Study*, International Policy Centre for Inclusive Growth, Brasilia, DF - Brazil.

Coradin, L., Beltrame, D.M.O. (2016) 'From dish to podium – It's about a lot more than just sports in Rio', Bioversity International, Aug 2016, Available at: www.bioversityinternational.org/news/detail/from-dish-to-podium-its-about-a-lot-more-than-just-sports-in-rio/ (Accessed 27 June 2019).

Coradin, L., Siminski, A., Reis, A. (eds) (2011) *Espécies Nativas da Flora Brasileira de Valor Econômico Atual ou Potencial. Plantas para o Futuro – Região Sul*, Ministerio do Meio Ambiente, Brasília, DF, Brazil.

da Cunha, M.C., Mace, G., Mooney, H. (eds) (2019) *Summary for policymakers of the global assessment report on biodiversity and ecosystem services of the Intergovernmental Science-Policy Platform on Biodiversity and Ecosystem Services*, Intergovernmental Science-Policy Platform on Biodiversity and Ecosystem Services. Available at: www.ipbes.net/news/ipbes-global-assessment-summary-policymakers-pdf

Dahanayake, N. (2015) 'Some neglected and underutilized fruit-crops in Sri Lanka', *International Journal of Scientific and Research Publications*, vol 5, no 2, pp 1–7.

Dernini, S., Meybeck, A., Burlingame, B., Gitz, V., Lacirignola, C., Debs, P., Capone, R., El Bilali, H. (2013) 'Developing a methodological approach for assessing the sustainability of diets: The Mediterranean diet as a case study', *New Medit*, vol 12, no 3, pp 28–36.

Dogan, Y. (2012) 'Traditionally used wild edible greens in the Aegean region of Turkey', *Acta Societatis Botanicorum Poloniae*, vol 81, no 4, pp 329–342. doi: 10.5586/asbp.2012.037.

Erem, C. (2015) 'Prevalence of overweight and obesity in Turkey', *IJC Metabolic & Endocrine*, vol 8, pp 38–41. doi: 10.1016/j.ijcme.2015.07.002.

Ertug, F. (2014) 'Yenen Bitkiler (Edible Plants)', in A. Güner, T., Ekim (eds) *Resimli Türkiye Florası (İllustrated Flora of Turkey). Vol. 1*, pp 345–380, Ali Nihat Gökyiğit Vakfi, Flora Araştırmaları Derneği and Türkiye İş bankası Kültür Yayınları, İstanbul.

Fanzo, J., Holmes, M., Junega, P., Musinguzi, E., Smith, I.F., Ekesa, B., Bergamini, N. (2011) *Improving nutrition with agricultural biodiversity. A manual on implementing food systems field projects to assess and improve dietary diversity, and nutrition and health outcomes*, Bioversity International, Rome, Italy.

FAO. (2010) *Expert consultation on nutrition indicators for biodiversity 2.* Food consumption. FAO, Rome, Italy.

FAO. (2012) 'Sustainable diets and biodiversity. directions and solutions for policy, research and action', in Burlingame, B., Dernini, S. (eds) *Proceedings of the International Scientific Symposium. Biodiversity and Sustainable Diets against Hunger*, 3–5 November 2010, FAO Headquarters, Rome. Available at: www.fao.org/3/a-i3004e.pdf

FAO. (2017) *The State of Brazil's biodiversity for food and agriculture*, Food and Agriculture Organization of the United Nations. Available at: www.fao.org/3/CA3475EN/ca3475en.pdf

FAO. (2019) 'FAO commission on genetic resources for food and agriculture assessments', in J. Bélanger, D. Pilling (eds) *The state of the world's biodiversity for food and agriculture*, p572. Rome. Available at: www.fao.org/3/CA3129EN/CA3129EN.pdf

FAO, Government of Kenya. (2018) 'Kenya food composition tables', Nairobi, Kenya. Available at: www.fao.org/3/I8897EN/i8897en.pdf

FAO, IFAD, UNICEF, WFP, WHO. (2019) *The state of food security and nutrition in the world 2019. Safeguarding against economic slowdowns and downturns*, Food and Agriculture Organization of the United Nations, Rome, Italy. Available at: https://docs.wfp.org/api/documents/WFP-0000106760/download/?_ga=2.207800402.1765058888.1569506848-1262155193.1566480304

FAO, INFOODS. (2012) '*FAO/INFOODS guidelines for checking food composition data prior to the publication of a user table/database-version 1.0*', Food and Agriculture Organization of the United States, Rome, Italy.

Flora do Brazil 2020. (2019) 'Flora do Brasil 2020 em construção', Jardim Botânico do Rio de Janeiro. Available at: http://floradobrasil.jbrj.gov.br/

FNDE. (2018) 'IQ COSAN – Indice de qualidade – Manual'. Fundo Nacional de Desenvolvimento da Educação. Available at: www.fnde.gov.br/acesso-a-informacao/institucional/legislacao/item/12142-iq-cosan

Fraundorfer, M. (2013) 'Zero hunger for the world – Brazil's global diffusion of its Zero Hunger strategy', *Austral: Brazilian Journal of Strategy & International Relations*, vol 2, no 4, Jul–Dec 2013, pp 91–116.

Frison, E.A., Cherfas, J., Hodgkin, T. (2011) 'Agricultural biodiversity is essential for a sustainable improvement in food and nutrition security', *Sustainability*, vol 3, pp 238–253. doi: 10.3390/su3010238.

Gelli, A., Neeser, K., Drake, L. (2010) *Home grown school feeding: Linking small holder agriculture to school food provision*. HGSF Working Paper Series #1.

Greenfield, H., Southgate, D.A.T. (2003) *Food composition data. Production management and use*, Food and Agriculture Organization of the United Nations, Rome, Italy.

Güner, A., Aslan, S., Ekim, T., Vural, M., Babaç, M.T. (eds) (2012). Türkiye Bitkileri Listesi (Damarlı Bitkiler). List of Turkish Vascular Plants. Nezahat Gökyiğit Botanik Bahçesi ve Flora Araştırmaları Derneği Yayını. İstanbul.

Güzelsoy, N.A., Ucurum, Ö., Tokat, E., Tan, A., Tuğrul Ay, S., Özbek, K. (2017) 'Nutritional properties of some wild edible species in Turkey', *ANADOLU, Journal of AARI*, vol 27, no 2, pp 39–45.

Hajjar, R., Jarvis, D., Gemmil-Herren, B. (2008) 'The utility of crop genetic diversity in maintaining ecosystem services', *Agriculture Ecosystems & Environment*, vol 123, no 4, pp 261–270. doi: 10.1016/j.agee.2007.08.003.

Hawkes, C., Fanzo, J. (2017) *Nourishing the SDGs: Global nutrition report 2017*, Development Initiatives Poverty Research Ltd., Bristo.

Hunter, D., Borelli, T., Beltrame, D.M.O., Oliveira, C.N.S., Coradin, L., Wasike, V.W., Wasilwa, L., Mwai, J., Manjella, A., Samarasinghe, G.W.L., Madhujith, T., Nadeeshani, H.V.H., Tan, A., Tuğrul Ay, S., Güzelsoy, N., Lauridsen, N., Gee, E., Tartanac, F. (2019) 'The potential of neglected and underutilized species for improving diets and nutrition', *Planta*, vol 250, no 3, pp 709–729. doi: 10.1007/s00425-019-03169-4.

Hunter, D., Heywood, V.H. (eds) (2011) *Crop wild relatives; a manual of in situ conservation*, issues in agricultural biodiversity, Earthscan, Abingdon.

Hunter, D., Özkan, I., Beltrame, D.M.O., Samarasinghe, W.L.G., Wasike, V.W., Charrondière, U.R., Borelli, T., Sokolow, J. (2016) 'Enabled or disabled: Is the environment right for using biodiversity to improve nutrition?' *Frontiers in nutrition*, vol 3, article 14, doi: 10.3389/fnut.2016.00014.

IBGE. (2006) *Censo Agropecuario 2006*, Fundação do Instituto Brasileiro de Geografia e Estatística, Rio de Janeiro, Brazil.

IBGE. (2010) *Pesquisa de Orçamentos Familiares 2008–2009: antropometria e estado nutricional de crianças, adolescentes e adultos no Brasil*, Fundação do Instituto Brasileiro de Geografia e Estatística, Coordenação de Trabalho e Rendimento, Rio de Janeiro, Brazil.

IBGE. (2011) *Pesquisa de Orçamentos Familiares 2008–2009: análise do consumo alimentar pessoal no Brasil*, Fundação do Instituto Brasileiro de Geografia e Estatística, Coordenação de Trabalho e Rendimento, Rio de Janeiro, Brazil.

IFPRI. (2016) 'Taking action: Progress and challenges in implementing nutrition policies and programs' in *Global nutrition report 2016: From promise to impact: Ending malnutrition by 2030*, pp 44–59, International Food Policy Research Institute.

IPBES. (2016) 'The assessment report of the Intergovernmental Science-Policy Platform on Biodiversity and Ecosystem Services on pollinators, pollination and food production', in S.G. Potts, V. L. Imperatriz-Fonseca, H. T. Ngo (eds) *Secretariat of the intergovernmental science-policy platform on biodiversity and ecosystem services*, p552. Bonn. Available at:

https://ipbes.net/sites/default/files/downloads/pdf/2017_pollination_full_report_book_v12_pages.pdf

IPBES. (2019) 'Summary for policymakers of the global assessment report on biodiversity and ecosystem services of the Intergovernmental Science-Policy Platform on Biodiversity and Ecosystem Services', in S. Díaz, J. Settele, E. S. Brondizio, H. T. Ngo, M. Guèze, J. Agard, A. Arneth, P. Balvanera, K. A. Brauman, S. H. M. Butchart, K. M. A. Chan, L. A. Garibaldi, K. Ichii, J. Liu, S. M. Subramanian, G. F. Midgley, P. Miloslavich, Z. Molnár, D. Obura, A. Pfaff, S. Polasky, A. Purvis, J. Razzaque, B. Reyers, R. Roy Chowdhury, Y. J. Shin, I. J. Visseren-Hamakers, K. J. Willis, C. N. Zayas (eds) IPBES secretariat, Bonn, Germany. https://ipbes.net/sites/default/files/inline/files/ipbes_global_assessment_report_summary_for_policymakers.pdf

Karabak, S. (2017) 'Economic and socio-cultural importance of edible wild species', *ANADOLU, Journal of AARI*, vol 27, no 2, pp 26–38.

Katulanda, P., Jayawardena, M.A.R., Sheriff, M.H.R., Constantine, G.R., Matthews, D. R. (2010) 'Prevalence of overweight and obesity in Sri Lankan adults', *Obesity Reviews*, vol 11, no 11, pp 751–756. doi: 10.1111/j.1467-789X.2010.00746.x

Kehlenbeck, K., Asaah, E., Jamnadass, R. (2013) 'Diversity of indigenous fruit trees and their contribution to nutrition and livelihoods in sub-Saharan Africa: Examples from Kenya and Cameroon', in J. Fanzo, D. Hunter, T. Borelli, F. Mattei (eds) *Diversifying food and diets: Using agricultural biodiversity to improve nutrition and health*, pp 257–269, Routledge, Abingdon.

Kenya National Bureau of Statistics. (2018) *Basic report on well-being in Kenya*, Nairobi, Kenya.

Kitaoka, K. (2018) 'The national school meal program in Brazil: A literature review', *The Japanese Journal of Nutrition and Dietetics*, vol 76, pp S115–S125. doi: 10.5264/eiyogakuzashi.76.S115

Kuhnlein, H.V. (2003) 'Micronutrient nutrition and traditional food systems of indigenous peoples', in FAO (ed) (2003) 32/2003 *Food, nutrition and agriculture*, pp 33–39, Food and Agriculture Organization of the United Nations, Rome, Italy. Available at: http://www.fao.org/tempref/docrep/fao/005/y8346m/y8346m00.pdf

Lartey, A. (2016) 'Linking agriculture with nutrition within SDG2: Making a case for a dietary diversity indicator', *Field Exchange*, vol 50, p 51. Available at: www.ennonline.net/fex/50/sdg2dietdiversityindicator

Levy-Costa, R.B., Sichieri, R., Pontes, N.S., Monteiro, C.A. (2005) 'Household food availability in Brazil: Distribution and trends (1974–2003)', *Revista Saúde Pública*, vol 39, pp 530–540.

Maluf, R.S., Burlandy, L., Santarelli, M., Schottz, V., Speranza, J.S. (2015) 'Nutrition-sensitive agriculture and the promotion of food and nutrition sovereignty and security in Brazil', *Cien Saude Colet*, vol 20, no 8, pp 2303–2312. doi: 10.1590/1413-81232015208.14032014.

Mburu, S.W., Koskey, G., Kimiti, J.M., Ombori, O., Maingi, J.M., Njeru, E.M. (2016) 'Agrobiodiversity conservation enhances food security in subsistence-based farming systems of Eastern Kenya', *Agriculture & Food Security*, vol 5, p 19. doi: 10.1186/s40066-016-0068-2.

MMA. (2017) 'National Biodiversity Strategy and Action Plan (NBSAP)', Ministério do Meio Ambiente, Brasília, DF, Brazil. Available at: www.cbd.int/doc/world/br/br-nbsap-v3-en.pdf

MMA. (2018) 'Diário oficial da união. portaria interministerial N° 284, de 30 de Maio de 2018', Ministério do Meio Ambiente.

MMDE. (2016) 'National Biodiversity Strategy and Action Plan NBSAP (2016–2022)', Ministry of Mahaweli Development and Environment. Available at: www.cbd.int/doc/world/lk/lk-nbsap-v2-en.pdf

MoH Brazil. (2008–2017) Ministry of Health) (2008–2017). Surveillance System for Risk Factors and Protection for Chronic Diseases by Telephone Inquiry (VIGITEL).

MoH Kenya. (2011) 'The Kenya National Micronutrient Survey', Ministry of Health. Available at: www.nutritionhealth.or.ke/wp-content/uploads/Downloads/The%20Kenya%20National%20Micronutrient%20Survey%202011.pdf

MoMD&E (2016) *National biodiversity strategic action plan 2016–2022*, Biodiversity Secretariat, Ministry of Mahaweli Development and Environment, Colombo, Sri Lanka. Available at: www.cbd.int/doc/world/lk/lk-nbsap-v2-en.pdf

Muthoni, J., Nyamongo, D.O., Silingi, M. (2010) 'Participatory characterization and evaluation of some African leafy vegetables in Lari, Kiambu West District, Central Kenya', *Journal of Horticulture and Forestry*, vol 2, no 2, pp 12–16.

Mwamodenyi, J.M., Omondi, S. (2012) *The state of the world's forest genetic resources. Country report: Kenya*, FAO, Rome, Italy. Available at: www.fao.org/3/a-i3825e.pdf

Myers, N., Mittermeier, R.A. Mittermeier, C.G., Da Fonseca, G.A., Kent, J. (2000) 'Biodiversity hotspots for conservation priorities', *Nature*, vol 403, pp 853–858.

Padulosi, S., Bala Ravi, S., Rojas, W., Valdivia, R., Jager, M., Polar, V., Gotor, E., Mal, B. (2013) 'Experiences and lessons learned in the framework of a global UN effort in support of neglected and underutilized species', *ISHS Acta Horticulturae*, vol 979, pp 517–531.

Padulosi, S., Hodgkin, T., Williams, J.T., Haq, N. (2002) 'Underutilised crops: Trends, challenges and opportunities in the 21st century', in J. Engels, V.R. Rao, M. Jackson (eds.) *Managing plant genetic diversity*, pp 323–338, CAB International, Wallingford.

PAR. (2018) *Assessing agrobiodiversity: A compendium of methods*, Platform for Agrobiodiversity Research, Rome, Italy.

Pereira, P.A.A., Martha, J.R. Geraldo, B. Santana, Carlos, A.M., Alves, E. (2012) 'The development of Brazilian agriculture: Future technological challenges and opportunities', *Agriculture & Food Security*, vol 1, pp 4.

Prakash, V., Jaiswal, N., Srivastava, M. (2017) 'A review on medicinal properties of *Centella asiatica*', *Asian Journal of Pharmaceutical and Clinical Research*, vol 10, pp 10. doi: 10.22159/ajpcr.2017.v10i10.20760.

Rajapaksha, U. (1998) *Traditional food plants in Sri Lanka*, Hector Kobbekaduwa Agrarian Research and Training Institute, Colombo, Sri Lanka.

Redin, E., Fialho, M.A.V. (2010) 'Política Agrícola Brasileira: Uma Análise Histórica sa Inserção da Agricultura Familiar', in Sociedade Brasileira de Economia, Administração e Sociologia Rural (eds) *Congresso da sociedade brasileira de economia, administração e sociologia rural*, vol 48, pp. 1–19, Campo Grande – MS.

Remans, R., Flynn, D.F., DeClerck, F., Diru, W., Fanzo, J., Gaynor, K., Lambrecht, I., Mudiope, J., Mutuo, P.K., Nkhoma, P., Siriri, D., Sullivan, C., Palm, C.A. (2011) 'Assessing nutritional diversity of cropping systems in African villages', *PLoS One*, vol 6, no 6, p e21235. doi: 10.1371/journal.pone.0021235.

Remans, R., Wood, S.A., Saha, N., Anderman, T.L., DeFries, R.S. (2014) 'Measuring nutritional diversity of national food supplies', *Global Food Security*, vol 3, no 3–4, pp 174–182. doi: 10.1016/j.gfs.2014.07.001.

Republic of Kenya. (2015) *Fifth national report to the conference of parties to the Convention on Biological Diversity*, Nairobi, Kenya.

Republic of Turkey. (2014) *Fifth National Report to the UN Convention on Biological Diversity.*

Ruel, M.T., Minot, N., Smith, L. (2005). Patterns and determinants of fruit and vegetable consumption in sub-Saharan Africa: A multi-country comparison. World Health

Organization, Geneva, Switzerland, and International Food Policy Research Institute, Washington D.C., US.

Sánchez-Mata, M.d.C., Tardío, J. (eds) (2016) *Wild edible plants: Ethnobotany and food composition tables part II: Nutrient and bioactive compounds in mediterranean*, Springer, New York.

Şekercioğlu, C.H., Anderson, S., Akçay, E., Bilgin, R., Can, O.E., Semiz, G., Tavşanoğlu, C., Yokeş, M.B., Soyumert, A., İpekdal, K., Sağlam, I.K., Yücel, M., Dalfesm, H.M. (2011) 'Turkey's globally important biodiversity in crisis', *Biological Conservation*, vol 44, no 12, pp 2752–2769. doi: 10.1016/j.biocon.2011.06.025.

SiBBr. (2019) 'Biodiversity and nutrition online food composition tables database', Sistema de Informação sobre a Biodiversidade Brasileira. Available at: https://ferramentas.sibbr.gov.br/ficha/bin/view/FN

Sidaner, E., Balaban, D., Burlandy, L. (2012) 'The Brazilian school feeding programme: example of an integrated programme in support of food and nutrition security', *Public Health Nutrition*, vol 16, pp 1–6. doi: 10.1017/S1368980012005101

Souza, A.M., Rosangela, I., Pereira, A., Yokoo, E.M., Levy, R.B., Sichieri, R. (2013) 'Most consumed foods in Brazil: National dietary survey 2008–2009', *Revista Saúde Pública*, vol 47, no suppl 1, pp 190S–9S.

Statista. (2019) 'Share of adult population who were considered overweight in Brazil from 2008 to 2018'. Available at: www.statista.com/statistics/781286/share-adult-people-overweight-brazil/ (Accessed September 2019).

Tan, A. (2009) *Second national report of Turkey on conservation and sustainable utilisation of plant genetic resources for food and agriculture*, Food and Agriculture Organization of the United Nations, Rome, Italy.

Tan, A., Adanacioğlu, N., Karabak, S., Aykas, L., Tas, N., Taylan, T. (2017) 'Biodiversity for food and nutrition: Edible wild plant species of the Aegean Region of Turkey', *ANADOLU, Journal of AARI*, vol 27, no 2, pp 1–8.

Turkstat. (2017) 'Turkey health survey 2016', Turkish Statistical Institute.

Ulloa, C., Acevedo-Rodríguez, P., Beck, S., Belgrano, M.J., Bernal, R., Berry, P.E., Brako, L., Celis, M., Davidse, G., Forzza, R.C., Gradstein, S.R., Hokche, O., León, B., León-Yánez, S., Magill, R.E., Neill, D.A., Nee, M., Raven, P.H., Stimmel, H., Strong, M.T., Villaseñor, J.L., Zarucchi, J.L., Zuloaga, F.O., Jørgensen, P.M. (2017) 'An integrated assessment of the vascular plant species of the Americas', *Science*, vol 358, no 6370, pp 1614–1617. doi: 10.1126/science.aao0398

UN. (2012) *Report of the United Nations Conference on Sustainable Development, Rio de Janeiro, Brazil, 20–22 June 2012*, United Nations, New York.

Valencia, V., Wittman, H., Blesh, J. (2019) 'Structuring markets for resilient farming systems', *Agronomy for Sustainable Development*, vol 35, p 25. doi: 10.1007/s13593-019-0572-4

Von Grebmer, K., Bernstein, J., Nabarro, D., Prasai, N., Amin, S., Yohannes, Y., Sonntag, A., Patterson, F., Towey, O., Thompson, J. (2016) *2016 global hunger index: Getting to zero hunger*, Welthungerhilfe, Bonn, International Food Policy Research Institute, Washington, DC and Concern Worldwide, Dublin.

WHO. (2003) *WHO technical report series 916: Diet, nutrition and the prevention of chronic diseases, report of a joint WHO/FAO expert consultation*, World Health Organization, Geneva, Switzerland. Available at: http://health.euroafrica.org/books/dietnutritionwho.pdf

Williams, J.T., Haq, N. (2002) *Global research on underutilized crops – An assessment of current activities and proposals for enhanced cooperation*, International Centre for Underutilized Crops, Southampton.

World Bank. (2015) 'Poverty & equity data portal – Kenya'. Available at: http://povertydata.worldbank.org/poverty/country/KEN (Accessed September 2019).

PART III

Future opportunities and directions

In this section:

- Further opportunities for biodiversity and ecosystem services to contribute to healthy and sustainable food systems in the context of urban-rural linkages.
- The key achievements, lessons learned and good practices of the BFN Project are considered in terms of strategic insights and widely recognized opportunities and leverage points that can contribute to food systems transformation.

6

BIODIVERSITY, FOOD SYSTEMS AND URBAN-RURAL LINKAGES

Florence Egal and Thomas Forster[1]

Introduction

Biodiversity is often addressed in the context of either the rural landscapes of soil organisms, birds, insects and numerous species across the spectrum of flora and fauna, or in the context of urbanization, in terms of blue and green corridors or natural habitats in urban and peri-urban spaces. This chapter focuses instead on the dynamic interactions of biodiversity and food systems *across* the urban-rural continuum. Each of these are important for biodiversity conservation, as all biodiversity is under enormous threat of extinction (IPBES, 2019). Agricultural biodiversity (or agrobiodiversity) is a vital subset of biodiversity, as it is the foundation for all anthropogenic food systems, without which our species would not survive. What is less well addressed in policy and research is the spatial and functional intersection of food systems and biodiversity in the flows that define Urban Rural Linkages (URLs) (UN-Habitat, 2019).

This chapter addresses current trends in the intersection of biodiversity, food systems and URLs. The field is developing rapidly in response to socio-economic and environmental drivers which are examined below. Global policy agendas in recent years have begun to address interconnected challenges and opportunities for integrating urban and rural areas, selectively reviewed in this chapter. Nonetheless, significant gaps exist between placed-based practices, the lack of research to generate practice-based evidence, and global policy addressing this intersection of biodiversity, food and URLs. The formal literature of peer-reviewed, rigorous analysis and synthesis is weak at this intersection. However, at the level of practice, there are innovations, investments and initiatives in all regions of the world that hold promise for both biodiversity conservation and sustainable urban and territorial (or city region) food systems. Where formal research literature is lacking, the co-authors cite informal and grey literature as well as personal experience with the issues and processes discussed.

Recently there has been attention in policy and research to challenges at the intersection of food systems and biodiversity from an environmental perspective, with increasing concerns about the impacts of industrial monocultures on natural resources and in particular on biodiversity, and more recently, the fact that food systems are failing in delivering healthy diets to people (GLOPAN, 2016). Due to the dual processes of urbanization and globalization, diets are changing all over the world, and agricultural policies which have determined to a great extent these shifts tend to prioritize a commodity approach and international trade. The consumption of high-calorie, nutrient-poor foods that are high in fats, sweeteners and salt is increasing in both rural and urban areas (Hawkes, 2007). Dynamically changing food systems, with significant impacts on diets and nutrition, are inseparable from the relationships between rural and urban areas, and deeply connected to the underlying natural resource foundation of food systems, including soil, water, forest and trees, marine and riverine resources. These linkages have been, by and large, overlooked by both food and agriculture policies and urban policies to date, and they have wide implications for biodiversity (in particular agrobiodiversity) and related ecosystem functions (UN-Habitat, 2019).

Despite the fact that the majority of the world's food supply comes from smallholder producers, the commodification of agriculture has led to the standardization of food chains and an erosion of biodiversity (IPBES, 2019). Out of 6,000 plant species cultivated for food today, only 200 make major contributions to food production globally, regionally or nationally. Only nine of these plant species account for 66% of total crop production, with rice, maize and wheat contributing to a disproportionate majority of human calorie intake (FAO, 2019a). Small family farms traditionally produced, gathered or collected a wide range of plant and animal products, all of which required substantial labour and appropriate environment management. Small-scale farmers have been encouraged for decades to abandon traditional food and farming practices that were risk-averse and low input, and instead to specialize in fewer crops, access new markets and engage in commercial agriculture (associated with land consolidation, mechanization, more costly inputs and usually higher risk). In addition, areas where wild foods were available have shrunk. These transformations have in turn led to declines in labour demands and increased need for land, water and agriculture inputs, pushing vulnerable rural households out of agriculture and reducing availability of wild foods. The consolidation and concentration of post-harvest processing activities and the adoption of new legal and regulatory measures have also presented barriers to micro, small and medium-size enterprises. Smallholder producers and food enterprises which traditionally fed the local population have been relegated to an informal economy that receives little support from government authorities, when not actively discouraged or even barred from the market (Berdegué et al., 2014; Proctor and Berdegué, 2016).

In many areas, reduced diversity and the adoption of 'modern' food practices dependent on external inputs and imported foods has led to increased exposure to shocks and stresses, further accelerating migration to urban areas on a transitory or permanent basis. Rural areas that supported livelihoods of the local population now provide few or no acceptable opportunities for youth. While in-demand services such as schools, health, business and cultural opportunities are decreasing or absent, the increasing age and feminization of the remaining population further aggravate the situation in rural territories. Perceived job availability and access to health and education in urban areas, together with the barriers to economic viability in rural areas, have served to drive rural out-migration in many regions while fuelling urban growth and the expansion of urban slums.

Impacts of urban-rural linkages on diets and biodiversity

Rural migrants bring their food habits to towns and cities, but they often cannot find or process the foods (and biodiversity) with which they are familiar. In addition to encountering different distribution systems and living conditions, they are frequently obliged to switch to foods that are cheaper, more convenient and readily available, and usually more highly processed. Global marketing strategies have also resulted in teenagers worldwide switching to junk foods and soft drinks. The diets of migrants therefore evolve, though in many cases they retain connections to their culture and communities of origin. Traditional diets, often perceived to be healthier, remain the basis of complementary and child feeding, and the solution to addressing diet-related diseases as well as a means of preserving cultural traditions and social networks (Bichard et al., 2005).

In many cases, an exchange of foods and related practices develops across the urban-rural continuum, with urban people sending rice, sugar and vegetable oil to their rural homes, in particular during the hungry season, and rural families sending fresh products, wild foods (including game) or fuel wood, especially in times of crisis. Rural households also adopt and produce new vegetables for consumption and sale[2]. This contributes to the hybridization of diets in rural areas, with families perceiving city foods as 'modern' and substituting traditional foods as status symbols, but also increasingly shifting to the hybridization of urban diets. In Burkina Faso, urban consumers have revisited 'babenda', a mix of wild leaves (substituted by sorrel) and cereals traditionally prepared during the hungry seasons and now sold by street food vendors. For example, in Niamey, moringa-based kopto sauce is also in increasing fashion across social groups. Such modernized traditional dishes are perceived by urban consumers as a link to culture and rural origins[3].

In many countries, food aid projects distributing wheat, maize or rice to vulnerable urban and rural households has led to a decrease in demand for (and therefore a decrease in both production and consumption of) local foods such as sorghum, roots and tubers, and in particular indigenous species. These processes have also led to a decrease in the cultural acceptability of such traditional foods. The notion of a 'national diet' increasingly determined by central planning, urbanization and food import, is undermining the development – and sustainable development – of sociologically and environmentally important bioregions.

Limited attention has been given so far to understanding and optimizing rural-urban linkages as a consciously planned component of urban food security. Too often, as cities become increasingly dependent on imported foods, rural areas, geared to export, progressively lose vitality as their cultures and local biodiversity continue to erode alongside traditional rural lifestyles and social networks. Rural development projects often do not take into account urban demand: food is produced, and value chains are set up without understanding and analysing what urban consumers want, leading to missed opportunities for family agriculture and local development[4].

Solutions that link biodiversity conservation, food security and nutrition in urban areas with their rural territories have emerged, but they are usually the outcomes of nongovernmental and civil society initiatives, and policy and research support are generally lacking. Academic and research institutions have begun to analyse the complex interactions between urban and rural areas in the context of the multiple economic, social and environmental challenges facing local and national governments (FAO, 2018b). Other drivers such as climate change, biodiversity loss and rural-to-urban migration have impacted global policy agendas, but without connections to emerging practice and early research in these interlinkages at the urban-rural level.

Policy frameworks integrating biodiversity, food systems and urban-rural linkages

The intersection of biodiversity, urbanization and food systems has gained global relevance as the nexus of biodiversity, food systems and URLs, and it is increasingly recognized in policy frameworks at different levels – even without the experience of place-based practitioners or evidence-based research. Normative policy agendas addressing some of these interlinkages have been evolving in the last decade through a succession of often parallel but synergistic political initiatives (Figure 6.1).

New understanding of the critical importance of strengthening URLs and overcoming the dichotomy of urban and rural has been introduced in policy and implementation initiatives at different levels in recent years. Some of the initiatives include reassessing the roles of local people as stewards and managers of diverse landscapes that supported traditional diets that need to be restored and recovered. It is urgent to document, revisit and protect the knowledge vested in multi-generational, traditional and indigenous peoples' communities. This knowledge provides a time-tested and low-input basis for sustainable territorial

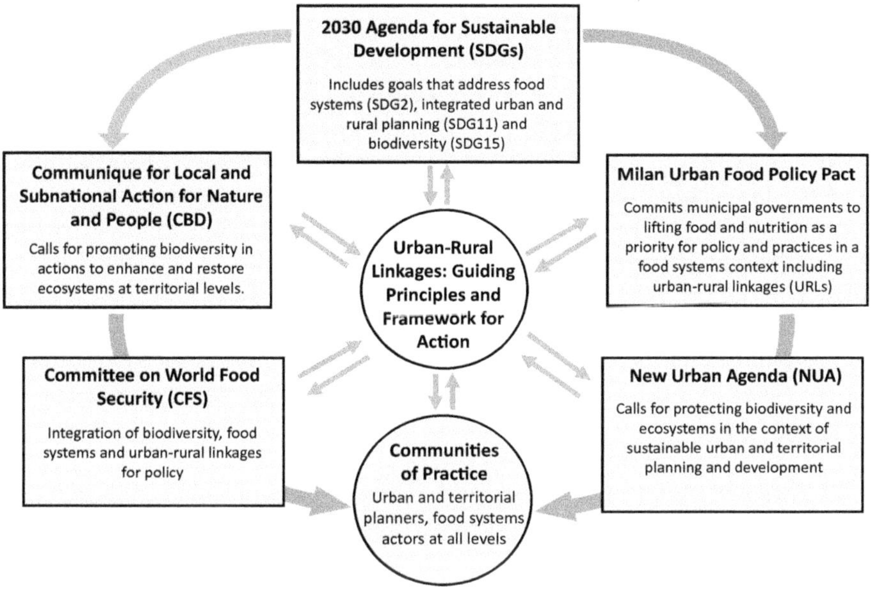

FIGURE 6.1 Linking normative agendas to biodiversity URLs

development as an essential dimension for protecting the ecological systems stewarded by local knowledge. It is also important to understand how traditional values adapt in urban contexts as an important dimension of urban demand and sustainable territorial development (Forster and Getz, 2014).

National attention to the importance of URLs was recognized in The Future We Want (UN, 2012) outcome document of the 2012 United Nations Conference on Sustainable Development (Rio+20), which in turn was the basis for debate and negotiation of the Sustainable Development Goals (SDGs) that succeeded the Millennium Development Goals (MDGs). The 2030 Agenda for Sustainable Development (UN, 2012) includes an SDG for ending hunger, improving nutrition and promoting sustainable agriculture (SDG 2); the conservation of biodiversity (SDG 15); and sustainable urbanization (SDG 11) with a target addressing urban-rural linkages (SDG 11.a). These targets call on national governments to 'support positive economic, social and environmental links between urban, peri-urban and rural areas by strengthening national and regional development planning'. The interlinkages of these and other SDGs are important but guidance for how these goals may be linked cannot be found in the 2030 Agenda. However, other parallel, internationally agreed upon normative frameworks and guidelines, addressed below, do begin to highlight the interlinkages.

At the subnational municipal policy level, cities from around the world negotiated in 2015, a mayoral protocol facilitated by a technical team convened by the City of Milan. This process occurred over the same period during which the SDGs and targets were being negotiated. The resulting

Milan Urban Food Policy Pact[5] adopts a sustainable food systems framework for urban food policy and planning. This framework draws upon cities' experiences in six areas: governance sustainable diets and nutrition, social and economic equity, food supply and distribution, food production and food waste. As of March 2020, signatory cities numbered around 210 small, intermediate and large cities from every continental region. After just three years, many Milan Pact Cities are exchanging and adopting practices and policies that link biodiversity, food systems and urban-rural linkages, as documented in the 2018 report, *The Role of Cities in the Transformation of Food Systems: Sharing Lessons of Milan Pact Cities* (FAO, 2018b).

The MUFPP explicitly refers to the links between sustainable food systems, biodiversity and urban-rural linkages. The preamble of the Milan Pact statement includes:

> acknowledging that urban and peri-urban agriculture offers opportunities to protect and integrate biodiversity into city region landscapes and food systems, thereby contributing to synergies across food and nutrition security, ecosystem services and human well-being; … recalling that cities have made commitments (…) to promote sustainable management of biodiversity through city biodiversity initiatives as part of the Convention on Biological Diversity.

Among recommended actions to operationalize these commitments, cities are encouraged to:

> Apply an ecosystem approach to guide holistic and integrated land use planning and management in collaboration with both urban and rural authorities and other natural resource managers by combining landscape features, for example with risk-minimizing strategies to enhance opportunities for agro-ecological production, conservation of biodiversity and farmland, climate change adaptation, tourism, leisure and other ecosystem services.
>
> *MUFPP Framework for Action: Food Production*

In the 2030 Agenda, there is no direct mention of food systems or biodiversity in SDG 11 on sustainable urbanization, as these issues are covered in other SDGs. However, the New Urban Agenda or NUA (UN-Habitat, 2016) adopted by UN member states at the *3rd United Nations Conference on Housing and Sustainable Urban Development* (Habitat III) in Quito, Ecuador, in 2016, gives specific attention to both food systems and biodiversity in the context of integrated urban and territorial planning and development.

The vision for sustainable urbanization calls to: 'Protect, conserve, restore and promote their ecosystems, water, natural habitats and biodiversity, minimize their environmental impact and change to sustainable consumption and production patterns' (para 13.h). The NUA also commits to: 'Ensure environmental sustainability by promoting … sustainable use of land and resources in urban development, by protecting ecosystems and biodiversity, including adopting healthy lifestyles in harmony with nature, by promoting sustainable consumption and production patterns' (para 14.c). In relation to food systems, the NUA states:

> We will promote the integration of food security and the nutritional needs of urban residents, particularly the urban poor, in urban and territorial planning, in order to end hunger and malnutrition. We will promote coordination of sustainable food security and agriculture policies across urban, peri-urban and rural areas to facilitate the production, storage, transport and marketing of food to consumers in adequate and affordable ways in order to reduce food losses and prevent and reuse food waste. We will further promote the coordination of food policies with energy, water, health, transport and waste policies, maintain the genetic diversity of seeds, reduce the use of hazardous chemicals and implement other policies in urban areas to maximize efficiencies and minimize waste.
>
> *para 123*

While biodiversity, food systems and URLs all are present in the 2030 and New Urban Agendas, they are not effectively tied together in ways that are operational for subnational policymakers and planners who have the task to operationalize these normative agendas.

The multi-actor global food governance policy platform of the Committee on World Food Security (CFS), however, did bring the intersection of biodiversity, food systems and URLs to a level of new integration and clarity. At its 43rd session in October 2016, the CFS held a *Forum on Urbanization, Rural Transformation and Implications for Food Security and Nutrition* in order to reach a better understanding of the issues at stake, and to identify key areas for policy attention and possible roles for CFS. The Committee then compiled experiences and effective policy approaches having URLs as a primary focus. The resulting document, 'Addressing Food Security and Nutrition in the Context of Changing Rural-Urban Dynamics: Experiences and Effective Policy Approaches', identified some key policy implications and knowledge gaps which were discussed at CFS 44 in October 2017 (CFS, 2017). Challenges related to biodiversity were brought up by almost all case studies.

> The interdependencies and synergies between urban and rural spaces and functions are further asserted through the economic dynamics, social links and environmental synergies. Ecology and biodiversity are part of these complex interrelations between rural and urban areas. Biodiversity and

> strong ecosystems are cornerstones to sustainable food systems, food security and nutrition. As cities and urban areas grow and food production becomes more industrial, there is an increased risk to the sustainability and destruction of the ecosystems that support food systems, and in particular the degradation of biodiversity.
>
> *p.21 para 39*

> Specific approaches based on territorial and food system characteristics are needed to ensure access to foods that are affordable and convenient, while generating income and employment of small-scale food producers and suppliers – in particular youth and women – and making sustainable use of local biodiversity.
>
> *p.33 para 46*

Sustainable food systems and healthy diets can be fostered through:

- Media campaigns which promote eating fresh, agro-ecological foods by linking urban consumers and rural smallholders;
- Short food supply chains and participatory approaches which establish an effective collaboration between consumers and producers and revive traditional food systems based on local biodiversity;
- Promoting nutrition education and ensuring access to culturally acceptable local foods in childcare centres;
- Public procurement that uses organic food in school canteens.

pp.33–34 following para 48: Key Messages

In recognising the significance of URLs in the context of food systems, each of these policy agendas has acknowledged the importance of biodiversity, and in particular agrobiodiversity, but only at a general level. Operational guidance and tools to help local and national governments integrate biodiversity/agrobiodiversity, food systems and URLs have only very recently become more concrete. One example of operational guidance developed for implementing governments and other actors at a global level is the *Urban-Rural Linkages: Guiding Principles and Framework for Action to Advance Integrated Territorial Development* (URL-GP), a multi-stakeholder initiative convened by UN-Habitat in 2018 (UN-Habitat, 2019). The initiative led to a set of agreed upon principles for strengthening URLs and a Framework for Action with practical entry points based on existing practice contributed by stakeholders. In this way, the URL-GP is structurally similar to the Milan Urban Food Policy Pact.

Guiding Principle 8 refers to the integration of territorial development with biodiversity and ecosystem services, stating:

> Among the other principles for strengthening integrated territorial development, investments along the urban-rural continuum should also prioritize

protecting, sustaining, and expanding areas important to biodiversity and ecosystem services in order to ensure an integrated urban-rural transition to resilient, resource efficient, low-carbon, and circular economies.

The URL-GP Framework for Action includes recommendations, for example in section J. Environmental Impact and Natural Resource and Land Management offer the following concrete provisions:

- **Conduct participatory multi-stakeholder assessments and establish monitoring systems** to identify environmental and health impacts of urbanization and/or migration across the urban-rural continuum involving different levels of government. The goal of such assessment should be to reduce risks and unequal burdens while promoting ecosystem-based approaches to planning and management for sustainable development. Finance mechanisms and incentives for equitable compensation for provision of ecosystem services between rural and urban areas must take into account the fact that the world's most fertile and productive lands are often found in the path of rapid urbanization. (para J.1)
- **Incorporate integrated landscape management into territorial development plans** and support primary producers and smallholders as the stewards of ecosystem services to protect natural resources and biodiversity (including agricultural biodiversity), and the reduction of waste. (para J.3)
- **Use systems approaches and circular economy frameworks** to link and prioritize interventions that integrate resilience to climate change, promote biodiversity and protect ecosystem functions and natural resources, and allow a successful transition to sustainable consumption and production. Recycling, reuse and remediation at the landscape level (including soil, water, plant and animal habitat), or waste and energy reuse across the urban-rural continuum are examples of urban-rural synergy in the circular economy. Integration of natural and built environments through nature-based solutions such as ecosystem-based adaptation to climate change will foster sustainability, liveability, productivity and resilience.
- **Establish service and product procurement systems that encourage and/or require sustainable use and management of natural resources** for multiple economic and ecosystem benefits. This should include fulfilling the demands from towns and cities that can be met by peri-urban and rural producers. Natural, resource-based production systems and their linkages to services and manufacturing are both casualties and causes of climate change. Policy interventions that encourage varied diets and emphasize plant-based sources of protein (especially legumes), for example, will not only be better for human nutrition but will also reduce the environmental costs of production (for example greenhouse gas emissions, groundwater pollution and land use).

Following the finalization and launch of the URL-GP, UN-Habitat and other partners have begun to develop participatory diagnostic and assessment tools for national and subnational application of both the principles and actions to strengthen urban-rural linkages in guidance policies, programmes and initiatives.

In late 2018, URLs were incorporated into the Global Biodiversity Summit for Local and Regional Governments, resulting in the 'Sharm El-Sheikh Communiqué for Local and Subnational Action for Nature and People' adopted by the *14th Conference of the Parties of the UN Convention on Biological Diversity* (COP 14). This outcome report welcomed the Urban-Rural Linkages Guiding Principles and Framework for Action to Advance Integrated Territorial Development (CBD, 2018). The Communiqué calls for 'aligning efforts for maximum impact to mainstream nature-based solutions, promote biodiversity actions and enhance and restore ecosystems in our territories'. Primary organizers of this summit, including the Local Governments for Sustainability (ICLEI) and the Network of Regional Governments for Sustainable Development (nrg4SD) have begun to include URLs and territorial approaches in their efforts to mainstream biodiversity in the context of climate change with new attention to both URLs and city region food systems.

Building on events such as the COP 14, the URL-GP has been recognized at other policy review events including the Global Landscape Forum (December 2018), COP 24 of UNFCCC (December 2018), President of the UN General Assembly (February 2019) and the UN-Habitat Assembly (May 2019). Nonetheless, there is an ongoing need to further mainstream URLs, linking to a variety of other policy fora and processes including:

- The **Sendai Framework for Disaster Risk Reduction**[6] 2015–2030, which mentions biodiversity as a key dimension of disaster risk governance and refers to the importance of urban planning and urban development, but no explicit connection is made between these topics.
- The **UN Decade of Action on Nutrition**[7] **2016–2025** acknowledges the importance of biodiversity for sustainable food systems and healthy diets (developed in other chapters in this publication). But here, again, an explicit urban-rural lens would contribute to further synergy and more effective operationalization (UNSCN, 2018). Demand of urban consumers for tasty and healthy local foods (including products of geographic origin) can contribute to sustainable territorial development (UNSCN, 2020).
- **Indigenous peoples' food systems** – closely related to local biodiversity – are gaining increased recognition as informants in the global debate on sustainability and climate resilience in the context of the 2030 Agenda and the United Nations Decade of Action on Nutrition 2016–2025. This is especially significant given the fact that indigenous peoples, often marginalized, have had to migrate to urban areas while retaining close linkages with their

cultures and communities of origin. They are therefore potential mediators for making the best of URLs and sustainable use of biodiversity.

- The Global Action Plan of the **UN Decade of Family Farming**[8] **2019–2028** (FAO and IFAD, 2019) aims to support family farmers in fulfilling their role as agents of change in achieving a more balanced and resilient planet by managing natural resources, protecting the environment and achieving sustainable development. This can build on the linkages established by the UN Decade on Biodiversity[9] and Decade on Nutrition (Tutwiler, 2016).
- The FAO's first assessment on **The State of the World's Biodiversity for Food and Agriculture** (FAO, 2019a) did not engage an explicit urban-rural lens, but it touches on various relevant issues including URLs and dynamics that are critical in the changing agriculture sector.
- The recently declared **UN Decade for Ecosystem Restoration**[10] **2021–2030** offers great opportunities for synergistic linkages and mainstreaming.
- Closely related to the GP-URL initiative, and led by some of the same organizations, is a **call to support 'living territories'**, or rural transformation in the context of urbanization that links food systems, agrobiodiversity and ecosystem services to address rural-urban inequality. These issues are at the heart of a recent paper, *Fostering Territorial Perspectives for Development: Towards a Wider Alliance* (CIRAD, 2017). The growing attention to bioregions could provide a framework for bringing together these different approaches.

Bridging initiatives with policy and research linking biodiversity, food systems and URLs

This chapter was introduced with a reference to the disconnections and gaps between emerging practices making the connections between biodiversity and food systems at place-based levels across the urban-rural continuum within both the policy arena and research community. Highlighted below are several examples of initiatives which aim to bridge this trichotomy of practice, research and policy. (The co-authors stress that, at the time of this writing, these bridges are being improved from each of the three entry points.)

- Between 1996 and 2004, Bioversity International and other partners worked with partners in Kenya to revive the interest of researchers, growers and consumers in African Leafy Vegetables (ALVs), resulting in increased production, consumption and marketing of a wide range of local leafy vegetables and an improvement in household well-being (Bioversity International, 2013). Rural production of biodiversity important foods was linked directly to the creation of new urban markets; in Nairobi, researchers found a ten-fold increase in peri-urban production, with 80% of households reporting consumption of ALVs from 1997–2007 (Bioversity International, 2013; see how this was built upon in Part II of this book).

- In South Korea, policy initiatives link cities with neighbouring provinces that signed the Milan Urban Food Policy Pact. Here URLs have taken on new importance in projects such as bringing rice paddy cultivation into urban corridors in Daegu or legislating urban-rural associations between Seoul neighbourhoods and agriculture production regions in rural provinces (FAO, 2018b).
- In Ecuador, the practitioner-initiated campaign *Qué rico es: comer sano y de nuestra tierra* aims to link 250,000 urban households to indigenous farmers to protect biodiversity and supply local markets. This initiative is supported by the national government's Ministry of Culture and Heritage and fulfils multiple goals and objectives at national and international policy levels, while addressing biodiversity and traditional foods across the urban-rural continuum[11].
- In Antananarivo, Madagascar, agro-ecologically trained smallholders are connected directly to consumers and markets via 'collectors', giving the farmers more agency through knowledge of market pricing, and building a market for environmentally safe and healthy food. This initiative connects practice, research and policy in another Milan Pact signatory city (FAO, 2018b).
- In Brazil's innovative school food procurement programme, schools supply the next generation with meals, knowledge, skills and values that shape their health and our future food systems. Such programmes are an important means for bridging the urban-rural gaps and raising awareness of children, teachers and parents of the importance of local biodiversity for healthy diets and sustainable territorial development. School feeding has been mandatory in primary and secondary schools in Brazil since 1955, and in 2009, a new policy was introduced to establish a 30% minimum quota of procuring organic agricultural products from local farmers. This has required inter-disciplinary cooperation across departments of education, health and agriculture to support local farmers in introducing organic farming and supplying schools with quality produce and nutritional food. This process empowers producers through increased income and the creation of cooperatives which often include marginalized producers, such as indigenous communities. The research community has also been engaged in the process of linking practice to policy in the case of Brazil. Together, this example demonstrates the multiple benefits of connecting structured urban markets with rural smallholders, for both diets and upstream biodiversity conservation (Valencia et al., 2019, see also Part II for more about the incorporation of food biodiversity into Brazil's school procurement policies).
- Understanding the importance of ethno-cultural foods to migrants, entrepreneurial practitioners are making the connections of biodiversity, food systems and URLs in both rich and poor countries. In Lima, Peru, plastic sealed semi-processed quinoa from producers working with business intermediaries is available in supermarkets. In rural areas near Toronto, Canada, 'world crops' are grown in order to meet the growing demand of Asian immigrants[12].

- Another opportunity for rural communities that has employment potential across the rural-urban continuum is the incorporation of wild foods, such as shea butter or néré seeds, in commercialization circuits. These foods traditionally played an important role in West Africa in local diets and food systems and are in high demand with urban consumers[13].
- Since 2004, the New Nordic Cuisine builds on local biodiversity and related food production for sustainable diets. Initiated by Nordic chefs, this framework links the policy goals of Nordic countries and cities with practitioners in food production and natural resource stewardship and sustainable diets (Nordic Council, 2019 see also Chapters 8 and 10).
- In Quelimane, Mozambique (also a Milan Pact city), a municipal waste system, managed by the city, farmers and community organizations, links organic food production and waste recovery to biodiversity conservation (FAO, 2018b).

While these efforts illustrate the potential of improved URLs to contribute to sustainable food systems and sustainable management of biodiversity, there are also negative effects and missed opportunities when these linkages are not given appropriate attention. For example:

- In 2001, in Polokwane, South Africa, urban dwellers expressed a demand for sorghum-based products that was not being met by the national agri-food industry. This could have been an opportunity for local food chains to meet the expectations of the urban market in terms of process and regarding the variety of sorghum bred by small scale farmers in Sekhukhuneland. Linked research and policy support could have led to other beneficial outcomes[14].
- Some practices in the private sector may have significant research and policy support, but do not accept or even recognize the consequences of their practices on biodiversity and sustainable food systems. For example, the worldwide promotion of fortified foods to improve the quality of diets – such as the insertion of micronutrients in foods – may problematically divert attention from micronutrient local biodiversity and further erode the livelihoods of small-scale food producers[15].

Concrete examples of experiences that bridge practice with policy and research that document the impact and benefits of linking biodiversity, food systems and urban-rural linkages, are only at a formative stage. More research on many existing initiatives and experiences is critical in order to exchange practices, scale up and inform policy at national and international levels. These experiences should feed back into policy review and implementation, for example with SDGs and the 2030 Agenda, the New Urban Agenda, as well as in other frameworks such as those related to indigenous people's food systems, the

Decade of Action on Nutrition, among others. In that perspective, FAO has developed an Urban Food Actions platform to pool and share information (for example academic research, case studies, policy document) related to urban food policies and programmes in a wide range of aspects: governance and planning, sustainable diets and nutrition, social and economic equity, food production and ecosystem management, food supply and distribution, food loss and waste (FAO, 2019b). Horizontal exchanges of practice, policy and research are essential, between city-to-city networks such as ICLEI, UCLG and nrg4SD at the global level, regional and national networks of local governments, or between academic and civil society actors, such as the Food for Cities Dgroup[16] hosted by FAO.

Way forward/recommendations

Thus far, from the evidence and analysis informing policy debates, there are several conclusions that could help inform a new consensus and policy narrative based on a fundamental agreement that challenges common to both urban and rural areas must be addressed in a more integrated way, transcending separate, binary urban and rural approaches. These recommendations include:

- Urban and territorial planning and development must be inclusive of and balanced between urban and rural actors and interests, and different size cities and towns;
- The economic engine and social development advantages of urban communities must benefit rural development as well as urban development;
- Food systems, encompassing urban and rural spaces and actors from production to consumption and waste management/recovery, are at the heart of urban-rural linkages;
- Consumer preferences reflected in public procurement and private purchasing can be powerful drivers of agrobiodiversity, incentivizing farmers to grow biodiverse foods;
- To improve resilience and adapt to climate change, biodiverse blue/green living landscapes should include food/forest/livestock/aquaculture components;
- Ecological production systems will increasingly need to be an important part of the planning and design of city-region spatial and functional mosaics;
- Municipalities, territorial authorities and national governments need to recognize that the nexus of biodiversity, food systems and URLs presents many opportunities to realize the interlinkages of the SDGs and the New Urban Agenda;
- Associations of municipalities and towns in functional territories or city systems with a priority of conserving ecological functions of waterways and landscapes that provision human settlements will become important governance mechanisms to meet needs and challenges across the urban-rural continuum.

An overall goal should be to build on exchanges of practice, research and policy to support an emerging policy consensus and a new unifying narrative for a nexus of biodiversity, food systems and URLs. In service of this goal there are some short and intermediate term pathways to maintain and increase the synergies that connect actors at both horizontal and vertical levels. A few of these pathways for addressing current gaps between practice, research and policy include:

- Find ways to continue to build on both 'vertical' or 'transversal' synergy between local, national and international practice and policy as indicated above;
- Unite institutional champions for the integration of biodiversity, food systems and URLs, for example among actors seeking to deepen the guidance coming from the GP-URLs in specific country/landscape settings;
- As part of the joining together of champion agencies, advance solidarity work with networks of indigenous peoples and allied social actors such as smallholder organizations defending rights/lands in specific geographies where positive solutions can offer model approaches;
- Lift the roles of women and youth as a priority in applying new technical and creative tools to the design and planning of environments that protect biodiversity while providing healthy foods and related ecosystem services.

Notes

1 The co-authors have been lead technical consultants on several of the processes described in this chapter including the inclusion of URLs in the New Urban Agenda, the Milan Urban Food Policy Pact and the Urban-Rural Linkages Guiding Principles and Framework for Action. Other contributors to this chapter include Ingrid Coetzee, Paul Currie, Eliot Gee, Danny Hunter and Annah MacKenzie.
2 Author's field experience.
3 Author's field experience.
4 Author's experience with Milan Pact signatory cities
5 Milan Urban Food Policy Pact (2015) www.milanurbanfoodpolicypact.org/wp-content/uploads/2018/10/CA0912EN.pdf
6 Sendai Framework for Disaster Risk Reduction (available at www.unisdr.org/we/coordinate/sendai-framework)
7 UN Decade of Action on Nutrition 2016–2025. www.unscn.org/en/topics/un-decade-of-action-on-nutrition
8 UN Decade of Family Farming 2019–2028. www.fao.org/family-farming-decade/en/
9 UN Decade for Biodiversity 2011–2020. www.cbd.int/doc/strategic-plan/UN-Decade-Biodiversity.pdf
10 UN Decade for Ecosystem Restoration 2021–2030. www.unwater.org/the-united-nations-general-assembly-declare-2021-2030-the-un-decade-on-ecosystem-restoration/
11 Author's field experience.
12 Author's field experience, see also: Campaña Qué rico es comer sano y de nuestra tierra! www.quericoes.org/
13 Author's field experience.

14 Author's field experience.
15 Author's field experience.
16 FAO's Food for Cities is an email-based discussion platform that has been operating since 2009.

References

Berdegué, J.A., Proctor, F.J., Cazzuffi, C. (2014) *Inclusive Rural–Urban Linkages. Working Paper Series N° 123*, Rimisp, Santiago, Chile. Available at: www.rimisp.org/wp-content/files_mf/files_mf/1421411559123_InclusiveRural_UrbanLinkages_edited.pdf

Bichard, A., Dury, S., Schönfeldt, H.C., Moroka, T., Motau, T., Bricas, N. (2005) 'Access to urban markets for small-scale producers of indigenous cereals: A qualitative study of consumption practices and potential demand among urban consumers in Polokwane', *Development Southern Africa*, vol 22, no 1, pp 125–141. doi: 10.1080/03768350500044560.

Bioversity International. (2013) *African Leafy Vegetables Come Out of the Shade*, Bioversity International. Available at: www.bioversityinternational.org/fileadmin/user_upload/online_library/publications/pdfs/African_leafy_vegetables_come_out_of_the_shade_2008.pdf

Cabannes, Y., Marocchino, C. (2018) *Integrating Food into Urban Planning*, UCL Press, London, UK, and The Food and Agriculture Organization of the United Nations, Rome, Italy. doi: 10.14324/111.9781787353763.

CBD. (2018) *Conference of the Parties to the Convention on Biological Diversity, Fourteenth Meeting Sharm El-Sheikh, Egypt, 17–29 November 2018: Sharm El-Sheikh Communiqué for Local and Subnational Action for Nature and People 2018*, Conference on Biological Diversity. Available at: www.cbd.int/doc/c/c745/007e/3ac98825a03a8073bf0d547d/cop-14-inf-48-en.pdf

CFS. (2017) *Forty-fourth Session – "Making a Difference in Food Security and Nutrition", Rome, Italy, 9–13 October 2017: Addressing food security and nutrition in the context of changing rural-urban dynamics: Experiences and effective policy approaches with draft decision*, Committee On World Food Security. Available at: www.fao.org/3/a-mu135e.pdf

CIRAD. (2017) *Fostering Territorial Perspectives for Development: Towards a Wider Alliance*, CIRAD, Montpelier, France. Available at: https://collaboratif.cirad.fr/alfresco/s/d/workspace/SpacesStore/6daa60e1-d89e-4a59-9bfd-ff5f66a93130/TP4D_vENG.pdf

FAO. (2018a) *High-level Expert Seminar on Indigenous Food Systems: Building on Traditional Knowledge to achieve Zero Hunger, 7–9 November 2018, FAO Headquarters, Rome, Italy*, Food and Agriculture Organization of the United Nations. Available at: www.fao.org/indigenous-peoples/ifs-seminar/en/

FAO. (2018b) *Role of Cities in the Transformation of Food Systems: Sharing Lessons from Milan Pact Cities*, Food and Agriculture Organization of the United Nations. Available at: www.milanurbanfoodpolicypact.org/2018/10/04/mpa-narrative-report/

FAO. (2019a) *The State of the World's Biodiversity for Food and Agriculture*, FAO Commission on Genetic Resources for Food and Agriculture Assessments, Rome, Italy, Available at: www.fao.org/3/CA3129EN/CA3129EN.pdf

FAO. (2019b) *Urban Food Actions Platform*, Food and Agriculture Organization of the United Nations. Available at: www.fao.org/urban-food-actions/knowledge-products/en/

FAO and IFAD. (2019) *United Nations Decade of Family Farming 2019–2028*, Global Action Plan. Rome. Licence: CC BY-NC-SA 3.0 IGO.

Forster, T. and Getz, A.E. (2014) *City Regions as Landscapes for Food, People and Nature*, EcoAgriculture Partners International. https://peoplefoodandnature.org/publication/city-regions-as-landscapes-for-people-food-and-nature/

GLOPAN. (2016) *Food systems and diets: Facing the challenges of the 21st century*, Global Panel on Agriculture and Food Systems for Nutrition, London. Available at: www.glopan.org/wp-content/uploads/2019/06/ForesightReport.pdf

Hawkes, C. (2007) 'Globalization and Nutrition Transition: A Case Study', in Pinstrup-Andersen, P., Cheng, F. (eds.) *Food Policy for Developing Countries: Case Studies*, pp 1–16, Cornell University, Ithaca, NY.

IBPES. (2019), *Global Assessment Summary for Policy Makers*, Science and Policy for People and Nature, Bonn. Available at: www.ipbes.net/global-assessment-report-biodiversity-ecosystem-services

MCP. (2019) *Que Rico es Comer Sano y de mi Tierra* Ministerio de Cultura y Patrimonio, Government of the Republic of Ecuador. Available at: www.culturaypatrimonio.gob.ec/que-rico-es-comer-sano-y-de-mi-tierra-documentales/

Nordic Council. (2019) *New Nordic Food*, The Nordic Council. Available at: www.norden.org/en/new-nordic-food

Proctor, F.J., Berdegué, J.A. (2016) *Food systems at the rural-urban interface. Working Paper series N° 194.* Rimisp, Santiago, Chile. Available at: http://rimisp.org/wp-content/files_mf/1467380890194_Felicity_Proctor_Julio_Berdegue.pdf

Tutwiler, A.M. (2016) *Bioversity Blog: Why governments should look to agricultural biodiversity to contribute to the Decade of Action on Nutrition*, Bioversity International. Available at: www.unscn.org/en/topics/un-decade-of-action-on-nutrition?idnews=1794

UN. (2012) *General Assembly Sixty-sixth session Agenda item 19: Resolution Adopted by the General Assembly on 27 July 2012 66/288 The future we want*, United Nations. Available at: www.un.org/ga/search/view_doc.asp?symbol=A/RES/66/288&Lang=E

UN-Habitat. (2016) New Urban Agenda, Outcome document from Habitat III, UN Conference on Housing and Sustainable Urban Development, Quito.

UN-Habitat. (2019) *Urban-Rural Linkages: Guiding Principles Framework for Action to Advance Integrated Territorial Development*, UN-Habitat, Nairobi, Kenya. Available at: https://urbanrurallinkages.files.wordpress.com/2019/04/url-gp.pdf

UNSCN. (2018) *Expert Group Meeting on Nutrition and the SDGs under Review in Preparation for the High-Level Political Forum Background Document*, United Nations System Standing Committee on Nutrition. Available at: www.unscn.org/uploads/web/news/EGM-Background-Document.pdf

UNSCN. (2020) *Urban-Rural Linkages for Nutrition: Territorial Approaches for Sustainable Development.* www.unscn.org/uploads/web/news/document/Urban-rural-linkages-for-nutrition-EN-WEB-OK-Feb.pdf

Valencia, V., Wittman, H., Blesh, J. (2019) 'Structuring Markets for Resilient Farming Systems', *Agronomy for Sustainable Development*, vol 35, no 25. doi: 10.1007/s13593-019-0572-4.

7

DIVERSIFYING PUBLIC FOOD PROCUREMENT AND SCHOOL FEEDING

Luana F. J. Swensson and Florence Tartanac

Introduction

Public food procurement programmes (PFPP)[1], including school feeding initiatives, are important policy instruments to deliver multiple benefits for multiple beneficiaries, including food consumers, food producers and local communities (Tartanac et al., 2019). These initiatives are supported by an increasing number of international regulatory frameworks that recognize that public institutions, when using their financial capacity and purchasing power to award contracts, can go beyond the immediate scope of responding to the state's procurement needs by addressing additional social, environment or economic needs that contribute to the overall public good.

These frameworks – such as the 2011 United Nations Commission on International Trade Law's (UNCITRAL) Model Law on public procurement – recognize the potential as well as the right for public entities to use procurement as a strategic instrument to pursue development goals. They are also endorsed by the United Nations (UN) Sustainable Development Goals, which identifies sustainable public procurement (which includes environmental, economic and social aspects) as a key prerequisite for achieving more sustainable patterns of consumption and production (Target 12.7).

One particularity of public food procurement – which distinguishes it from other types of public procurement – is its potential to have a direct impact on both suppliers and consumers. The PFPP – and related policy and legal frameworks – can determine not only the way food is procured, but, in particular (i) what food will be purchased (such as local, diverse, nutritious, healthy, culturally adequate, environmental-friendly); (ii) from whom (for example from local and/or family farming producers, small and medium food enterprises, women, youth and/or other vulnerable groups); and (iii) from which type of production (for example from agricultural production that ensures environmental sustainability as well as biodiversity)

(Tartanac et al., 2019). Considering the weight of public sector demands and based on how those choices are made, PFPP has the potential to influence both food consumption and food production patterns (Bontrager Yoder et al., 2014; Fitch and Santo, 2016; Foodlinks, 2013; IPES, 2016; Tartanac et al., 2019, Valencia et al., 2019).

Furthermore, one important characteristic of PFPP is that by creating a demand for certain types of food products (such as biodiverse, from local and smallholder farming, and from agroecological production), governments have the power to set positive trends. Public sector institutions – such as schools, hospitals, universities, prisons, armed forces, care homes and canteens in government buildings – represent a significant part of the procurement of any national food economy. They can send strong signals about government ambitions on the future directions of the food system, which has the power to incentivize supply chain actors to align their values and practices accordingly, fostering a transition towards sustainable food consumption and production (De Schutter, 2014; Foodlinks, 2013; IPES-Food, 2016; Sambuichi et al., 2013; Tartanac et al., 2019; UNSCN, 2017).

This potential is particularly relevant in the school feeding context. This is due to not only the predictable and stable demand that school feeding initiatives can provide, especially for local and smallholder food producers, but also due to the specificities of its direct beneficiaries; i.e. the children. School feeding programmes, especially when combined with sound nutrition education initiatives, also have the potential to influence consumption patterns of children and lifetime dietary behaviours, contributing to the formation of future educated consumers (Fitch and Santo, 2016; UNSCN, 2017).

It is in this context that this chapter argues that – depending on the choice of what, from who and from which type of production, food shall be purchased – PFPP, particularly school feeding programmes, constitute one important instrument with the potential to deliver multiple dividends to a multiplicity of beneficiaries, with contributions crossing all three areas of sustainability: social, economic and environmental. These include the conservation and sustainable use of biodiversity. Nevertheless, this chapter also recognizes that these choices are not free and will be dependent on a series of conditions. These conditions include conducive regulatory frameworks, which, despite their relevance, are still very often overlooked both by the literature (Stefani et al., 2017) as well as by policymakers (Brooks et al., 2014; Swensson, 2018).

Building on the existing literature on public food procurement and on two countries' experiences, this chapter aims to explore the multiple benefits that PFPP can contribute, and the role of the regulatory framework in supporting this. The chapter is organized into three sections. Section one will provide an overview of the literature on the multiple potential benefits and beneficiaries of PFPP, including the promotion of biodiversity. Section two will focus on country experiences and provide examples of good practices from the Brazilian food purchase programme (Programa de Aquisição de Alimentos – PAA), the National School Feeding

Programme (Programa Nacional de Alimentação Escolar – PNAE) and the Danish Organic Policy and Action Plan. These examples complement the experiences provided in this publication within the context of the BFN Project and in particular the Brazilian and Kenya examples presented in Part II of this book. The last section is dedicated to concluding remarks, with a discussion of enabling and constraining conditions, focusing on the role of a conducive regulatory framework.

Multiple benefits and beneficiaries

There is a large range of literature and policy documents that recognize the multiple dividends that PFPPs have the potential to achieve and the contributions that public procurement of food can make to sustainability and its three core components (De Schutter, 2014, 2015; Espejo et al., 2009; FAO, 2015; FAO and WFP, 2018; Fitch and Santo, 2016; Foodlinks, 2013; Gelli et al., 2010; Global Panel, 2015; Morgan and Sonnino, 2008; Smith et al., 2016). The literature also recognizes that public procurement has the potential to benefit those who receive the food through the public institutions (food consumers), those who supply the food (food producers) and the wider community in general.

Depending on the choice of **from who** food should be purchased (for example from local and smallholder farmers), PFPP can become an instrument to support local and smallholder agriculture production and stimulate community economic development. Studies demonstrate that schools and other public institutions that require regular and predictable demand targeting smallholder farmers' products can encourage, facilitate and reduce the risk of investments for farmers to increase and diversify their agriculture production; ultimately this can contribute to increasing their incomes and access to formal markets (Drake et al., 2016; FAO, 2015; FAO and WFP, 2018; Global Panel, 2015; IPC and WFP, 2013; Kelly and Swensson, 2017; Sumberg and Sabates-Wheeler, 2010; UNSCN, 2017). PFPP can also provide an important market opportunity for small processors and micro, small and medium-sized food enterprises, which may supply schools and other public institutions with nutritious processed food products such as bread, biscuits, fruit juice, as in the case of the school feeding programme in Brazil (FAO and WFP, 2018; Global Panel, 2015). As women and/or youth often own these enterprises, PFPP and school feeding initiatives have the potential also to contribute to increased youth empowerment and gender equity as well as to job creation (FAO and WFP, 2018; UNSCN, 2017).

In the same way, by defining **what food to be purchased** (such as fresh, healthy and biodiverse – including underutilized, nutrient-rich species) PFPP and, in particular school feeding initiatives, can become a powerful instrument to enhance food security and nutrition of children as well as of smallholder farmers and their communities, through increased and diversified food production, consumption and incomes. Diversified school feeding programmes can, indeed, have a direct effect on schoolchildren's food consumption, dietary diversity and nutrition status (FAO and WFP, 2018). Studies from countries like Brazil, Kenya and Ghana show that properly designed school feeding can lead to an increase in the variety and quantity of healthy foods served in schools, and to improvements of the dietary diversity of schoolchildren (Niebylski et al., 2014; Sidaner et al., 2012; UNSCN, 2017).

Considering the weight of public sector contracts, a diversification of the demand – especially when linked to the purchase from local and smallholder producers – has the potential to stimulate production diversification by these actors. As recently assessed by Valencia et al. (2019) in the Brazilian context, this production diversification can promote a higher level of agrobiodiversity in terms of an increase in diversity in crop and livestock varieties and species as well as on the area of farmland under diversified production. This production diversification may lead – according to the context – to an increase in the consumption by smallholder farmers of, for instance, fresh, healthy and biodiverse products, contributing to household dietary diversity (Bioversity International, 2016; HLPE, 2017; IPES-Food, 2016; Niebylski et al., 2014; Valencia et al., 2019). At a community level, by supporting higher levels of agrobiodiversity, an increase in the production diversification may also lead to an increase in the availability of these products in local markets, contributing to the nutritional security of non-farming populations and, ultimately, to the conservation and use of food biodiversity (Valencia et al., 2019).

In this sense, depending on the choice of what products to purchase, school feeding programmes have the potential to diversify diets of children, farmers and local communities, by promoting production of, and access to, a wider variety of foods. This is particularly relevant considering that homogenous diets, limited food access and poorly developed markets for biodiverse species are one of the key causes for the continued increase in malnutrition and poverty (Bioversity International, 2016; Burlingame and Dernini 2012; FAO, 2019). The promotion of agricultural production diversification can be considered, therefore, as a key element to explore agricultural biodiversity, rural development and sustainable diets (Tartanac et al., 2019, Valencia et al., 2019).

PFPP can also target **food that is produced** in a specific way, and, therefore, use public purchasing power to support and promote forms of agricultural production that ensure environmental sustainability as well as biodiversity. This includes, for instance, the purchase of food from low-impact production systems, such as organic production, biodiversity attuned practices and reduced carbon inputs and greenhouse gas emission. (Fitch and Santo, 2016; Foodlinks, 2013; IPES-Food, 2016) A recent study of the Brazilian experience mentioned above shows, for instance, how the participation in the school feeding programme played a direct role in farmers' decision to shift their household's primary economic focus from low agrobiodiversity, input-intensive farming systems to a more diversified, low external input system one (Valencia et al., 2019). It also shows how for a subset of farmers – who were already market-oriented horticultural producers – the price premium offered for certified agroecological cultivation incentivized a transition to organic production practices (Valencia et al., 2019). Furthermore, PFPPs have the potential to promote environmental benefits in terms of reduced packaging; food waste; and lower food miles (Foodlinks, 2013). Within this context, PFPPs have been recognized also with great potential to influence positively water and land use, and climate change, as well as the conservation and sustainable use of biodiversity (Beltrame et al., 2016; Fitch and Santo, 2016; Foodlinks, 2013; IPES-Food, 2016)

The decision on what type of food to be procured, from who, and from what type of production – and consequently on which of the multiple potential benefits and beneficiaries to focus – will depend on decision-makers' choices according to government priorities and on programme design. Policymakers can focus on single components, on specific beneficiaries or benefits; they can also adopt a multiple approach, favouring synergy (FAO and WFP, 2018; Hunter et al., 2016;). These decisions will be dependent, however, on a series of conditions. These conditions are linked to supply and demand side factors as well as to the policy, legal and institutional enabling environment (Kelly and Swensson, 2017; Smith et al., 2016), particularly a conducive public procurement regulatory framework (Swensson, 2018).

One key characteristic of PFPP is that, just like any type of public purchase, PFPPs are operationalized and regulated by specific and detailed rules. These rules govern the entire procurement process, shaping and limiting the choices available to governments regarding (i) what food to purchase; (ii) how to purchase it; and (iii) from whom to purchase. As a result, the objectives and implementation of any public food procurement initiative is intrinsically linked to the existing public procurement regulatory framework (Swensson, 2018).[2] This includes the objective of supporting the promotion and conservation of biodiversity through public food procurement, which can be directly supported – or constrained – by specific regulatory instruments.

Brazil and Denmark provide good examples on the different choices and possibilities of using PFPP to achieve multiple policy goals. These experiences also provide insights on some key elements that may support or constrain these choices and their implementation, including a conducive regulatory framework. They have both been awarded the 2018 'Future Policy Award' from the World Future Council, dedicated to policies that scale up agroecology, contribute to the protection of biodiversity, life and livelihoods of small-scale food producers, ensure sustainable food production systems and implement climate resilient agricultural practices.

Examples of good practices

Brazil

Brazil exemplifies the political choice of using public food procurement and school feeding as instruments to achieve a multiplicity of benefits including the promotion of biodiversity conservation and sustainable use. On defining what food shall be purchased (such as diverse, healthy, safe and culturally adequate in the case of school feeding), from whom (family farmers and family rural entrepreneurs)[3] and giving priority to food that is produced in a specific way (such as from biodiverse attuned practices, organic and agroecological), Brazilian programmes are intentionally designed as multifaceted interventions, with objectives that go beyond the simple response to the state food needs at the lowest price.

The Brazilian Food Purchase Programme (PAA) (see also Part II, Box 5.B6) purchases food directly from family farmers, rural entrepreneurs, and their organizations to meet the state's various food needs, with nine goals that clearly articulate the multifaceted nature of the programme[4]. These goals include: (i) support family farming production by promoting its economic and social inclusion, and fostering income generation, sustainable food production, processing and industrialization; (ii) support the consumption and valorization of food produced by family farming; (iii) promote access to food, in the quantity, quality and regularity necessary for people with food and nutritional insecurity; (iv) promote and enhance biodiversity, organic and agroecological food production and encourage healthy eating habits at local and regional level; and, (v) stimulate the development of cooperatives and associations (Law No. 10.696/2003, Law No. 12.512/2011, Decree No. 7775/2012).

The reformulation of the Brazilian National School Feeding Programme (PNAE) in 2009 broadened the programme's traditional objective of responding to student's nutritional needs during their time in the classroom. Currently PNAE has among its directives: (i) the adoption of healthy and adequate food, including the use of diversified and safe food products which respect culture, tradition and healthy eating habits; and (ii) the support of local and sustainable development, fostering the purchase of diversified and locally produced food from family farming producers (Law No. 11.947/2009). The programme requires that at least 30% of the federal budget allocated for the purchase of food for school feeding must be reserved for contracts with local family farming producers, giving priority and a price premium to, among others, organic and/or agroecological products.

The PNAE and the PAA and related objectives are aligned with the aforementioned international frameworks and national public procurement law that recognize the key objectives of public procurement and the pursuit of sustainable development (Law n. 8.666/1993). Furthermore, both Brazilian programmes were created through legislation which provides specific tools aimed at aligning the regulatory framework to the policy objectives. These legal instruments allow, among other issues: (i) the reservation of procurement opportunities to specific target beneficiaries (i.e. reservation schemes); (ii) the utilization of innovative award criteria that acknowledge the socio, economic and environmental quality of the food products offered; (iii) the adoption of alternative and simplified procurement methods which substitute the standard open tender procedures; and (iv) the application of tailored participation requirements. These instruments are crucial to enable and support the implementation of all the aforementioned policy objectives (see Swensson, 2018).

The possibility of selecting suppliers based on awarding criteria that goes beyond the selection of the best price (which is traditionally the bottom line applied for public sector selection of suppliers) is an important enabling factor for the multifaceted nature of both programmes. Although prices cannot be higher than market prices, when selecting suppliers (among the eligible family farming producers), procuring entities shall apply specific awarding criteria aimed at reaching the following specific policy objectives:

- Support the poorest and most vulnerable producers. To foster the social inclusion of the most vulnerable and marginalized populations, priority is given to these producers in the selection process. These groups include, among the eligible family farming producers, the land reform settlers, members of traditional communities (indigenous people and *quilombolas*[5], see also Part II, Box 5.C3) and, in the case of PAA, beneficiaries of social programmes (such as the *Bolsa Família* social welfare programme) and women.
- Support agroecological and organic production. Eligible suppliers of agroecological and organic production take precedence over conventional produce, with up to a 30% differential in the price margin allowed.
- Strengthen the development of collective actions. While individuals can access and sell their products to the programmes, farmer organizations, cooperatives or associations, including informal ones, take priority in the selection process over individual access (see Swensson, 2015).
- Support local production. The locality of production, giving preference to the nearest locality of production based on the administrative division of the country (municipality, neighbouring municipalities, other municipalities within the same state, neighbouring states and other states within the national territory, in this order) is recognized as a priority selection criterion and prevails over all other criteria in the case of PNAE.

Furthermore, in both programmes, the potential, as well as the political choice, of using public food procurement as an instrument to achieve different types of beneficiaries is very clear. Both programmes have two distinct beneficiaries: food consumers and food producers. The first include schoolchildren (in the case of PNAE) as well as people with food and nutritional insecurity supported by the government social assistance network (in the case of PAA). The second focuses on family farming producers.

Regarding the PAA, in the case of Brazil these target beneficiaries are defined by specific legislation (Law n. 11.326/2006) which establishes the criteria with which rural actors must comply in order to be classified as family farming producers. In the case of Brazil's PFPP, these beneficiaries include not only farmers, but also family rural entrepreneurs, foresters, aquaculturists, extractivists, fisherfolks, indigenous people and members of traditional communities. This legal instrument, combined with a registration system, is of great relevance in targeting, implementing and monitoring the programme (Swensson, 2018). It also supports the compliance with public procurement principles of transparency and equal treatment which requires that, when some advantage is given to a defined category of suppliers, such as the reservation of the contractual opportunities to family farming producers, it is essential that the criteria for category membership for those receiving that advantage is well defined.

Although implementation challenges still exist, we can affirm that the Brazilian experience distinguishes itself not only by the policy choice of using PFPP as a policy instrument to achieve a multiplicity of benefits and beneficiaries with benefits across all the three pillars of development, but also by the conducive policy and, particularly, legal enabling environment, which supported and made these choices possible. Despite the lack of an overall impact evaluation, qualitative studies demonstrate and acknowledge the multiplicity of benefits of the Brazilian approach (see Sambuichi et al., 2013 and Swensson, 2015 for an overview). These include its potential in promoting positives changes in the food system, such as increased production diversification[6], that can support both the conservation and sustainability of biodiversity (Valencia et al., 2019).

Denmark

Denmark provides another interesting example of a policy choice of using public food procurement as an instrument to achieve development goals. By targeting food that is produced in a specific way (i.e. organic), Denmark has tailored public food procurement as an instrument to support and trigger the consumption and production of organic products in the country. Although this initiative is not based on specific legislation – as in the case of PAA and PNAE programmes in Brazil – it is supported by a conducive regulatory framework which, together with other policy and institutional factors, have been a key enabling condition for its development (Spyridon and Mikkelsen, 2018; Mikkelsen and Lundø, 2016; Sørensen et al., 2015).

Policy support to organic farming is not new in Denmark, but, until recently, the focus was mainly on supporting agriculture. In 2011, however, the Danish government launched the new Organic Action Plan (OAP), which adopted a unique approach, using public food procurement – together with other strategies – as a key instrument to foster organic production in the country (Sørensen et al., 2015).

With approximately half a million public meals produced per day at schools, hospitals and other public institutions, the Danish government recognizes the potential of public food procurement to increase demand and motivate farmers to switch from conventional to organic food production. Furthermore, public food procurement is essential to achieving the OAP overall goal of doubling the organically cultivated area in the country from the 2007 level by 2020 (Sørensen et al., 2015). The key action adopted for the implementation of the strategy was to set the goal to increase, by up to 60%, the procurement of organic food in all public kitchens by 2020.

Although the focus of the Danish OAP is on increasing organic production in the country, the purchase of organic food by public institutions has also been recognized as an instrument to deliver multiple public goods simultaneously (Berg, 2017; Spyridon and Mikkelsen, 2018; Sørensen et al., 2015). From an environmental perspective, organic production is recognized to be more sustainable compared with conventional methods, delivering benefits in terms of energy utilization, soil quality maintenance, water conservation, pest control and biodiversity improvements (Sørensen et al., 2015). Regarding biodiversity improvements, comparative studies demonstrate that organic farming potentially increases biodiversity in fields and adjacent habitats by 30% relative to conventional farms (Hole et al., 2005; Jespersen et al., 2017). Although there is large variation of this average depending on organism group, crop, landscape and duration of the organic management, this result has been assessed as robust in Denmark over the last 30 years (Jespersen et al., 2017).

Potential benefits also include health (through a decrease in consumer's intake of chemical fertilizer and pesticide residues) and nutritional outcomes. Although the Danish strategy does not target directly 'what food to purchase', by integrating higher levels of organic food within their existing budget, public kitchens have been adopting strategies which include buying more local and seasonal foods, less processed products, limiting meat consumption as well as reducing food waste (Sørensen et al., 2015). Similarly, although not targeting directly 'from who' food should be purchased, the OAP is believed to have the potential to go hand in hand with local and regional sourcing strategies and also support the local agriculture production, stimulating community economic development (Mikkelsen and Lundø, 2016). This enhanced demand for more local, seasonal and less processed food (see also Chapter 10, Box 10.3), combined with a local and regional sourcing strategy, may have the potential to stimulate the diversification of production, contributing theoretically to increased agrobiodiversity in terms of diversity in crop and livestock varieties and species, as accessed in the Brazilian case (Valencia et al., 2019). Nevertheless, specific studies exploring this hypothesis in the Danish context could not be found.

The Danish OAP and the objectives to be reached through public food procurement are aligned with, and supported by, the public procurement regulatory frameworks, both at the national and European level. These frameworks (European Directive 2014/24/EU enacted by Danish Public Procurement Act n. 1564/2015) recognize that public procurement should be used strategically as a 'market-based instrument to achieve smart, sustainable and inclusive growth while ensuring the most efficient use of public funds' (EU Directive) and provide instruments in order to enable food procurers to make better use of public procurement in support of common societal goals.

Among these instruments, similarly to the Brazilian case, there is the possibility of using special awarding criteria that go beyond the lowest price, including both environmental and social aspects. This instrument is particularly relevant considering that over-emphasis on price as the awarding criterion (to the detriment of quality and other environmental and socio-economic ones) is considered as one of the key bottlenecks for the implementation of sustainable public food procurement initiatives (De Schutter, 2014, 2015; Morgan and Sonnino, 2008; Swensson, 2018).

The objective of targeting organic food products in public purchases is also supported by the national Organic Law (Law No. 1657/2015). Based on the EU legislation on organic farming, this Law defines clear eligibility criteria for organic products and producers and sets out the standards and rules for the certification scheme, i.e. the Danish stated-owned organic Ø-label. Denmark was one of the first countries in the world to adopt a stated-owned organic label back in 1990. Currently the label is not only familiar to (98%) but also highly trusted by (81%) the Danish consumers (Frederiksen, 2016).

This instrument, similarly to the Brazilian case, facilitates the targeting and implementation of the food procurement initiative and the compliance with public procurement principles of transparency and equal treatment of suppliers. It also provides the baseline for monitoring and evaluating the results and impacts of the initiative and related policies – a factor that is also considered of key importance for the sustainable development of public procurement initiatives that pursue broader development goals (Adell and Schaefer, 2016; UNEP, 2017). In the case of Denmark, in fact, great effort is dedicated to the monitoring of the OAP, including a specific scheme (the Organic Cuisine Label) which, building on the organic certification scheme, allows the measurement and monitoring of organic food procurement by public kitchens (Sørensen et al., 2015).

It is interesting to note, as reported by Berg (2017), in order to support the conversion to organic production, efforts were made also to make the rules on how to become an organic producer easier to understand. This goal has been achieved partly through informing farmers via different channels about the requirements for becoming an organic farmer and also by making the rules on how to become an organic producer clear and flexible enough to make the transition easier (Berg, 2017). These instruments, together with other factors, have been considered as key enabling factors of the Danish strategy (Spyridon and Mikkelsen, 2018; Mikkelsen and Lundø, 2016)

Currently, Denmark has the highest market share of organic products in the world, with almost 80% of Danes purchasing organic food. The OAP was very successful in motivating public kitchens to augment their percentage of organic food purchases. Already in its first two years of implementation, the percentage of participating public kitchens that reached the target of 60% of organic food purchase doubled from 31% to 62% (Sørensen et al., 2016). A significant example is the city of Copenhagen that met the goal of 90% organic food in public kitchens in 2015. Also, thanks to the increase in demand, the OAP has amply met its original goal of doubling organic land compared to a 2007 baseline (Berg, 2017).

The policy choices as well as the achieved results of the Danish OAP are a good illustration of the potential of PFPP – when properly supported by a conducive policy and regulatory frameworks – to influence both production and consumption patterns and promote positive changes in the food system.

Concluding remarks

The literature and the country experiences presented demonstrate the multifaceted nature of PFPP and the multiplicity of benefits and beneficiaries that PFPPs have the potential to achieve. They also demonstrate how PFPPs can be tailored to the pursuit of different policy objectives and influence both production and consumption patterns and ultimately drive more sustainable food systems.

Deciding which of the many potential benefits to focus on is primarily a policy choice and can differ from country to country according to government priorities. They may include the support of local and smallholder producers, as well as of agricultural practices that ensure environmental sustainability and biodiversity conservation, as shown by the examples presented in this as well as in other chapters of this book. These choices, however, are dependent on a series of conditions, which include, with particular relevance, a conducive and aligned regulatory framework.

One key characteristic of PFPPs which distinguishes them from private parties is that government institutions are not completely free to choose (i) what food to purchase; (ii) how to purchase; and (iii) from whom to purchase. These choices are shaped and limited by the public procurement regulatory framework. As a result, the objectives and implementation of any food procurement initiative will be intrinsically linked to the existing public procurement regulatory framework, its alignment with the policy objectives and the possibility of amending it. Despite this strong linkage between PFPPs and public procurement regulatory frameworks, the importance of the latter as a key enabling factor as well as the need for alignments and reforms is often overlooked both by the literature and by policymakers.

The experiences of Brazil and Denmark analysed in this chapter are good examples that demonstrate the role of public food procurement and related regulatory frameworks in supporting policymakers' choices on the use of public food procurement as an instrument to pursue multiple development goals, including the promotion of biodiversity. These include, among others, the recognition of sustainable development as one of the objectives of public procurement; awarding criteria that goes beyond the lowest price; legal instruments that provide clear criteria for target beneficiaries and; adapted procurement procedures for facilitating the access (especially in the case of smallholder farmers) to public contractual opportunities. These instruments – together with other elements such as appropriate funding, institutional capacities, and strong political support – are key enabling factors of these policy choices.

The findings of this study highlight the need for more research in this area as well as the importance of raising awareness of policymakers on the need for accompanying PFPP with regulatory reforms and appropriate instruments to allow the use of public food procurement as an instrument to reach multiple benefits and beneficiaries. This includes its use for promoting the conservation and sustainable use of biodiversity.

Notes

1 These programmes are often referred also as Institutional Food Procurement Programmes (IFPP).
2 The term 'regulatory framework' used in this publication comprises all public procurement laws and regulations, legal texts of general application, and administrative rulings made in connection with public procurement.
3 The terms 'smallholder' and 'family' farmers are used interchangeably in this study.
4 For an overview of the Brazilian PFPP see, among many, Swensson, 2015.
5 According to Brazilian legislation (Decree No. 4.887/2003), the remaining members of *quilombola* communities are ethnic and racial groups with their own historical past, characterized by specific territorial relations and with the assumption of black ancestry, related to the resistance of historical oppression.
6 In a survey reported by Sambuichi et al. (2013), based on 29 case studies in different regions of Brazil, diversification in production was assessed as the most common impact of PAA in 72% of cases. Similarly, Valencia et al (2019) reports the increasing production of diversified foods as one of the most important effects of PNAE.

References

Adell, A., Schaefer, B. (2016) *Monitoring Sustainable Public Procurement Implementation: Recommendations and Case Studies*. United Nations Environment Programme (UNEP), 10-Year Framework of Programmes on Sustainable Consumption and Production Patterns (10-YFP), Nairobi/Kenya.

Beltrame, D.M.O., Oliveira, C.N.S., Borelli, T., Santiago, R.A.C., Monego, E.S., Rosso, V.V., Coradin, L., Hunter, D. (2016) 'Diversifying institutional food procurement – opportunities and barriers for integrating biodiversity for food and nutrition in Brazil', *Raízes*, vol 36, no 2, pp 56–69.

Berg, T. (2017) *The Danish Organic Action Plan*. www.oneplanetnetwork.org/initiative/danish-organic-action-plan, Accessed 11 October 2018.

Bioversity International. (2016) *Mainstreaming Agrobiodiversity in Sustainable Food Systems: Scientific Foundations for an Agrobiodiversity Index*, Bioversity International, Rome, Italy.

Bontrager Yoder, A.B., Liebhart, J.L., McCarty, D.J., Meinen, A., Schoeller, D., Vargas, C., LaRowe, T. (2014) 'Farm to elementary school programming increases access to fruits and vegetables and increases their consumption among those with low intake', *Journal of Nutrition Education and Behavior*, vol 46, no 5, pp 341–349. doi: 10.1016/j.jneb.2014.04.297.

Brooks, J., Commandeur, D., Vera, E. (2014) *Inclusive Procurement and Transparency Connecting Smallholder Farmers to School Feeding Procurement Governance for Home Grown School Feeding Project Learning Series #3*. SNV. Available at: www.snv.org/public/cms/sites/default/files/explore/download/snv_pg-hgsf_learning_document_-_inclusive_procurement_and_transparency.pdf

Burlingame, B., Dernini, S. (eds) (2012) *Sustainable diets and biodiversity: Directions and solutions for policy, research and action*, Food and Agriculture Organization of the United Nations, Bioversity International, Rome, Italy. Available at: www.fao.org/3/a-i3004e.pdf

De Schutter, O. (2014) *Briefing Note 08 - April 2014 The power of procurement: Public purchasing in the service of realizing the right to food*, United Nations Special Rapporteur on the Right to Food. Available at: www.srfood.org/images/stories/pdf/otherdocuments/20140514_procurement_en.pdf

De Schutter, O. (2015) 'Institutional food purchasing as a tool for food system reform', in Global Alliance for the Future of Food (ed), *Advancing Health and Well-being in Food Systems: Strategic Opportunities for Funders*. Available at: https://futureoffood.org/wp-content/uploads/2016/09/Global-Alliance-Advancing-Health-Wellbeing-Compendium-April-2015.pdf

Drake, L., Woolnough, A., Burbano, C., Bundy, D. (2016) *Global school feeding sourcebook: Lessons from 14 countries*, Imperial College Press, London, UK.

Espejo, F., Burbano, C., Galliano, E. (2009) *Home Grown School Feeding: A Framework to Link School Feeding with Local Agricultural Production*. World Food Programme, Rome, Italy. Available at: https://documents.wfp.org/stellent/groups/public/documents/newsroom/wfp204291.pdf

FAO. (2014) *School Feeding and Possibilities for Direct Purchases from Family Farming: Case Studies from Eight Countries*, Food and Agriculture Organization of the United Nations, Santiago, Chile.

FAO. (2015). *Las compras públicas a la agricultura familiar y la seguridad alimentaria y nutricional en América Latina y el Caribe: Lecciones aprendidas y experiencias*. FAO, Santiago, Chile.

FAO. (2019) *The State of the World's Biodiversity for Food and Agriculture*, Bélanger, J., Pilling, D. (eds) FAO Commission on Genetic Resources for Food and Agriculture Assessments, Rome, Italy.

FAO and WFP. (2018) *Home-Grown School Feeding. Resource Framework. Technical Document*, The Food and Agriculture Organization of the United Nations and the World Food Programme, Rome, Italy.

Fitch, C., Santo, R. (2016) *Instituting Change: An Overview of Institutional Food Procurement and Recommendations for Improvement*, The Johns Hopkins Center for a Livable Future, Baltimore, USA.

Foodlinks. (2013) *Revaluing Public Sector Food Procurement in Europe: An Action Plan for Sustainability*, The Foodlinks Project.

Frederiksen, A.C. (2016) 'The Danish organic label'. Available at: http://organicdenmark.com/organics-in-denmark/the-danish-organic-label, accessed 5 November 2018.

Gelli, A., Neeser, K., Drake, L. (2010) *HGSF Working Paper No 1 Linking Smallholder Agriculture to School Food Provision*, Home Grown School Feeding Initiative and the Partnership for Child Development, Rome.

Global Panel. (2015) *Healthy Meals in Schools: Policy Innovations Linking Agriculture, Food Systems and Nutrition Policy Brief no. 3*, Global Panel on Agriculture and Food Systems for Nutrition, London.

HLPE. (2017) *Nutrition and food systems: A report by The High Level Panel of Experts on Food Security and Nutrition September 2017*, Committee on World Food Security, High Level Panel of Experts, Rome.

Hole, D.G., Perkins, A.J., Wilson, J.D., Alexander, I.H., Grice, P.V., Evans, A.D. (2005) 'Does organic farming benefit biodiversity?', *Biological Conservation*, vol 122, no 1, pp 113–130. doi: 10.1016/j.biocon.2004.07.018.

Hunter, D., Beltrame, D., Wasike, V. (2016) 'The school food revolution: Can local farmers and food biodiversity be part of it?' Presented at: The Secure, Safe, Sustainable Food Systems: Safe Today, Optimal for the Future Workshop University of Sydney, Sydney, Australia, 30 Nov.–2 Dec. 2016.

IPC and WFP. (2013) *Structured Demand and Smallholder Farmers in Brazil: The Case of PAA and PNAE*, International Policy Centre for Inclusive Growth and the World Food Programme Centre of Excellence against Hunger, Brazil.

IPES-Food. (2016) *From uniformity to diversity: A paradigm shift from industrial agriculture to diversified agroecological systems*, International Panel of Experts on Sustainable Food Systems, Brussels.

Jespersen, L.M., Baggesen, D.L., Fog, E., Halsnaes, K., Hermansen, J.E., Andreasen, L., Strandberg, B., Sørensen, J.T., Halberg, N. (2017) 'Contribution of organic farming to public goods in Denmark', *Organic Agriculture*, vol 7, no 3, pp 243–266. doi: 10.1007/s13165-017-0193-7.

Kelly, S., Swensson, L.F.J. (2017) *Leveraging institutional food procurement for linking small farmers to markets: Findings from WFP's Purchase for Progress initiative and Brazil's food procurement programmes*, The Food and Agriculture Organization of the United Nations and the World Food Programme, Rome, Italy.

Mikkelsen, B.E., Lundø, M. (2016) 'Role of public organic procurement policies (POPP's) in the implementation of organic food and farming strategies: Lessons learnt from Denmark', *Raízes*, vol 36, no 2, pp 45–54.

Morgan, K., Sonnino, R. (2008) *The School Food Revolution: Public Food and the Challenge of Sustainable Development*, Earthscan, London.

Niebylski, M.L., Lu, T., Campbell, N.R.C., Arcand, J., Schermel, A., Hua, D., Yeates, K. E., Tobe, S.W., Twohig, P.A., L'Abbé, M.R., Liu, P.P. (2014) 'Healthy food procurement policies and their impact', *International Journal of Environmental Research and Public Health*, vol 11, no 3, pp 2608–2627. doi: 10.3390/ijerph110302608.

Sambuichi, R.H.R., Galindo, E.P., Oliveira, M.A.C.d. and Magalhães, A.M.d. (2013) *Compras públicas sustentáveis e agricultura familiar: a experiência do Programa de Aquisição de Alimentos (PAA) e do Programa Nacional de Alimentação Escolar (PNAE)*. Mimeo, Brasilia.

Sidaner, E., Balaban, D., Burlandy, L. (2012) 'The Brazilian school feeding programme: An example of an integrated programme in support of food and nutrition security', *Public Health Nutrition*, vol 16, no 6, pp 989–994. doi: 10.1017/S1368980012005101.

Smith, J., Andersson, G., Gourlay, R., Karner, S., Mikkelsen, B.E., Sonnino, R., Barling, D. (2016) 'Balancing competing policy demands: The case of sustainable public sector food procurement', *Journal of Cleaner Production*, vol 112, no 1, pp 249–256. doi: 10.1016/j.jclepro.2015.07.065.

Sørensen, N.N., Lassen, A.D., Løje, H., Tetens, I. (2015) 'The Danish organic action plan 2020: Assessment method and baseline status of organic procurement in public kitchens', *Public Health Nutrition*, vol 18, no 13, pp 2350–2357. doi: 10.1017/S1368980015001421.

Sørensen, N.N., Tetens, I., Løje, H., Lassen, A.D. (2016) 'The effectiveness of the Danish Organic Action Plan 2020 to increase the level of organic public procurement in Danish public kitchens', *Public Health Nutrition*, vol 19, no 18, pp 3428–3435. doi: 10.1017/S1368980016001737.

Spyridon, F., Mikkelsen, B.E. (2018) *More organic food on the public plate – need for a New Public Governance? WG 2. Institutional food procurement and school feeding programmes: Exploring the benefits, challenges and opportunities*, Presented at: Third International Conference: Agriculture and Food in an Urbanizing Society, Porto Alegre, Brazil.

Stefani, G., Tiberti, M., Lombardi, G.V., Cei, L., Sacchi, G. (2017) 'Public food procurement: A systematic literature review', *International Journal on Food System Dynamics*, vol 8, no 4, pp 270–283. doi: 10.18461/ijfsd.v8i4.842.

Sumberg, J., Sabates-Wheeler, R. (2010) 'Linking Agricultural Development to School Feeding in sub-Saharan Africa: Theoretical perspectives', *Food Policy*, vol 36, no 3, pp 341–349. doi: 10.1016/j.foodpol.2011.03.001.

Swensson, L.F.J. (2015) 'Institutional Procurement of food from smallholder farmers: The case of Brazil', Rome, FAO. http://www.fao.org/3/a-bc569e.pdf

Swensson, L.F.J. (2018) *Working Paper no. 177: Aligning policy and legal frameworks for supporting smallholder farming through public food procurement: The case of home-grown school feeding programmes*, The Food and Agriculture Organization of the United Nations and International Policy Centre for Inclusive Growth.

Tartanac, F., Swensson, L.F.J., Galante, A.P., Hunter, D. (2019) 'Institutional food procurement for promoting sustainable diets', in Burlingame, B., Dernini, S. (eds) *Sustainable diets: The transdisciplinary imperative*, pp 240–247, CABI, Wallingford.

UNEP. (2017) *Global Review of Sustainable Public Procurement.* United Nations Environmental Programme, Nairobi, Kenya.

UNSCN. (2017) *Schools as a System to Improve Nutrition: A New Statement for School-based Food and Nutrition Interventions.* United Nations System Standing Committee on Nutrition, Rome, Italy.

Valencia, V., Wittman, H., Blesh, J. (2019) 'Structuring markets for resilient farming systems'. *Agronomy for Sustainable Development*, vol 39, no 2, pp 25.

8

FOOD BIODIVERSITY, LOCAL SOURCING AND CONSUMERS

Gastronomy as a critical interface

Willy Legrand, Clare Hindley and Gabriel C. M. Laeis

Introduction

The General Assembly of the United Nations adopted Resolution 71/246 designating 18 June as Sustainable Gastronomy Day[1]. This decision clearly indicates a perceived need to review and redefine the processes of and behind 21st century gastronomy. For decades, and in response to the globalisation and industrialisation of our food manufacturing and gastronomy sectors, some stakeholders, including non-governmental organisations, consumers and chefs, have, with mixed results, advocated for a change in how we eat. The consequence has been a myriad of short-lived food fads and a few long-term food trends. In this chapter, by analysing the role and history of both food and gastronomy, we argue that the gastronomic sector is a critical interface in which consumers and chefs negotiate what is eaten and consequently what food is sourced. The impact of local and organic food on the gastronomy sector is critically reviewed along with the concept of 'sustainable gastronomy', with an emphasis on the challenges associated with 'sustainable dining options'. Finally, we examine the crucial linkages between food biodiversity, gastronomy and local sourcing. The chapter concludes that consumers, as well as gastronomic 'producers', have a role to play in the type and location of food grown. However, current gastronomic food trends, such as local and organic, are unlikely to have a substantial impact on our food production systems without recognition that food choices are not only the concern of restaurateurs and chefs but necessitate significant support from policymakers.

Food to gastronomy

Throughout the history of human development, food has always had a special significance. For thousands of years, food resources were everyone's primary preoccupation, if not obsession. The practice of hunting and gathering, the discovery of farming cereals, the science of stock breeding, the consumption of salt, sugar and spices from far off places, have all changed the foundations of many cultures' cuisines (Kittler, Sucher and Nahikian-Nelms, 2017; Mazoyer and Roudart, 2006). This has had a profound impact on the development of societies and cultures as well as on the ecological condition of our planet. Some species have been chosen by humankind as the major sources of food and thus bred and propagated in large quantities, whereas others have been neglected (Diamond, 1999). This process of choosing certain species for consumption and breeding, or even genetically engineering them according to our needs, continues to dramatically change the way we eat as well as our global ecosystems. Chefs, restaurateurs and consumers are key stakeholders in the negotiation of this process.

From a social perspective, food is a critical factor in physical well-being and its preparation a major expression of culture. However, food, or rather its availability and quality, is also of great concern for governments and consumers. Many feel a genuine disconnection with the processes around agriculture and food preparation, something which is also associated with inequality of food distribution around the world. Today, it is estimated that about 821 million people are suffering from chronic undernourishment, while another 672 million people are obese (FAO, IFAD, UNICEF, WFP, WHO, 2018; WHO, 2018a). Paradoxically, this even occurs simultaneously in many countries. Counting calories is at times the currency of purchasing decisions, yet little effect can be seen on the obesity epidemic (Friedrich, 2017; The GDB 2015 Obesity Collaborators, 2017).

Food fads come and go in developed countries where consumers are in a constant quest for ever more exotic and fashionable new products. There is now an increased interest in food from other cultures and geographical areas. Affordable travel options and international tourism as well as a globalised commodity market render access to 'world foods' and a 'globalised' gastronomy easier and easier (Mak, Lumbers and Eves, 2012; Hwang et al., 2018). As outlined below, a number of the fads and trends seem to show the consumers' concern with the wider implications of their food consumption. However, the environmentalist Wendell Berry believed that in these countries most individuals have rather become 'ideal industrial food consumer[s] [who] would be strapped to a table with a tube running from the food factory directly into his or her stomach' (Berry, 1990, p. 146).

Many countries, whether developed or developing economies, have experienced an increase in industrialised, manufactured and convenience food at the expense of a diverse selection of traditional and locally adapted food. Moreover, studies have outlined that the global food system, despite an increase in calorie output in recent decades, does not operate for the benefit of all people equally (McIntyre et al., 2009; Pretty, 2001). One of the key reasons for this situation lies within the structure of the agribusiness sector, as Constance, Hendrickson and Howard (2014) illustrate; since the 1990s, the food industry has seen an increasing market power concentration and consolidation of multinational corporations through horizontal and vertical integration. In many instances procurement and distribution policies are determined by a few global players, such as JBS (the world's largest beef packer) or Cargill (the world's largest agribusiness company), to the disadvantage of consumers and producers (Constance et al., 2014). Consequently, if one aspect of everyone's lives has radically changed in recent decades, it is certainly food including, but not limited to, its production, sourcing, preparation and cooking techniques and nutritive value. This, in turn, dramatically impacts the issue of biodiversity (see below).

'Gastronomy' has been very diversely and broadly defined, originally being attributed to 4th century BC Greek poet and philosopher Archestratus in his fragmented work known as *The Life of Pleasure, The Science of Dining* or *The Art of Cooking* (Gordon, 2012). Archestratus' description of 'cooking as an art' provides the basis for the modern definition of gastronomy. The Oxford Dictionary defines gastronomy as 'the practice or art of choosing, cooking, and eating good food'[2], alluding to the modern-day founder of gastronomic science, Brillat-Savarin. He argued that gastronomy was the science of 'good eating', predominantly concerned with the culture of consumption and taste (Brillat-Savarin, [1825] 1941), including concepts such as the combination of flavours, aromas, taste and the visual appearance of a meal.

Today, cooking and gastronomy continue to play a significant role in the culture of food. Cooking food has as much to do with pleasure and well-being, sharing moments, experiencing culture and the social fabric of society as it has to do with ecology and biodiversity (Godfray, 2011; Hjalager and Richards, 2011; Tscharntke et al., 2012). However, today's gastronomy is also at the heart of a host of transformations arising from changing consumer behaviour, agro-industrial food production and arising environmental changes. While it is not a goal of this chapter to provide a detailed description and analysis of these transformations, a number of key issues arise including the relationship of gastronomy and biodiversity; the role of gastronomy in sustainable development; consumer-driven 'sustainable' dining options and the role of consumers and chefs in sustainable gastronomy.

Biodiversity to food

Jacques Weber, anthropologist and economist at the French Agricultural Research Centre for International Development (CIRAD, *Centre de coopération internationale en recherche agronomique pour le développement*) summarises the importance of biodiversity in gastronomy and in our daily lives in these words:

> If this morning, you had the chance to have breakfast, you only consumed products derived from biodiversity. All of your clothes come from the biodiversity. And if you have consumed elements of biodiversity this morning and have

> properly digested them, that's thanks to the collaboration of 1.5 kilos of intestinal bacteria on average. […]. We are a part of the living world; we are not next to it: we participate in these interactions between all organisms.
>
> *Rovillé, Courchamp and Bunle, 2009, p. 23*

The world of gastronomy depends entirely on biodiversity. The World Health Organization (2018a, para. 6) advocates that 'healthy local diets, with adequate average levels of nutrients intake, necessitates maintenance of high biodiversity levels'. A well-prepared meal may be based on the diversity of ingredients (fruits, cereals, animal products such as milk, eggs etc.) or products transformed by microscopic biodiversity (cheese, bread, wine, etc.).

Bioversity International discusses how biodiversity contributes to the way we eat in five ways. Firstly, the biodiversity of food 'provides a wide range of nutritious, colourful options to choose from that meet diverse cultural and taste preferences' (Kennedy and Stoian, 2017, para. 3). This has a particular relevance to gastronomy and the way chefs interpret and prepare meals that are pleasing to the eye and flavourful to the palate (Klosse, 2014). Secondly, food biodiversity 'takes advantage of important and significant nutritional differences *between* species' (Kennedy and Stoian, 2017, para. 4). The nutrition component of food is high on the radar of consumers and food writers alike, craving the latest 'super fruit' or 'super veggie', but the nutritional component of food should be an essential aspect in designing balanced menus (Salais, 2015). Thirdly, food biodiversity 'takes advantage of important and significant nutritional differences *within* species' (Kennedy and Stoian, 2017, para. 5). The authors give the example of the orange-fleshed bananas from the Pacific which contain higher levels of beta-carotene compared to the typical white bananas found in supermarkets around the globe (Kennedy and Stoian, 2017). Similarly, the World Health Organization reports (2018b, para. 6), '[The] nutritional composition between foods and among varieties/cultivars/breeds of the same food can differ dramatically, affecting micronutrient availability in the diet' (see also Part II of this book). Fourthly, food biodiversity 'provides the ingredients for a portfolio of nutritious foods in different seasons of the year' (Kennedy and Stoian, 2017, para. 6). The argument is made that lesser-known plants and animal species tend to be well-adapted to local conditions thus requiring fewer farming inputs. Many of those traditional, indigenous foods have been forgotten with a clear shift towards modern commercial foods with, at times, catastrophic consequences on health including the likes of 'obesity, diabetes, heart disease, high blood pressure and other diet-related disorders' (FAO, 2018, para. 3). Finally, food biodiversity should offer food options that are 'available in sufficient quantities year-round, accessible, affordable, and acceptable for people to eat it' (Kennedy and Stoian, 2017, para. 7). This may be the crux on the biodiversity front: how do we recover and strengthen local food systems to ensure that 'we can leverage long-standing knowledge and practices to preserve biodiversity and benefit from it today and in the future?' (FAO, 2018, para. 5).

One frequently highlighted way to tackle this question is to opt for organic agricultural practices. Among a number of positive environmental and social corollaries, supporting organic agriculture has a positive impact on biodiversity generally, and on healthy insect populations specifically. This is mainly due to particular farming practices, such as employing frequent crop rotations, allowing for green fallow and wide field margins, using a higher crop diversity and banning all chemical herbicides, pesticides, fungicides and fertilisers. Studies suggest that this can lead to 30% higher species richness on organic farms as opposed to conventionally managed farms. Even though results vary, birds, insects and plants especially fare much better under this environmentally friendly farming system (Bengtsson, Ahnström and Weibull, 2005). Insects are particularly relevant to the food chain by playing a 'central role in a variety of processes, including pollination, herbivory and detrivory, nutrient cycling, and providing a food source for higher trophic levels such as birds, mammals and amphibians' (Hallmann et al., 2017, para. 2). It is estimated that insects are responsible for 80% of wild plants' pollination (Ollerton, Winfree and Tarrant, 2011) and are the main food source to 60% of birds (Morse, 1971, cited in Hallmann et al., 2017, para. 2). In a recent *New York Times* article, Jarvis (2018, para. 48) summarises the important role of insects:

> By eating and being eaten, insects turn plants into protein and power the growth of all the uncountable species – including freshwater fish and a majority of birds – that rely on them for food, not to mention all the creatures that eat those creatures. We worry about saving the grizzly bear, says the insect ecologist Scott Hoffman Black, but where is the grizzly without the bee that pollinates the berries it eats or the flies that sustain baby salmon? Where, for that matter, are we?

The global 'loss of insect diversity and abundance is expected to provoke cascading effects on food webs and to jeopardize ecosystem services' (Hallmann et al., 2017, para. 1) and 'what we're losing is not just the *diversity* part of biodiversity, but the *bio* part: life in sheer quantity' (Jarvis, 2018, p. 41). Insects fare better in environmentally friendly farming systems and are essential to the biodiversity which is in turn critical to our food supply. Biodiversity plays a vital role in *what we eat, when we eat it* and, *how we consume it.* Therefore, the support of a diet that supports a rich biodiversity should be a priority for any sustainable gastronomy.

Sustainable gastronomy

While there is no accepted definition of 'sustainable gastronomy', Scarpato (2002) suggests that it is 'about producing food that is environmentally sensitive and preparing and eating it so that it nourishes both mind and body' (p. 139). Sustainable gastronomy implies 'the survival of local food production […], the transmission of culinary knowledge [… and …], the right to pleasure and diversity' (Scarpato, 2002, p. 140) as well as aspects surrounding the preparation of food at home, the creation of authentic experiences and the well-being of individuals and communities. The United Nations (UN) declared June 18th World Sustainable Gastronomy Day, thus recognising that 'sustainable gastronomy' is crucial, has a responsibility for the three dimensions of sustainable development (economic, environmental and social) and in achieving the Sustainable Development Goals adopted in 2015 (United Nations, 2017). In addition, sustainable gastronomy 'can play a role promoting agricultural development, food security, nutrition, sustainable food production and the conservation of biodiversity[3]'.

In some sections of gastronomy, such as restaurants, there is a general trend towards a more natural and 'fresher' kitchen. Some chefs develop a privileged relationship with local and artisan-like farmers helping to create an identity around the quality of products and supporting the local community (Legrand, Sloan, Fett and Manten, 2019). The use of rare, traditional varieties (also referred to as *heirloom* or *heritage*) of fruits, vegetables and grains gives this type of gastronomy a dimension that adds to its ecologic, economic and social sustainability (Pröbstl-Haider, Hochwarter, and Schrank, 2015). A sustainable gastronomy also means a commitment and responsibility of all institutions, organisations, businesses and communities in relation to the production and consumption of food. The UN specifically mentions that all actors, whether the public or private sectors, producers, businesses, industry professionals or travel institutions 'should support the consumption of sustainable food and create links with local producers'[4].

Sustainable gastronomy thus involves the components of 'sustainable consumption' within a local context, via the usage of locally-grown ingredients and the support for local farmers and communities. This is emphasised within the UN's website dedicated to Sustainable Gastronomy Day where consumers follow the 'Think Globally – Eat Locally' *mantra*. As much as biodiversity plays a critical role in *what we eat, when we eat* and, *how we consume* – as discussed in the previous section – in sustainable gastronomy, consumers are urged to consider the following when making food choices: Is the product in season, where does it come from and how was it grown?

Local and sustainable trends

'Local' and so-called 'sustainable' food has for years been omnipresent on the agendas of gastronomic trend surveys. In the 2018 US National Restaurant Association's (NRA) annual report on gastronomic trends, kitchen chefs reported that 'hyper-local sourcing' (i.e. food grown on the premises of the restaurant), 'locally sourced meats and seafood' and 'locally sourced produce' were amongst the top eight restaurant concept trends. This is by no means a new development. For about a decade local food has been amongst the top trends (NRA, 2018). Other food trends in the 2018 report also underlined consumers' interests in presumably sustainable dining options. 'Natural ingredients', 'environmental sustainability' and 'simplicity/back to basics' were arguably calls for food that is produced and traded with a minimum environmental impact. Chefs also forecasted food trends to evolve mainly around 'new cuts of meat', 'house-made' and 'farm-branded' menu items and a variety of ethnic foods (NRA, 2018, p. 1). The BBC (2018) '15 food trends for 2018' are a little lighter on the local side, with hyper-local sourcing only on place six, and more focused on health-minded choices such as 'gut-friendly food', 'booze-free beverages', 'specialised teas' and 'plant-based protein'. Similarly, the chefs of New Zealand's Restaurant Association assert that customers are especially looking for local, sustainable and plant-based options (NZ Herald, 2018). Germany's Future Institute proposes that, among others, flexitarians, urban gardening and cradle-to-cradle food concepts will be major determinants of the country's future foodscape (Zukunftsinstitut, 2018). Scholarly research further corroborates the notion that chefs and restaurateurs are very interested in concepts around local food (Roy, Hall and Ballantine, 2016). It seems that any restaurateur is well-advised to put the local and sustainable on their menu as soon as possible.

The challenges of sustainable gastronomy

Restaurateurs need to think carefully about questions such as what 'local' or 'ethnic' food really means, what exactly makes for an environmentally sustainable menu and to what extent their restaurant can tackle such issues. Finding answers to such questions is problematic because the situation is aggravated by the ambiguity of many trends. For instance, the wish for more 'sustainable' dining options is hard to reconcile with consumers' interest in 'new cuts of meat' and 'sustainable seafood', themselves problematic concepts. The production of meat is a significant greenhouse gas contributor and has a number of detrimental environmental impacts along its value chain (Hedenus, Wirsenius and Johansson, 2014). Looking at the current state of the world's oceans, scientists argue that there may not be a 'sustainable seafood' option any longer, due to the overexploitation of marine species (Pauly et al., 2002), a situation exacerbated with the introduction of micro-plastics in the food chain. A particularly pervasive term in the debate around sustainable food provisioning is 'local' (for more on procurement refer to the previous chapter, 7). It has become 'something of a chimera' which successfully evades definition and can potentially acquire a variety of meanings (DeLind, 2011, p. 273; see also Roy, Hall and Ballantine, 2016, in a restaurant context). For restaurateurs, despite the ongoing interest, the concept of local food may be as much a challenge as it is an opportunity.

The challenges of local gastronomy

The above-described interest in local food comes largely from consumer concerns over the current industrialised food production system and its social and environmental implications. 'Locavorism', the interest of consumers to limit their diet to food from a certain geographic area, arose from such concerns. This is not considered only a food movement, but as Beer (2015) states, it is linked to 'a well-established mantra […] that local is good (whatever local is)' (p. 47). This definition of local is certainly challenged by growing urbanisation on a global scale. The UN predicts that by 2050, 66% of the global population will live in urban centres, the fastest move from rural to urban areas the world has yet seen (United Nations, 2018). Bauman (2008, p. 226) argues that 'sooner or later, Kant warned, there will be no empty space left into which those of us who found the already populated places too cramped – or too inconvenient, awkward, and uncomfortable – can venture'. Agricultural food production is minimal in an urban environment and thus the notion of 'local food' may change, or already has changed, from direct vicinity to regional to national or beyond.

This open interpretation of 'local' also raises the question of who or what benefits from the purchase of 'local'. Locavores have the desire to eat local foods because they think it may taste better, is better for the farmers, the local community or for the environment in general. However, Noll (2014, p. 220) states that 'if corporations gain the majority of the profit from selling local goods, then this money does not go back into the community'. Additionally, 'local' may have a different meaning depending on where one lives and the term, similar to 'regional', is often overused. For example, Noll (2014, p. 220) explains that '200 miles away may be local for Wal-Mart, but in actuality, the goods are being sold in a different community or even a different state'. Consequently, if large corporations have the freedom to define the concept of 'local' 'into a commodity attribute […] then, in essence, the term "local" loses its meaning and connection to a particular place' (Noll, 2014, p. 220). Thus, it is crucial to separate the concept of geographic 'local' from its business-related use since 'through universalizing, the local food movement may no longer support the local community and this could go against the systems focused direction' (Noll, 2014, p. 220).

Despite all the ambiguity associated with 'local food', locally produced and sourced food has, for many years, been seen as a gastronomic trend. Recently this trend has been revived with the arrival of hyper-local sourcing concepts (De Chabert-Rios and Deale, 2018). The use of local food in restaurants is arguably a concept that can promote the culinary heritage of entire regions, it is an essential component of food tourism and can support rural economic development and biodiversity through direct farm-to-table linkages (Bessière, 1998; Broadway, 2017; Timothy and Ron, 2013). Local food is therefore closely linked to what Gössling and Hall (2013) have termed 'sustainable culinary systems'. From a broader perspective, local food systems claim to be more equitable, just and socially inclusive than the current agro-industrial system (Pretty, 2001). Unfortunately, the localisation of food systems has often become an end goal in itself, rather than a sustainable development strategy or process. To this end, Born and Purcell (2006) argued that there is nothing inherently good about 'the local'. It depends on how and according to whose agenda food systems are localised; all too often the term becomes conflated with 'sustainable', which is not necessarily the case (Duell, 2013; Laeis and Lemke, 2016).

Equally critical, yet not well researched, is the notion that local food may be a cornucopia of culinary creativity and business opportunities for restaurants, as the current trends outlined above and successful restaurant concepts, such as *noma* (Copenhagen, Denmark) might imply. Case studies from Ireland and South Africa showed that chefs and restaurant owners face challenges in defining local food and dealing with seasonal fluctuations of local and hyper-local supply, while maintaining authentic menus and satisfying guest demands (Duram and Cawley, 2012; Laeis, 2016). On a practical level, Wellton, Jonsson and Walter (2016) concluded from case studies in Sweden that the organisational implications of using local foods can collide with the daily practices of restaurants. Kitchens are pressure-driven and complex environments. If personnel are not well-trained in restaurant organisation, cooking skills and menu planning, then including local food into menus poses a significant challenge to the restaurant's success. This is largely due to the need for chefs to react promptly to the often-varying quality, quantity and availability of local food supply. Another constraint that emerges from the daily practices of restaurants was found by Roy, Hall and Ballantine (2016) in Vancouver, Canada. Half of the chefs and restaurant managers surveyed did not source from farmers' markets because they lacked the time to do so or farmers did not offer delivery services. The majority of respondents sourced from local and regional distributors. Overall, logistical issues in the delivery processes of local food was also cited by Green and Dougherty (2008).

The consumer perspective

The consumer perspective on sustainable food and local food in particular is also challenging. Since the turn of the century, there is an increasing interest in local food and consumers are willing to pay a premium for its credence qualities (Feldmann and Hamm, 2015). Consumers that are ethnocentric, environmentally concerned, quality-focused or favour local production from 'their' region seem to be most likely to opt for local choices (Aprile, Caputo and Nayga, 2016). Brown, Dury and Holdsworth (2009) found in a study on local and organic box schemes that they were favoured predominantly by comparatively wealthy and educated consumers. Interestingly, restaurant patrons find dishes made from local produce more enticing, if their price is marginally higher than other non-local options (Alfnes and Sharma, 2010). Seemingly, there needs to be an air of exclusivity, perhaps to appeal to the lifestyle of well-educated and affluent guests. Other scholars found that local food might not be quite as enticing to customers as the trends above seem to imply. Herzog and Murray (2013) found that patrons did not choose to dine at a particular restaurant because of its local dishes, even though they were a prominent part of the menu; main courses were chosen according to their expected taste rather than their local origin. Local food is only seen as attractive if other quality attributes are present, such as freshness, visual appeal and pleasant taste.

Undoubtedly, food is a major aspect of societal development. Public concerns in developed countries regarding environmental and social sustainability have started to shape culinary trends and may offer significant opportunities for gastronomic entrepreneurs. A study from the US showed that consumers are even willing to pay a premium for sustainable restaurant practices (Namkung and Jang, 2017). Guests have moved away from particular menu items and are looking for restaurant concepts that represent their lifestyle choices. For example, the LOHAS (lifestyle of health and sustainability) consumer group might be a valuable target market. In fact, restaurants in Vancouver, Canada, reported that selling locally produced food was profitable for their establishments (Roy, Hall and Ballantine, 2016). Nevertheless, statistics highlight that meat consumption in the US, Europe and New Zealand is still unsustainably high (OECD, 2018). This underlines the attitude-behaviour gap amongst many consumers (Vermeir and Verbeke, 2006). Even though restaurant guests might claim interest in local food, their actual purchasing behaviour tells a different story. Despite the apparent interest of some consumer groups in sustainable options when eating out, restaurateurs need to bear in mind that mere attitudes are poor predictors of actual behaviour (Ajzen, 2001). So-called food trends need to be taken with a pinch of salt.

This fragmented body of research might suggest two things. Firstly, the interest in local and sustainable food is predominantly present amongst well-off and educated consumers in developed countries. Secondly, 'local' as such does not make for an attractive menu, if the food is otherwise not appealing. In other words: the success of a highly-acclaimed restaurant such as *noma* might be primarily based on the craftsmanship and creativity of its chefs and the fact that its menu prices demand a high disposable income. From this point of view, it appears that the fact that most of what *noma* serves could be defined as local is just the cherry on top. It is doubtful whether the 'local' trend will ever become mainstream or add significantly to a change of the current food system. To this end it is interesting to look at the past organic food trend which started in the early 2000s. In Europe, more than a decade later, the share of farm land under organic management was just 2.4%, meanwhile in Germany, Europe's largest organic food market, the share of organic retail sales was under 5% (IFOAM, 2016). Even though organic produce has become a fixed part of many developed countries' foodscapes, especially the US and Europe, it has failed to become anything near mainstream. Limited data is available on the sales of 'local' food, probably mainly due to the term not being strictly defined. However, rising numbers of farmers' markets (USDA, 2010), for instance, further corroborate the consumers' interest in local food.

Looking to the future

Even though local and organic foods appear to be rather long-lasting gastronomic trends and there is evidence that their importance is increasing overall, it also seems likely that they will remain a niche phenomenon. Local and organic food appears to be positively associated with supporting biodiversity in agriculture and diet, but without widespread endorsement from consumers and producers – including gastronomic 'producers' such as chefs and restaurateurs – they will not have a meaningful impact on how we produce food. However, a greater effort by policy-makers concerned with agriculture and food production might change this outlook.

Food biodiversity is a critical component of a rich and diverse gastronomy through the diversity between various food groups (for example grains, leafy vegetables, fruits), within a food group (for example apple, pears, peaches) or within a species (for example the traditional varieties of cultivated apples, bananas). Food biodiversity also provides gastronomy with the potential to work with numerous, perhaps lesser-known varieties which are well-adapted to local conditions, making farming less dependent on external inputs. While the local food movement has received attention in recent decades, there are still inherent limitations in its (i) definition, (ii) size and scope, (iii) impact mitigation potential and (iv) acceptance. It might be advisable for the gastronomy sector to consider 'sustainable diets' rather than focusing on micro-trends such as local or hyper-local. Contrary to what one may first think, sustainable diet is 'more than just the food'. Sustainable diets are 'healthy diets that also optimise the environment, food quality, nutrition, socio-cultural values, the economy and governance' (Mattioni and Caraher, 2018, p. 430).

Without wishing to claim that individual food choices and chefs' menu planning decisions are inconsequential, the above brief discussion shows that food choices are part of other life decisions. If we wish these decisions to be consequential, they need to be addressed not only by individuals, but to be part of a wider political movement. The culinary arts, catering and restaurant industry with the brigade of cooks, entrepreneurs, artisan chefs and innovators all have a role to play in what is being offered to consumers and have a direct influence on individual food choices. We certainly cannot be expected to dramatically change the field of gastronomy in the short term. However, food decisions do not stand alone and need to take place in the political arena, not just in the kitchen or dining room.

Notes

1 https://www.un.org/en/events/sustainablegastronomy/background.shtml
2 https://en.oxforddictionaries.com/definition/gastronomy
3 https://www.un.org/en/events/sustainablegastronomy/background.shtml
4 Ibid.

References

Ajzen, I. (2001) 'Nature and operation of attitudes', *Annual Review of Psychology*, vol 52, pp 27–58. doi: 10.1146/annurev.psych.52.1.27.

Alfnes, F., Sharma, A. (2010) 'Locally produced food in restaurants: Are the customers willing to pay a premium and why?', *International Journal of Revenue Management*, vol 4, no 3–4, pp 238–258. doi: 10.1504/IJRM.2010.035955.

Aprile, M.C., Caputo, V., Nayga, R.M., Jr. (2016) 'Consumers' preferences and attitudes toward local food products', *Journal of Food Products Marketing*, vol 22, no 1, pp 19–42. doi: 10.1080/10454446.2014.949990.

Bauman, Z. (2008) *Does ethics have a chance in a world of consumers?* Harvard University Press, Cambridge, MA.

BBC. (2018) '15 food trends for 2018', *BBC Good Food*. Available at: www.bbcgoodfood.com/howto/guide/15-food-trends-2018

Beer, S. (2015) 'Does the pursuit of local food destroy our environment? Questions of authenticity and sustainability', in Sloan, P., Legrand, W., Hindley, C. (eds) *The Routledge handbook of sustainable food and gastronomy*, pp 47–56, Routledge, London. doi: 10.4324/9780203795699.

Bengtsson, J., Ahnström, J., Weibull, A.C. (2005) 'The effects of organic agriculture on biodiversity and abundance: A meta-analysis', *Journal of Applied Ecology*, vol 42, no 2, pp 261–269. doi: 10.1111/j.1365-2664.2005.01005.x.

Berry, W. (1990) *What People Are for?* Counterpoint Press, Berkley, CA.

Bessière, J. (1998) 'Local development and heritage: Traditional food and cuisine as tourist attractions in rural areas', *Sociologia Ruralis*, vol 38, no 1, pp 21–34. doi: 10.1111/1467-9523.00061.

Born, B., Purcell, M. (2006) 'Avoiding the local trap: Scale and food systems in planning research', *Journal of Planning Education and Research*, vol 26, no 2, pp 195–207. doi: 10.1177/0739456X06291389.

Brillat-Savarin, J.A. ([1825] 1941) *The physiology of taste*. (M.F.K. Fisher, Trans.), Alfred A. Knopf, New York.

Broadway, M. J. (2017) '"Putting place on a plate" along the West Cork food trail', *Tourism Geographies*, vol 19, no 3, pp 467–482. doi: 10.1080/14616688.2016.1276615.

Brown, E., Dury, S., Holdsworth, M. (2009) 'Motivations of consumers that use local, organic fruit and vegetable box schemes in Central England and Southern France', *Appetite*, vol 53, no 2, pp 183–188. doi: 10.1016/j.appet.2009.06.006.

Constance, D.H., Hendrickson, M., Howard, P.H. (2014) 'Agribusiness concentration: Globalization', in W. D. Schanbacher (ed) *The global food system*, pp 31–58, Praeger, Santa Barbara, CA.

De Chabert-Rios, J., Deale, C.S. (2018) 'Taking the local food movement one step further: An exploratory case study of hyper-local restaurants', *Tourism and Hospitality Research*, vol 18, no 3, pp 388–399. doi: 10.1177/1467358416666137.

DeLind, L.B. (2011) 'Are local food and the local food movement taking us where we want to go? Or are we hitching our wagons to the wrong stars?' *Agriculture and Human Values*, vol 28, no 2, pp 273–283. doi: 10.1007/s10460-010-9263-0.

Diamond, J.M. (1999) *Guns, germs and steel: The fates of human societies*, W. W. Norton, New York.

Duell, R. (2013) 'Is "local food" sustainable? Localism, social justice, equity and sustainable food futures', *New Zealand Sociology*, vol 28, no 4, pp 123–144.

Duram, L.A., Cawley, M. (2012) 'Irish chefs and restaurants in the geography of "local" food value chains', *The Open Geography Journal*, vol 5, pp 16–25. doi: 10.2174/1874923201205010016.

FAO. (2018) *Biodiversity reveals the treasures all around us: How local varieties can support nutrition and food security*, The Food and Agriculture Organization of the United Nations, Rome, Italy. Available at: www.fao.org/fao-stories/article/en/c/1133432/

FAO, IFAD, UNICEF, WFP, WHO. (2018) *The State of Food Security and Nutrition in the World 2018. Building climate resilience for food security and nutrition*, The Food and Agriculture Organization of the United Nations, Rome, Italy. Available at: www.fao.org/3/i9553en/i9553en.pdf

Feldmann, C., Hamm, U. (2015) 'Consumers' perceptions and preferences for local food: A review', *Food Quality and Preference*, vol 40. part A, pp 152–164. doi: 10.1016/j.foodqual.2014.09.014.

Friedrich, M.J. (2017) 'Global obesity epidemic worsening', *JAMA*, vol 318, no 7. doi: 10.1001/jama.2017.10693.

The GDB 2015 Obesity Collaborators. (2017) 'Health Effects of overweight and obesity in 195 countries over 25 Years', *The New England Journal of Medicine*, vol 377, pp 13–27. doi: 10.1056/NEJMoa1614362.

Godfray, H.C.J. (2011) 'Food and biodiversity', *Science*, vol 333, no 6047, pp 1231–1232. doi: 10.1126/science.1211815.

Gordon, P. (2012) *The invention and gendering of Epicurus*, University of Michigan Press, Ann Arbor, MI. doi: 10.3998/mpub.1826277.

Gössling, S., Hall, C.M. (2013) 'Sustainable culinary systems: An introduction', in Gössling, S., Hall, C.M. (eds) *Sustainable culinary systems: Local foods, innovation, tourism and hospitality*, pp 3–44, Routledge, Abingdon.

Green, G.P., Dougherty, M.L. (2008) 'Localizing linkages for food and tourism: Culinary tourism as a community development strategy', *Community Development*, vol 39, no 3, pp 148–158. doi: 10.1080/15575330809489674.

Hallmann, C.A., Sorg, M., Jongejans, E., Siepel, H., Hofland, N., Schwan, H., Stenmans, W., Müller, A., Sumser, H., Hörren, T., Goulson, D., de Kroon, H. (2017) 'More than 75 percent decline over 27 years in total flying insect biomass in protected areas', *PLoS ONE*, vol 12, no 10. doi: 10.1371/journal.pone.0185809.

Hedenus, F., Wirsenius, S., Johansson, D.J. (2014) 'The importance of reduced meat and dairy consumption for meeting stringent climate change targets', *Climatic Change*, vol 124, no 1–2, pp 79–91. doi: 10.1007/s10584-014-1104-5.

Herzog, C., Murray, I.P. (2013) 'Is "local" just a hot menu trend? Exploring restaurant patrons' menu choices when encountering local food options', in Gössling, S., Hall, C. M. (eds) *Sustainable culinary systems: Local foods, innovation, tourism and hospitality*, pp 122–134, Routledge, Abingdon.

Hjalager, A.-M., Richards, G. (2011) *Tourism and gastronomy*, Routledge, Abingdon.

Hwang, J., Kim, S., Choe, J. Y., Chung, C.-H. (2018) 'Exploration of the successful glocalization of ethnic food: A case of Korean food', *International Journal of Contemporary Hospitality Management*, vol 30, no 12, pp 3656–3676. doi: 10.1108/IJCHM-07-2017-0452.

IFOAM. (2016) *Organic in Europe: Prospects and developments 2016*, International Federation of Organic Agriculture Movements, Brussels. Available at: www.ifoam-eu.org/sites/default/files/ifoameu_organic_in_europe_2016.pdf

Jarvis, B. (2018) 'The insect apocalypse is here: What does it mean for the rest of life on earth?', The *New York Times Magazine*. Available at: www.nytimes.com/2018/11/27/magazine/insect-apocalypse.html

Kennedy, G., Stoian, D. (2017) *Five ways food biodiversity contributes to healthier diets*, Biodiversity International. Available at: www.bioversityinternational.org/news/detail/five-ways-food-biodiversity-contributes-to-healthier-diets/

Kittler, P.G., Sucher, K.P., Nahikian-Nelms, M. (2017) *Food and culture*, *Seventh* Edition, Cengage Learning, Boston, MA.

Klosse, P. (2014) *The essence of gastronomy: Understanding the flavor of foods and beverages*, CRC Press, Boca Raton, FL.

Laeis, G.C.M. (2016) 'Agriculture and food sourcing', in Sloan, P., Legrand W., Chen J.S. (eds) *Sustainability in the hospitality industry: Principles of sustainable operation, Third* Edition, pp 197–226, Routledge, Abingdon.

Laeis, G.C.M., Lemke, S. (2016) 'Social entrepreneurship in tourism: Applying sustainable livelihoods approaches', *International Journal of Contemporary Hospitality Management*, vol 28, no 6, pp 1076–1093. doi: 10.1108/IJCHM-05-2014-0235.

Legrand, W., Sloan, P., Fett, M., Manten, T. (2019) 'The roles of terroir, food and gastronomy in destination authenticity', in Dixit S.K. (ed) *The Routledge handbook of gastronomic tourism*, pp 260–271, Routledge, Abingdon.

Mak, A.H.N., Lumbers, M., Eves, A. (2012) 'Globalisation and food consumption in tourism', *Annals of Tourism Research*, vol 39, no 1, pp 171–196. doi: 10.1016/j.annals.2011.05.010.

Mattioni, D. and Caraher, M. (2018) 'Moving towards ecologically sustainable diets: Lessons from an Italian box delivery scheme', *International Journal of Consumer Studies*, vol 42, no 4, pp 430–439. doi: 10.1111/ijcs.12437.

Mazoyer, M., Roudart, L. (2006) *A history of world agriculture: From the Neolithic Age to the Current Crisis*, translated by Membrez, J.H., Earthscan, London.

McIntyre, B.D., Herren, H.R., Wakhungu, J., Watson, R.T. (eds) (2009) *Synthesis Report: A Synthesis of the Global and Sub-Global IAASTD Reports*, International Assessment of Agricultural Knowledge, Science and Technology for Development (IAASTD), Island Press, Washington DC.

Namkung, Y., Jang, S. (2017) 'Are consumers willing to pay more for green practices at restaurants?', *Journal of Hospitality & Tourism Research*, vol 41, no 3, pp 329–356. doi: 10.1177/1096348014525632.

Noll, S. (2014) 'Liberalism and the two directions of the local food movement', *Journal of Agricultural and Environmental Economics*, vol 27, no 2, pp 211–224. doi: 10.1007/s10806-013-9460-0.

NRA. (2018) *What's hot 2018 culinary forecast*, National Restaurant Association. Available at: www.restaurant.org/Restaurant/media/Restaurant/SiteImages/News%20and%20Research/Whats%20Hot/Whats_Hot_Culinary_Forecast_2018.pdf

NZ Herald. (2018) *New Zealand restaurants' biggest food trends for 2018*. Available at: www.nzherald.co.nz/lifestyle/news/article.cfm?c_id=6&objectid=12028888

OECD. (2018) 'Meat consumption', *OECD Agriculture Statistics: OECD-FAO Agricultural Outlook (Edition 2019)*. Available at: https://data.oecd.org/agroutput/meat-consumption.htm

Ollerton, J., Winfree, R., Tarrant, S. (2011) 'How many flowering plants are pollinated by animals?', *Oikos*, vol 120, no 3, pp 321–326. doi: 10.1111/j.1600-0706.2010.18644.x.

Pauly, D., Christensen, V., Guénette, S., Pitcher, T. J., Sumaila, U. R., Walters, C. J., Watson, R., Zeller, D. (2002) 'Towards sustainability in world fisheries', *Nature*, vol 418, no 6898, pp 689–695. doi: 10.1038/nature01017.

Pretty, J. (2001) *Some benefits and drawbacks of local food systems, Briefing Note for TVU/Sustain AgriFood Network, November 2nd 2001*, University of Essex. Available at: https://www.sustainweb.org/pdf/afn_m1_p2.pdf

Pröbstl-Haider, U., Hochwarter, E., Schrank, J. (2015) 'The Autumn-Pear: A symbol for local identity, local specialities, biodiversity and collaborative park management, an Austrian case study', in Sloan, P., Legrand, W., Hindley, C. (eds) *The routledge handbook of sustainable food and gastronomy*, pp 326–337, Routledge, London. doi: 10.4324/9780203795699.

Rovillé, M., Courchamp, F., Bunle, A. (2009) 'La biodiversité, comprendre pour mieux agir', *Association Française des Petits Débrouillards, Association des Petits Débrouillards de*

Lorraine, CNRS. Available at: www.lespetitsdebrouillardspaca.org/IMG/pdf/livret_biodiversite.pdf

Roy, H., Hall, C.M., Ballantine, P. (2016) 'Barriers and constraints in the use of local foods in the hospitality sector', in Hall, C.M., Gössling, S. (eds) *Food tourism and regional development: Networks, products and trajectories*, pp 255–272, Routledge, Abingdon.

Salais, L. (2015) 'Foodservice, health and nutrition: Responsibility, strategies and perspectives', in Sloan, P., Legrand, W., Hindley, C. (eds) *The routledge handbook of sustainable food and gastronomy*, pp 253–266, Routledge, London. doi: 10.4324/9780203795699.

Scarpato, R. (2002) 'Sustainable gastronomy as a tourist product', in Hjalager, A.-M. and Richards, G. (eds) *Tourism and gastronomy*, pp 132–152, Routledge, Abingdon.

Timothy, D.J., Ron, A.S. (2013) 'Heritage cuisines, regional identity and sustainable tourism', in Gössling, S., Hall, C.M. (eds) *Sustainable culinary systems: Local foods, innovation, tourism and hospitality*, pp 275–290, Routledge, Abingdon.

Tscharntke, T., Clough, Y., Wanger, T.C., Jackson, L., Motzke, I., Perfecto, I., Vandemeer, J., Whitbread, A. (2012) 'Global food security, biodiversity conservation and the future of agricultural intensification', *Biological Conservation*, vol 151, no 1, pp 53–59. doi: 10.1016/j.biocon.2012.01.068.

UN. (2017) *Resolution adopted by the General Assembly on 21 December 2016 71/246. Sustainable Gastronomy Day*, United Nations General Assembly Seventy-first session. Available at: https://undocs.org/a/res/71/246

UN. (2018) *68% of the world population projected to live in urban areas by 2050, says UN*, United Nations Department of Economic and Social Affairs. Available at: www.un.org/development/desa/en/news/population/2018-revision-of-world-urbanization-prospects.html

USDA. (2010) 'Local Food Research & Development', US Department of Agriculture. Available at: www.ams.usda.gov/AMSv1.0/FARMERSMARKETS

Vermeir, I., Verbeke, W. (2006) 'Sustainable food consumption: Exploring the consumer "attitude-behavioral intention" gap', *Journal of Agriculture and Environmental Ethics*, vol 19, no 2, pp 169–194. doi: 10.1007/s10806-005-5485-3.

Wellton, L., Jonsson, I.M., Walter, U. (2016) 'Culinary collisions: The vision of local food use collides with daily restaurant practice', in Hall, C.M., Gössling, S. (eds) *Food tourism and regional development: Networks, products and trajectories*, pp 273–284, Routledge, Abingdon.

WHO. (2018a) *Obesity and overweight*, World Health Organization. Available at: www.who.int/news-room/fact-sheets/detail/obesity-and-overweight

WHO. (2018b) *Biodiversity*, World Health Organization. Available at: www.who.int/globalchange/ecosystems/biodiversity/en/

Zukunftsinstitut. (2018) *Wie wir morgen essen werden*, Zukunftsinstitut, Available at: www.zukunftsinstitut.de/artikel/wie-wir-morgen-essen-werden/

9

LINKING FOOD BIODIVERSITY AND FOOD TRADITIONS TO FOOD TOURISM IN SMALL ISLAND DEVELOPING STATES (SIDS)

Tracy Berno

Tourism: food for thought

One of the most significant economic, social and cultural phenomena since the mid-1900s has been the growth of international tourism. Over the past 68 years, tourism has increased exponentially, growing from 25 million international tourist arrivals in 1950 to nearly 1.4 billion international tourists in 2018 (United Nations World Tourism Organisation, 2019). Some of these millions of tourists will engage in food-tourism-related activities as part of their journey, others will travel specifically to have culinary encounters, yet for others, food is no more than a transactional part of their tourist experience. Regardless of their travel and food motivations, every single one of these tourists must eat.

As an everyday activity, most tourists put some degree of thought into their food consumption. However, despite eating being an integral part of tourism, most tourists are unaware that their food-related decisions can have far-reaching consequences. Their choices can enhance a destination's well-being, or they can contribute to a broad range of negative impacts, both locally and globally. Estimating the number of meals consumed in international tourism is methodologically fraught; however, Gössling and Peeters (2015) have estimated that approximately 75 billion meals a year (in excess of 200 million meals per day) might be consumed. In a further study assessing food consumption in a resort hotel setting, Gössling (2015) found that 75% of participants reported that they had 'eaten more than at home' (p. 240), which was reflected in the total weight of food consumed per tourist, per day, being 3.1kg. A large proportion of the food consumed comprised of protein-based foods (on average, 385g of meat, 139g of seafood and fish and 294g of dairy products and eggs), which in itself has implications for the types of impact tourist food choices may have. Although these data only represented a single resort, Gössling (2001) confirmed that the results were consistent with previous studies of tourist food intake.

Ironically, despite the sheer volume of food consumed in tourism and the well-accepted need for tourism to be sustainable, it has only been relatively recently that attention has been paid to the relationship between food production systems, culinary traditions, tourist consumption and sustainability. These relationships, however, can be critical to sustainability (Berno et al., 2014; Rinaldi, 2017). This chapter will examine the relationship between food, biodiversity, cuisine and tourism with a focus on agriculture–tourism linkages, local food systems, community tourism development, and tourism and gastronomy. In particular, examples from the South Pacific will highlight the importance of food biodiversity and local food systems in tourism in Small Island Developing States (SIDS); specifically, the Fiji Islands. A case study of an agritourism venture in Fiji provides an example of how tourism can be used as a tool to conserve biodiversity, in this case with specific reference to taro.

Flour, sugar and spam: culinary colonialism and contemporary Pacific cuisine

Tourism is by no means the only factor associated with changes in cuisine, agriculture and biodiversity in the South Pacific. The cuisines of the Pacific have been subject to 'culinary colonialism' for centuries (Koa Dunsford, 2010–11). Socio-cultural influences on food preferences such as colonisation, urbanisation, migration and globalisation have all contributed to changes in cuisine and food consumption patterns in the region. As a result, many Pacific countries are experiencing the loss of their traditional foodways. This transformation places them at risk of losing both their unique food heritage as well as the diversity of crops that support it.

Although many view the influx of imported processed foods into the Pacific as a contemporary phenomenon, it was 19th century traders and missionaries who first started the transition away from traditional foods by introducing imported products to the region. These seemingly exotic ingredients such as sugar, flour and tinned meats were quicker and easier to prepare and cook and were less perishable than traditional foods. Pacific Islanders rapidly acquired a taste for these processed ingredients and incorporated them into their diets (Owens, 1999). Sexton (1991, cited in Owens, 1999) argued that foreign foods not only provided a welcome variation to an essentially monotonous diet, but were also favoured because of their association with the wealth and power of Westerners. By the 1900s, widespread colonisation throughout the Pacific and the overseas military presence during the World Wars also contributed to dramatic increases in the availability of imported foods in the region. Globalisation and the integration of the Pacific into the world economy further exacerbated food colonisation by increasing the availability of imported processed food and enabling multinational fast-food corporations to enter the region. More recently, trade liberalisation between Pacific island countries and larger economies have led to the region becoming the 'dumping ground' for low-quality, high-fat products such as mutton flaps and turkey tails (Berno, 2017a). As an outcome of this culinary colonialism, Pacific Islanders have experienced a loss of biodiversity, and they now have some of the highest rates of diet-related non-communicable diseases in the world, something almost unheard of when a more traditional diet was consumed (Thaman, 2002).

Tourism, as the most recent wave of colonisation in the Pacific, has also affected food production and consumption through socio-economic changes to the lifestyles of local communities. Over 35 years ago, Bolabola (1981) observed that hotel workers in Fijian resorts were becoming more reliant on processed and tinned foods as opposed to locally sourced food. This change was a result of changes in livelihood activities, particularly for women who were becoming increasingly engaged in shift work at the hotels. As a consequence of the shift work, women were spending less time on, and giving less emphasis to, traditional food collection and production. Similarly, Movono et al. (2018) found that Fijian villages with stronger purchasing power, received as a result of regular income from employment in tourism, had less motivation to engage in other (more traditional) forms of livelihood activities such as agriculture and food preparation. Beyond impacting food consumption patterns, this transition has eroded the traditional synergies between socio-cultural values and ecological custodianship. The subsequent loss of intergenerational knowledge about agriculture, hunting, fishing, foraging, food preparation and preservation methods has resulted in the redundancy of traditional knowledge and activities with deleterious effects on Fijian villagers' lifestyles and health (Movono et al., 2018).

Everybody eats: tourism, agriculture, culture and cuisine in the South Pacific

Tourism development is only one of several factors affecting biodiversity, local food production systems and cuisine in the South Pacific; however, its impact should not be underestimated in contributing to socio-economic (livelihoods) and socio-ecological changes for Pacific communities. Tourists' food preferences have also played a significant role.

In 2017, there were 2,137,538 international tourist arrivals to the South Pacific (South Pacific Tourism Organisation [SPTO], 2018). In global terms, this represents less than 1% of the world's international tourist arrivals, nonetheless this small number is enough for tourism to be the mainstay of the region's economy. For many small SIDS, tourism is one of the only (if not the only) economic sectors with sustainable growth. The SIDS of the South Pacific offer a stereotypical tropical island tourist idyll. However, some of the very attributes that make them popular tourism destinations pose challenges to food production. For the most part in the Pacific, the main purpose of agriculture is for subsistence and, to a lesser extent, trade in domestic and international markets (Barnett, 2010; Milne et al., 2017). The high cost of agricultural production, the cumbersome and perishable nature of agricultural goods along with limited technology, access to customary land, poor logistical links and high transportation costs to external markets, pose significant barriers to export-led agricultural development in the region and make it difficult for Pacific SIDS to compete and trade on a global scale (Chen et al., 2014; Duncan and Nakagawa, 2006; Milne et al., 2017; Pye-Smith, 2017).

Potentially, tourism can address some of these barriers by providing a proximate 'export' market for agricultural products in the Pacific. However, variable product quality and consistency, poor infrastructure, lack of training, complex supply chains and tourist food preferences (Berno, 2011) mean that it is often easier to offer tourists an international menu that relies on imported products – so much so that some South Pacific resorts' menus comprise up to 90% imported goods (Oliver and Berno, 2016), although there are notable exceptions. For the most part, when Pacific dishes are offered in touristic settings, they are often a parody of local cuisine presented at themed island-nights (the 'tikification' of Pacific cuisine, Berno, 2017a: 111; or international dishes with a 'Pacific flavour', Berno et al., 2016). Even in 1946, Victor Bergeron recognised this when he observed,

> Whenever commerce and shipping have taken over island cities and ports, and the further our Caucasian civilization spreads the less you see of true native life. The beloved tourist, with his pocket of loose change, however, has helped to bitch up the native way of life as much as anything else.
>
> *Bergeron, 1946: 135*

Further compounding tourism food-related issues in the South Pacific is the lack of a clear 'culinary identity' in the region. As with many tropical island locations, when one thinks of the South Pacific it is inevitably a sun, sea, sand image that comes to mind, rather than the cuisine (Berno, 2011; Pratt, 2013). Indeed, in defence of the 'Americanisation' of the Pacific foods on his menus, Victor Bergeron, the founder of the famous, Polynesian-themed Trader Vic's restaurant chain, said, 'The real, native South Seas food is lousy. You can't eat it' (cited in Reddinger, 2010: 208).

Recent research by Gibson and Berno (2019) found that tourists in Fiji had a low familiarity with Pacific cuisine. When asked to name Fijian dishes, the majority of tourists surveyed suggested items such as 'fruits', 'seafood' and 'coconuts' – products that are ubiquitous to tropical SIDS worldwide and are not unique to the South Pacific or Fiji. Berno has referred to this phenomenon as the 'same-same zone' when describing the ubiquitous nature of ingredients that are found throughout the coastal tropical world (Berno, 2017d). Although these products are key components of Fijian cuisine, they are ingredients, not Fijian cuisine in and of themselves.

Culture and cuisine have a symbiotic relationship: culture can be represented in cuisine, and cuisine can be an expression of individual and collective cultural identity. With holiday-makers who prefer burgers, chips and pizzas (Laeis, 2019b; Berno and Gibson, in preparation) the deeper cultural meaning of cuisine is mostly absent from tourist menus in the South Pacific.

This lack of culinary identity was also reflected by Pratt (2013) who suggested that the island nations of the South Pacific are perceived as being 'same-same', meaning that they:

> Share many of the same attributes indicating that there is a sameness to these destinations – something that has been reinforced over time [... all the Pacific islands] are perceived to have good weather, people known for their hospitality, and being a good place to relax.
>
> *2013: 607*

In other words, imagery, including that of the cuisine, of South Pacific destinations has become strongly stereotyped in the minds of tourists (Berno et al., 2016). Cuisine, when it did feature as a motivation for travellers to visit the Pacific, was also subject to 'same-same' with only the food of the French territories of New Caledonia and French Polynesia serving as a tourist motivator, but with French (not Polynesian) cuisine as the driver (Pratt, 2013). This lack of awareness of Pacific cuisine was a key driver behind Oliver, Berno and Ram's publications *Me'a Kai: The Food and Flavours of the South Pacific* (2010) and *Mea'ai Samoa: Recipes and Stories from the Heart of Polynesia* (2013). Similarly, the television programme *Real Pasifik*, inspired by Oliver et al.'s books and hosted by Robert Oliver, also sought to heighten awareness of Pacific cuisine as a salient attribute of the Pacific Islands' tourism sectors[1].

As discussed above, food, cuisine and culinary traditions are among the most foundational elements of culture (Berno, 2017b; Timothy and Ron, 2013). Significantly, Timothy and Ron (2013) suggested that cuisine, as a part of cultural heritage, is an important marker of social identity. They argue that traditional gastronomy, as part of the tourist experience, can empower communities, support them to achieve their goals for sustainability, and help them cope with tourism and achieve its potential in more responsible ways. Despite the significant impacts of colonisation, trade liberalisation and the broader influences of western lifestyle on food habits, local ingredients and traditional cuisine are still highly regarded in the Pacific, not only as a means for sustenance but as the foundation for meaningful exchanges and expressions of culture. Traditional dishes are still a part of everyday island life and can be found in private spaces and at almost all significant holidays and occasions (Berno, 2015). Despite several initiatives to increase local cuisine in tourism, and hence the use of locally grown products (see, for example Berno, 2017c; Oliver et al., 2010, 2013; Parkinson, 1989), Pacific cuisine is still mostly absent from tourists' tables. When asked why this was the case, local chefs in Fiji explained their belief that the only acceptable food to present to tourists is European-style cuisine (Laeis, 2019a, 2018). In essence, local chefs believed that their local cuisine was not good enough for the tourists.

The reasons for which Pacific chefs do not celebrate their cuisine through tourism are complex. In part, it is deeply rooted in the cultural and culinary changes discussed above. Another factor is the tourists, who have little familiarity with, or interest in, Pacific cuisines prefer a ubiquitous Western-style tourist menu instead. As a result, local chefs in the South Pacific have grown up in a tourism environment that does not appreciate their local cuisine. As Laeis (2019a) observed, 'In essence, tourists do not come to [the Pacific] for its cuisine and local chefs are socialised in an environment that degrades their culinary heritage'. In the Pacific, however, what tourists choose to eat is a critical driver of agricultural production and a potential contributor to more sustainable food systems. As chef Robert Oliver (2013: np) stated, 'In tourism-led economies [of the Pacific], the menus are the business plans of the nation'.

This idea that chefs are gatekeepers and that 'where the cuisine goes, the agriculture will follow' has been consolidated into a programme called 'Chefs for Development'. This programme is a collective initiative coordinated by the Technical Centre for Agricultural and Rural Cooperation with the Inter-American Institute for Cooperation on Agriculture, the Pacific Islands Private Sector Organisation, the Pacific Community (SPC), the South Pacific Tourism Organisation (SPTO), the International Fund for Agricultural Development, Women in Business Development Inc., farmers' organisations, agribusinesses and chefs. Its aim is to promote stronger linkages across the agriculture–tourism value chain, and to enhance the contribution made by healthier, locally sourced food and agrifood products to tourism menus across SIDS in the Caribbean, South Pacific and Indian Ocean (Chefs for Development, n.d.).

In the touristic settings of the South Pacific, the subsequent impact of tourists' food choices on agriculture is that the Western-centric tourist cuisine encounters a traditional agricultural sector that is largely unable to supply demand (Laeis, 2019a; Viet, 2007). The reliance on Western-style foods and imported food products in tourism have resulted in SIDS of the South Pacific developing agricultural agendas to match the requirements of tourists' food preferences. This has had a range of social and environmental consequences (Berno, 2015; Laeis, 2019a). In essence, in tourism, local food has become a problem to be overcome, rather than an opportunity to revitalise and value cultural and ecological attributes. This is consistent with Harrison and Pratt (2015) who suggested that increasing the use of local food in the tourism industry of the South Pacific requires both cultural and structural change. Despite this somewhat pessimistic outlook, Fiji, the most visited of the countries in the South Pacific, provides an interesting example of how tourism can be used as a conduit to promote food biodiversity, local food systems and cuisine with a focus on one of the most important crops in the region – taro.

Of Fiji, taro and tourists

The Pacific is not alone in experiencing changing agriculture, food production and culinary practices. There is a displacement of traditional foods that is occurring globally. As a consequence, the world's industrialised food-supply system is based on a narrow plant and animal base, which is continuing to narrow even further, despite agricultural biodiversity representing one of humanity's greatest resources (Hunter et al., 2017; Sage, 2012). This loss of biodiversity, and the genetic diversity that it represents, increases the risks for food insecurity and threatens traditional cultural and culinary practices. The foods, skills and food-related cultural practices of indigenous peoples are slowly vanishing, as this happens, culture is also lost. Kuhnlein et al. (2006: 1013) found that indigenous peoples' knowledge about their ecosystems, is:

> a treasure of knowledge that is typically overlooked and undervalued; which has potential benefits for the well-being and health not only of indigenous peoples themselves, but also industrialised and other populations' and that efforts should work 'towards making local cultural food resources sources of community pride, pleasure and responsibility, thereby ensuring local determination and sustainability of intervention effort.

These issues of food, biodiversity, conservation and sustainable food systems are becoming increasingly important in tourism. Just as tourism has contributed to changes in agricultural and culinary practices, equally it can play a role in rejuvenating and revaluing traditional indigenous knowledge about agriculture, biodiversity and cuisine by providing a market for local farmers, encouraging more local production and establishing value-chains from which the local communities can benefit (Pye-Smith, 2017). One example of how tourism can be used as a tool to connect food biodiversity, conservation and sustainability is an emerging agritourism venture that highlights the important role of taro in Fijian culture and cuisine.

Taro: an 'orphan crop' of immense importance

Taro is one of the world's oldest food crops, dating back over 9,000 years. Often seen as an 'orphan crop' (a crop that is of local significance but one that is not traded as an international commodity, and attracts limited attention for research), taro's global importance manifests most commonly at the local level through its role in traditional food systems. It is the farmers and communities who are the guardians of taro (Rao et al., 2010).

One of the reasons behind taro's importance at the community level is that, with the exception of the skin, all parts of taro plants are edible (see Figure 9.1):

FIGURE 9.1 Fijian taro (*Colocasia esculenta*) ready for market
Source: Shiri Ram (used with permission)

> Its corms are baked, roasted, or boiled and the leaves are frequently eaten as a vegetable and represent an important source of vitamins, especially folic acid. The blades and petioles of leaves can be preserved or dried and are an important food in times of scarcity. Petioles and stolons are also eaten fried or pickled. The inflorescence (a flowering stalk) is a delicacy in some food cultures of Asia and the Pacific. The corms and leaves are also used for medicinal purposes.
>
> *Rao et al., 2010: 1*

Along with its importance as a source of food, in many cultures, taro is a sacred plant with high prestige and strong cultural and symbolic importance. In cultures such as those of the South Pacific, taro is intrinsic to cultural identity. In the Pacific, taro is presented on formal occasions, in domestic or agricultural rituals, and as part of religious celebrations. Taro is also seen in local legends, stories, chants and proverbs (Rao et al., 2010). Similar to the situation in South East Asian countries where 'rice is life' (Berno et al., 2019), Pacific Island life without taro is almost inconceivable.

There are four main types of taro in the South Pacific: *Colocasia esculenta* is the most commonly grown type and is usually grown in rain-fed, dry land, though some varieties can also be grown in irrigated terraces or swamps. Giant swamp taro (*Cyrtosperma merkusii*) is the main root crop of the low-lying atolls in the Pacific as it can grow in sandy saline soils and can withstand high winds. Cyrtosperma is often grown in pits, and some varieties can be left in the pits for 10–15 years, forming an important source of food in times of floods or cyclones. Both Colocasia and Cyrtosperma taro play important roles in local custom. Giant taro (*Alocasia macrorrhizos*) is a hardy variety that grows in a range of soil types. The final type of taro in the Pacific is *Xanthosoma sagittifolium*, which is a more recent introduction, having been brought from tropical America just over 100 years ago. Its performance in dry and less fertile soils is superior to Colocasia. These different types and varieties of taro vary considerably in appearance, taste, use and other properties (SPC, 1999, 2006).

Taro corms form an important source of carbohydrates, fibre, essential minerals, calcium and iron for Pacific Islanders. Some varieties of Cyrtosperma are also a rich source of zinc – one of the few non-animal sources of this important trace mineral. Some varieties of yellow-fleshed Cyrtosperma also contain significant amounts of provitamin A carotenoids. Consuming two cups of this varietal provides 100% of the estimated daily requirements of vitamin A. The leaves, stems and other parts of taro plants also provide valuable nutrition. The leaves of Colocasia, for example, are an excellent source of provitamin A carotenoids, calcium, fibre and vitamin C, and are enjoyed widely across the Pacific (SPC, 2006). The stems of the taro plant, however, are only eaten in a few areas in the Pacific, baseisei (Fijian taro stem salad) being one example.

As is the case in other Pacific Island countries, taro (known as 'dalo' in the Fijian language) is a staple of the Fijian diet (see Figure 9.2), but even this mainstay of island cuisine is at risk. There are known to be 125 varieties of dalo in Fiji, of which at least 70 were grown and consumed by Fijians before the arrival of other settlers and before commercialisation of the root crop (NatureFiji-MareqetiViti, n.d.). Although there are currently 82 varieties of dalo recognised in Fiji (Ministry of Agriculture, 2018), emphasis on commercial production and marketability of the crop has seen a reduction to only a few (approximately 12) varietals being farmed commercially. Many of these 12 are hybrids, which in itself introduces significant vulnerabilities to things such as climate change, disease risk (for example the devastation of the taro leaf blight outbreak in Samoa in 1993), food and nutrition security, and economic livelihoods (SPC, 2019; Ministry of Agriculture, 2018; D. Hunter, pers. comm.). This shift

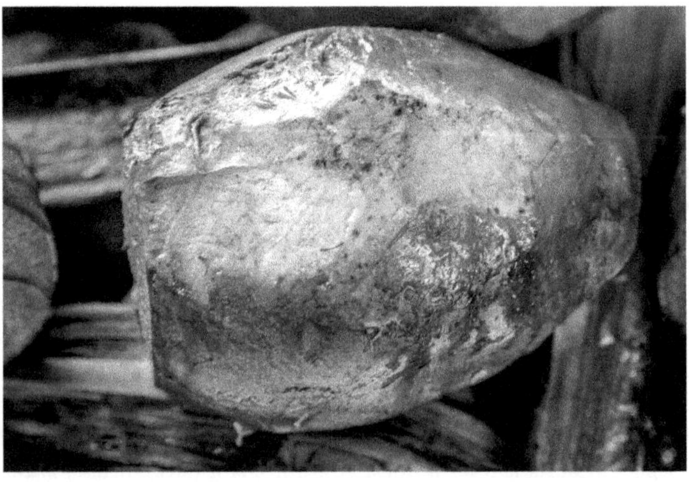

FIGURE 9.2 A taro corm that has been cooked in a Fijian lovo (earth oven)

Source: Shiri Ram (used with permission)

has meant a loss (and continuing loss) of the traditional varieties which many Fijians considered tastier but grow much slower and are therefore less desirable for commercial production. Commercially successful varietals have overwhelmed traditional ones to the point where most Fijians, particularly those in urban areas, now lack knowledge of them (NatureFiji-MareqetiViti, n.d.).

Food production structures and social structures are inextricably linked (Taylor et al., 2010). Thaman (2002) suggests that the loss of traditional knowledge about the uses, beliefs, management systems and language related to Pacific biodiversity is possibly the most serious obstacle to successful biodiversity conservation in the region. As such, the ongoing loss of traditional Fijian varieties of dalo is also the loss of cultural values and heritage. The urgent need to address the narrowing genetic base of taro across the Pacific has received considerable attention through the Taro Genetic Resources and Conservation Project ('TaroGen'; Hunter and Taylor, 2007) and, more recently, through the work of the Pacific's only regional gene bank, the Pacific Centre for Crops and Trees, which is a core part of the SPC's Land Resources Division. The sense of urgency to conserve dalo in the Pacific, however, is largely limited to specialist circles within ministries of agriculture and forests in the region, and with partner institutions and governments (SPC, 2019). This is compounded by a lack of awareness amongst Fijians of the significance of the ongoing loss of traditional varieties of this important food crop:

> Most farmers that had grown the traditional varieties and still know the traditional methods of growing dalo are getting older. If their knowledge is not documented or not passed onto the next generation (which may not happen because of the shift towards the commercial crop varieties), then our knowledge of traditional dalo varieties is lost, and our future generations may never have the opportunity to taste such a chiefly crop.
>
> *NatureFiji-MareqetiViti, n.d.: n.p.; Taylor et al., 2010*

Many of these now rare types of dalo are linked to traditional Fijian nutrition, health and healing, and food security, as different types have different properties. There is an urgent need to mainstream this food biodiversity into agricultural, health and nutrition, education, and as argued in this chapter, sustainable cuisine and tourism sectors. As Suliana Siwatibau, a prominent Fijian ethno-botanist indicated, 'If you lose the crop, you don't just lose the food, you lose the knowledge' (personal communication, 1 May 2019). As suggested by Heywood (2011: 2), 'ethnopharmacology, biodiversity, agriculture, health, food and nutrition are all inextricably interconnected'.

Conserving Fiji's food plants: an agritourism case study

The Global Action Programme on Food Security and Nutrition in Small Island Developing States (Food and Agricultural Organisation of the United Nations, 2017) suggests that, as part of developing inclusive and efficient nutrition-sensitive value chains, strengthening links between agriculture and tourism presents opportunities for agricultural development in SIDS, and that this offers a range of benefits. Specifically, tourism and agriculture can be linked in three main ways to biodiversity and ecosystems services: agritourism (agriculture as a tourism attraction); agriculture–tourism value chains; and tourism as a land-use alternative to intensive or non-sustainable agriculture (Morrison, 2016). Specifically, Morrison suggests that the protection of habitats and ecosystems in ecotourism can indirectly lead to the protection of species important to local communities for food, medical and ornamental uses, as well as communities' health and well-being. This is consistent with, and reinforces, what is suggested in the Pacific Organic Tourism and Hospitality Standard, that there is a synergistic relationship between sustainable tourism and organic production (SPC, 2017).

Biodiversity in a Pacific context is not just about scientific, economic, recreational or ecological value. Rather, biodiversity is a form of intergenerational capital inheritance. In essence, it is the 'capital' needed for development and maintenance of Pacific societies; it is the foundation of their cultures (Thaman, 1993: 2). Taro gardens are an integral part of this intergenerational cultural capital, and the guardians of it are local communities (Thaman, 2000). Most varieties of dalo in Fiji, other than the 12 that dominate commercial production, are found on small-scale farms and in-home gardens; thus, these are important sites for the conservation of the rich genetic diversity of dalo in Fiji (Taylor et al., 2010). As Thaman and Faulalo (2016) point out, it is the local farm (and home garden) 'where actions happen' (p. 126) that can support farmers to conserve biodiversity and ecosystem services in their farming practices, which can then form the foundation for environmental, food, health and livelihood security. Thaman (1993: 2) refers to this as 'community-based biodiversity conservation'.

However, it is not just about preserving a plant (in this case, taro), the diversity of culinary knowledge associated with the plant must also be preserved. Preserving traditional culinary knowledge is as important as preserving genetic diversity in taro: 'without the one, the other may lose relevance for people' (Matthews, 2004: 55). In essence, the protection of plant genetic diversity requires respect and support for communities' cultural diversity (Matthews, 2004). For example, in Japan, research and writing on taro is extensive. Recipes for taro, both old and new, are published across a range of media, often coinciding with the autumn taro-harvesting period. Food is a popular topic in Japan and taro commonly features in recipes. For example, in the popular *Oishinbo* series – the longest-running cooking manga – the 'Vegetables' issue contains a reference to a simple preparation of grilled taro. The taro was prepared in homage to Minakami Tsutomu's, *Days of Eating Earth: My Twelve Months of Devotion* (1978 cited in Tran, 2017), which is a book about Buddhist Shojin cooking – asking the garden what is good to eat, and then preparing it in a pure and simple way. Themes of local foods, organic farming and traditional cuisine run through all *Oishinbo* issues and the mention of the book *Days of Eating Earth* grounds those themes in religious tradition (Kariya, 2006). Rao et al. (2010)

suggest that the continuing popularity of taro in Japan can be attributed to a high regard for traditional foods, the relatively low price of taro compared to other starchy foods and the 'continuous dissemination' (Rao et al., 2010: 14) of old and new cooking methods across a broad range of popular media, such as *Oishinbo*.

In the South Pacific, on the other hand, there is little written about the culinary uses of taro (the works of the late Fiji-based nutritionist Susan Parkinson are notable exceptions). Yet there is clear potential to utilise the links between traditional crop production and cuisine to enhance biodiversity and nutrition in the region. Frison et al. (2011) suggest that a strategy for agriculture and biodiversity should include:

- An evidence-based approach to nutrition and health and sustainable agriculture by small-scale farmers.
- The evaluation and use of local foods and their varieties.
- Traditional cuisines.
- Culturally sensitive methods.
- Nutrition education.
- Research on novel and improved methods of food storage and processing.
- Greater attention to marketing.

The success of the 'Go Local' campaign in the Federated States of Micronesia is an example of how these strategies can lead to significant change in a community. The 'Go Local' campaign, and the affiliated 'Go Yellow' campaign, which promoted consumption of the rare karat and other yellow-fleshed bananas, used a variety of approaches including education, training, agriculture and social marketing. The 'Go Local' message promoted the consumption of bananas, giant swamp taro, breadfruit, pandanus, green leafy vegetables and fruits, while also incorporating concepts of local culture, environmental sustainability and long-term food security by using the acronym 'CHEEF' (Culture, Health, Environment, Economic and Food security). Results of the post-intervention assessment indicated increased (110%) provitamin A carotenoid intake; increased frequency of consumption of local banana (53%), giant swamp taro (475%) and local vegetables (130%); and increased dietary diversity from local food (Englberger et al., 2013; Englberger and Johnson, 2013; Kaufer et al., 2010).

It has been suggested that the approaches discussed above for increasing the use of local, traditional foods can also be holistically utilised in tourism for community development and to address food security issues (Mnguni and Giampiccoli, 2015). One example of how tourism can be used as a tool to connect food biodiversity, conservation and sustainability is an emerging agritourism venture in Fiji. Situated in a peri-urban region north of Suva, Fiji's capital city, an almost 12-acre farm is in the process of conversion by its owner from a lifestyle property to an agritourism enterprise. This, however, is no ordinary farm. The owner has a deep interest in traditional crops, not just because they are cultivated in the customary gardens of indigenous inhabitants of Fiji, but also because these crops contribute to a large gene pool (and crop features) that can be adapted for climate change mitigation and in response to changing consumer demands. Described as a living museum, the farm has been collecting traditional crop cultivars and is currently home to over ten traditional varieties of dalo (taro), including two highly regarded varieties known as *vavai vula* and *vavai loa*. Other varieties include the *rourou* (leaf), *dalo* (*Colocasia esculenta*) and *tausala* which is usually grown for the export market. *Tausala* can be found as *ni Samoa* and *ni Viti*, the latter in the indigenous garden collection. As part of the customary collection, the varieties known locally as *dalo ni Samoa* and *dalo in Tonga* are also being grown; legend has it that these varieties originally came from Samoa and Tonga respectively. Another variety of interest is the popular *uronivonu*, a type of *Colocasia esculenta* which has a dark purple colour.

Apart from dalo, the owner is also collecting varieties of a range of other crops including breadfruit (*Artocarpus altilis*), bananas, vudi (plantain – *Musa paradisiaca*), coconuts, yams (*Dioscorea spp.*), wild yams (*Dioscorea bulbifera*), via (giant taro – *Alocasia macrorrhiza*) and vutu (cutnut – *Barringtonia edulis*). Other customary crops on the farm include voivoi (pandanus – *Pandanus caricosus*), sugar cane, citrus, kavika (Malay apple – *Eugenia spp.* [*Myrtaceae*]), wi (Pacific apple – *Spondias dulcis*), dawa (Pacific lychee – *Pometia pinnata* [*Sapindaceae*]), pineapples, turmeric, cagolaya (wild ginger), cassava, kumala (sweet potatoes – *Batatas edulis* [*Convolvulaceae*]), dalo-ni-tana (*Xanthosoma sagittifolium*), bele (*Abelmoschus manihot*), ota (*Asplenium esculentum* [*Filices*]), as well as more modern crops like vanilla, star fruit and lemon grass. As is the practice in customary food gardens, the farm will collect other crop varieties as they are found.

When the agritourism enterprise becomes operational, visitors will tour the farm to learn about its biodiversity, the traditional (including medicinal) uses of plants and weeds, and local agricultural practices. They will also be taken through a native forest along the farm periphery and shown a range of forest plants that once furnished a wider range of the needs of customary communities in the olden days. Following the tour, visitors will be offered refreshments produced from ingredients produced largely on the farm. There will be a range of locally made cosmetic items for sale sourced from farm and forest-based ingredients such as beeswax, coconut oil and herbal extracts and infusions. There are also plans to produce books on medicinal and other uses of plants on the farm which will also be available for sale on site.

The experiential and educational aspects of the agritourism enterprise include more than just tourists. Biodiversity conservation programmes will only succeed in the long run when receptive locals take responsibility and run the programmes. As such, the farm also aims to engage in community development through involving a wide range of locals in its various activities. Women and girls will be trained in the production of the range of items for sale from cultivation of the crops to harvest, process and package. School teachers will be sourced from surrounding villages and their skills upgraded as necessary. Men and male youths likewise will be trained in carving small items for sale from locally grown wood so that they value the range of native trees in their surroundings instead of cutting them wantonly as they now tend to do. Local school children will be encouraged and welcome to visit the farm to learn of its philosophy and observe its activities. Using agritourism as a conduit for these broader activities and linkages has the potential to reinforce inter-generational knowledge transmission as well as to promote healthy diets and healthy eating. This in turn can help address the region-wide issues of diet-related non-communicable diseases and food security in the South Pacific.

Conclusion

As discussed in this chapter, tourism is one of the most significant economic, social and cultural phenomena in contemporary society. For many SIDS such as those in the South Pacific, tourism represents one of the only means for sustainable economic development. In theory, this should present opportunities for intersectoral linkages, such as with agriculture, but in the SIDS of the Pacific, this is a complex relationship. On one level, salaried work in tourism has contributed to changing livelihood patterns and, hence, changes in food production and consumption. This has posed threats to traditional ecological knowledge and biodiversity, as well as to health and nutrition. On the other side, complex interactions between historical and contemporary acculturative processes, tourist food preferences and local food production and culinary systems, have resulted in tourism-led agriculture mainly to support the exogenous cuisines preferred by tourists, with local cuisine and cultural knowledge being undervalued by both locals and tourists alike. This also has potentially negative outcomes for local knowledge and biodiversity.

Strengthening the relationships between food production systems, culinary traditions, tourist consumption and sustainability by stimulating demand for a sustainable Pacific cuisine based on locally grown and sourced products has the potential to enhance biodiversity and sustainability (Kambuou et al., 2007). Laeis (2019b) and Berno (2011) both found that this 'sustainable cuisine' (Berno, 2006: 209; 2015: 346) approach was more likely to be implemented successfully in smaller-scale operations that attract a niche tourist market. This suggests that agritourism enterprises, such as that highlighted in the case study, present opportunities to prioritise and positively re-evaluate indigenous foods in tourism. In this way, biodiversity can be fostered, nutrition improved, and indigenous knowledge and agriculture enhanced (Giampiccoli and Kalis, 2012).

For dalo in particular, the survival of its genetic diversity will depend on the personal interest of producers and consumers. Highlighting this important staple crop within tourism can stimulate and retain traditional knowledge, which in turn helps to establish value chains that benefit local communities economically, environmentally, socially and culturally. By specifically highlighting dalo as part of traditional culture and cuisine, a 'virtuous cycle' in which the preservation of culinary knowledge enhances local genetic diversity can be achieved.

Note

1 Real Pasifik www.imdb.com/title/tt4696898/

References

Barnett, J. (2010) 'Dangerous climate change in the Pacific Island: Food production and food security', *Regional Environmental Change*, vol 11, no 1, pp 229–237. doi: 10.1007/s10113-010-0160-2.

Bergeron, V. (1946) *Trader Vic's Book of Food and Drink*, Random House Value Publishing, New York.

Berno, T. (2006) 'Bridging sustainable agriculture and sustainable tourism to enhance sustainability', in Mudacumura, G.M., Mebratu, D., Haque M.S. (eds) *Sustainable Development Policy and Administration*, pp 208–223, CRC Press, Boca Raton, FL.

Berno, T. (2011) 'Sustainability on a plate: Linking agriculture and food in the Fiji Islands tourism industry', in Torres, R.M., Momsen, J. (eds) *Tourism and Agriculture: New Geographies of Consumption, Production and Rural Restructuring*, pp 87–103, Routledge, Abingdon.

Berno, T. (2015) 'Tourism, food traditions and supporting communities in Samoa: The Mea'ai Project', in Sloan, P., Legrand, W., Hindley, C. (eds) *The Routledge Handbook of Sustainable Food and Gastronomy*, pp 338–347, Routledge, Abingdon.

Berno, T. (2017a) 'Will the real Pasifika cuisine please stand up?', in Johnson E. (ed) *Kai and culture: Food stories from Aotearoa*, pp 110–114, Freerange Press, Christchurch.

Berno, T. (2017b) 'Putting the culture back into agriculture', in Burlingame, B., Massey, C. (eds) *The New Zealand Land & Food Annual: No free lunch Can New Zealand feed the world sustainably?*, pp 181–192, Massey University Press, Auckland.

Berno, T. (2017c) 'Fiji and Samoa: Food tourism, agriculture and intangible heritage', in S. Slocum, K. Curtis (eds) *Food and Agricultural Tourism: Theory and Best Practice*, pp 38–41, Routledge, Abingdon.

Berno, T. (2017d) 'Sun, sea, sand and cuisine: Putting the "cook" into the Cook Islands', Seminar presented at the New Zealand Tourism Research Institute, Auckland University of Technology, 27 October 2017.

Berno, T., Dentice, G., Wisansing, J.J. (2019) 'Kin kao laew reu young ("Have You Eaten Rice Yet")?: A New Perspective on Food and Tourism in Thailand', in Park, E., Kim, S., Yeoman, I. (eds) *Food Tourism in Asia*, pp 17–30, Springer Nature Singapore, Singapore.

Berno, T., Devlin, N., Ezaki, A, Wilson, D., Wolf, E. (2014) 'Sustainability in food and drink tourism', in E. Wolf, W. Lange-Faria (eds) *Have Fork Will Travel: Handbook for Food Tourism*, pp 309–322, World Food Travel Association, Portland.

Berno, T., Gibson, D. (in preparation) 'Assessing the demand for local food in the tourism industry: A Fiji case study'.

Berno, T., Milne, S., Thorburn, E., Sun, M., Oksanan, N., Pladdett, A. (2016) 'Sun, seas, sand, surf and sustenance: Food and tourism in the Cook Islands', paper presented at the CAUTHE conference, Sydney, 8–11 February.

Bolabola, C. (1981) 'Does tourism cause malnutrition: They seem to be connected', *Pacific Perspective*, vol 10, no 1, pp 72–77.

Chefs for Development. (n.d.) *Chefs for Development*. [online] Available at: http://chefs4dev.org/

Chen, H., Rauqeuqe, L., Singh, S.R., Wu, Y., Yang, Y. (2014) *Pacific Island Countries: In Search of a Trade Strategy*, IMF Working Paper 14/158, International Monetary Fund. Available at: www.imf.org/external/pubs/ft/wp/2014/wp14158.pdf

Duncan, R., Nakagawa, H. (2006) *Obstacles to Economic Growth in Six Pacific Island Countries*, World Bank, Washington DC.

Englberger, L., Johnson, E. (2013) 'Case study 1 – Traditional foods of the Pacific: Go local, a case study in Pohnpei, Federated States of Micronesia', in Fanzo, J., Hunter, D., Borelli, T., Mattei, F. (eds) *Diversifying Food and Diets: Using Agricultural Biodiversity to Improve Health and Nutrition*, pp 231–241, Routledge, Abingdon.

Englberger, L., Lorens, A., Pedrus, P., Albert, K., Levendusky, A., Hagilmai, W., Paul, Y., Moses, P., Jim, R., Jose, S., Nelber, D., Santos, G., Kaufer, L., Larsen, K., Pretrick, M. E., Kuhnlein, H.V. (2013) in Kuhnlein, H.V., Erasmus, B., Spigelski, D., Burlingame, B. (eds) *Indigenous Peoples' Food Systems& Well-Being Interventions & Policies for Healthy Communities*, pp 191–220, The Food and Agriculture Organization of the United Nations and The Centre for Indigenous Peoples' Nutrition and Environment, Rome, Italy, Available at: www.fao.org/3/a-i3144e.pdf

FAO. (2017) *Global Action Programme on Food Security and Nutrition in Small Island Developing States: Supporting the Implementation of the Samoa Pathway*, The Food and Agriculture Organization of the United Nations, Rome.

Frison, E.A., Cherfas, J., Hodgkin, T. (2011) 'Agricultural Biodiversity Is Essential for a Sustainable Improvement in Food and Nutrition Security', *Sustainability*, vol 3, no 1, pp 238–253. doi: 10.3390/su3010238.

Giampiccoli, A., Kalis, J.H. (2012) 'Tourism, food, and culture: Community-based tourism, local food, and community development in Mpondoland', *Culture, Agriculture, Food and Environment*, vol 34, no 2, pp 101–123. doi: 10.1111/j.2153-9561.2012.01071.x.

Gibson, D., Berno, T. (2019) 'Tourist Food Preferences and the Sustainable Development Goals: A Fiji case study', Presented at: Tourism and Sustainable Development Goals Conference, 24–25 January, 2019, Auckland, New Zealand.

Gössling, S. (2001) 'Tourism, economic transition and ecosystem degradation: Interacting processes in a Tanzanian coastal community', *Tourism Geographies*, vol 3, no 4, pp 430–453. doi: 10.1080/14616680110070504.

Gössling, S. (2015) 'New performance indicators for water management in tourism', *Tourism Management*, vol 46, pp 233–244. doi: 10.1016/j.tourman.2014.06.018.

Gössling, S., Peeters, P. (2015) 'Assessing tourism's global environmental impact 1900–2050', *Journal of Sustainable Tourism*, vol 23, no 5, pp 639–659. doi: 10.1080/09669582.2015.1008500.

Harrison, D., Pratt, S. (2015) 'Tourism in Pacific Island countries: Current issues and future challenges', in Harrison, D., Pratt, S. (eds) *Tourism in Pacific Islands: Current Issues and Future Challenges*, pp 3–35, Routledge, Abingdon.

Heywood, V.H. (2011) 'Ethnopharmacology, food production, nutrition and biodiversity conservation: Towards a sustainable future for indigenous peoples', *Journal of Ethnopharmacology*, vol 137, no 1, pp 1–15. doi: 10.1016/j.jep.2011.05.027.

Hunter, D., Guarino, L., Spillane, C., McKeown, P. (2017) 'Introduction: Agricultural biodiversity, the key to sustainable food systems in the 21st century', in Hunter, D., Guarino, L., Spillane, C., McKeown, P.C. (eds) *Routledge Handbook of Agricultural Biodiversity*, pp 1–12, Routledge, Abingdon.

Hunter, D., Taylor, M. (2007) *Networking and Learning Together for Taro Conservation and Improvement*, The Taro-Gen Project. Available at: www.capitol.hawaii.gov/session2010/testimony/HB1226_TESTIMONY_AGR_03-04-09_TaroGenPublications_.pdf

Kambuou, R., Gwabu, C., Taylor, M. (2007) 'The contribution of crop-plant genetic diversity to economic development and sustainable rural livelihoods in the Pacific region', *Pacific Economic Bulletin*, vol 22, no 3, pp 152–157.

Kariya, T. (2006) *Oishinbo: Vegetables: A la Carte*, VIZ Media, San Francisco.

Kaufer, L., Englberger, L., Cue, R., Lorens, A., Albert, K., Pedrus, P., Kuhnlein, H.V. (2010) 'Evaluation of a "traditional food for health" intervention in Pohnpei, Federated States of Micronesia', *Pacific Health Dialog*, vol 16, no 1, pp 61–73.

Koa Dunsford, C. (2010–11) 'Kaitiakitanga: Protecting our oceans, islands and skies by inspiring a climate change of consciousness', *Dreadlocks: Islands, Oceans and Skies*, vol 6/7, pp 1–22.

Kuhnlein, H., Erasmus, B., Creed-Kanashiro, H., Englberger, L., Okeke, C., Turner, N., Allen, L., Bhattacharjee, L. (2006) 'Indigenous peoples' food systems for health: Finding interventions that work', *Public Health Nutrition*, vol 9, no 8, pp 1013–1019. doi: 10.1017/S1368980006009876.

Laeis, G. (2018). 'What's on the Menu? How the cuisine of large-scale, upmarket tourist resorts shapes agricultural development in Fiji', PhD thesis. Massey University, Palmerston North.

Laeis, G. (2019a) 'Cuisine: A new concept for analysing tourism-agriculture linkages?', manuscript in preparation.

Laeis, G. (2019b) 'What's on the menu? How the cuisine of large-scale, upmarket tourist resorts shapes agricultural development in Fiji', PhD thesis, Massey University, Auckland.

Matthews, P.J. (2004) 'Genetic diversity in taro and the preservation of culinary knowledge', *Ethnobotany Research and Application*, vol 2, pp 55–71. doi: 10.17348/era.2.0.55-71.

Milne, S., Deuchar, C., Berno, T., Taumoepeau, T., Pusinelli, M., Raymond, J. (2017) *Private Sector Investment in the Pacific: Final Report – September 2017*, New Zealand Institute for Pacific Research.

Ministry of Agriculture. (2018) *Fiji Dalo Quality Manual*. Fiji Ministry of Agriculture. Available at: http://phama.com.au/wp-content/uploads/2018/09/Dalo_Manual_Final_e-copy.pdf

Mnguni, E.M., Giampiccoli, A. (2015) 'Indigenous food and tourism for community well-being: A possible contributing way forward', *Mediterranean Journal of Social Sciences*, vol 6, no 3 S2, pp 24–34. doi: 10.5901/mjss.2015.v6n3s2p24.

Morrison, C. (2016) 'Ecosystem services of ecotourism in agricultural zones', *Mainstreaming Ecosystem Services and Biodiversity into Agricultural Production and Management in the Pacific Islands Technical Guidance Document*, pp 91–109, The Food and Agriculture Organization of the United Nations, Rome, Italy. Available at: www.fao.org/3/a-i6505e.pdf.

Movono, A., Dahles, H., Becken, S. (2018) 'Fijian culture and the environment: A focus on the ecological and social interconnectedness of tourism development', *Journal of Sustainable Tourism*, vol 26, no 3, pp 451–469. doi: 10.1080/09669582.2017.1359280.

NatureFiji-MareqetiViti. (n.d.) *Dalo (Colocasia esculenta)*, Available at: https://naturefiji.org/colocasia-esculenta-dalo/

Oliver, R. (2013) 'Recipes for development', TEDX-Auckland, Available at: www.youtube.com/watch?v=orfB1UkoKdQ

Oliver, R., Berno, T. (2016) *Chefs for Development: The Role of Chefs in Linking Agriculture to Tourism in the South Pacific*, EU Technical Centre for Agricultural and Rural Cooperation (CTA), Wageningen, The Netherlands.

Oliver, R., Berno, T., Ram, S. (2010) *Me'a Kai: The Food and Flavours of the South Pacific*, Random House, Auckland.

Oliver, R., Berno, T., Ram, S. (2013) *Mea'ai Samoa: Stories and Recipes from the Heart of Polynesia*, Random House, Auckland.

Owens, K. (1999) What do we know of consumers preferences and food choices in the islands of the South Pacific' in *43rd Annual Agricultural and Resource Economics Society Conference*, Available at: https://pdfs.semanticscholar.org/1b56/0d781746e985fa9ac0af2e1c60695f4180fc.pdf

Parkinson, S. (1989) *Cooking the South Pacific Way! A Professional Guide to Fiji Produce*, Tourism Council of the South Pacific, Fiji.

Pratt, S. (2013) 'Same, same but different: Perceptions of South Pacific destinations among Australian travelers', *Journal of Travel and Tourism Marketing*, vol 30, no 6, pp 595–609. doi: 10.1080/10548408.2013.810997.

Pye-Smith, C. (2017) *Transforming Food Systems in the Pacific*, EU Technical Centre for Agricultural and Rural Cooperation (CTA), Wageningen, The Netherlands.

Rao, V.R., Hunter, D., Eyzaguirre, P.B., Matthews, P.J. (2010) 'Ethnobotany and global diversity of taro', In Rao, V.R., Hunter, D., Eyzaguirre, P.B., Matthews, P.J. (eds) *The Global Diversity of Taro Ethnobotany and Conservation*, pp 1–5, Bioversity International, Rome, Italy.

Reddinger, A. (2010) 'Pineapple Glaze and Backyard Luaus: Cold War cookbooks and the Fiftieth State', In Barnhisel, G., Turner, C. (eds) *Pressing the Fight: Print, Propaganda, and the Cold War*, pp 193–208, University of Massachusetts Press, Boston, MA.

Rinaldi, C. (2017). 'Food and gastronomy for sustainable place development: A multidisciplinary analysis of different theoretical approaches', *Sustainability*, vol 9, no 10, p1748.

Sage, C. (2012) *Environment and Food*, Routledge, Abingdon.

SPC. (1999) *The staples we eat*, SPC handbook number 35. Secretariat of the Pacific Community, Noumea, New Caledonia.

SPC. (2006) *Taro Pacific food leaflet number 5*. Secretariat of the Pacific Community, Noumea, New Caledonia. Available at: www.pacificcancer.org/pacp-resources/food-nutrition-physical-activity/spc-taro.pdf

SPC. (2017) *Pacific Organic Tourism and Hospitality Standard*. Secretariat of the Pacific Community, Suva, Fiji. Available at: www.organicpasifika.com/poetcom/wp-content/uploads/sites/2/2018/05/POTHS_Draft.pdf

SPC. (2019) 'Genetic resources for food and nutrition security, from the Pacific to the world', Secretariat of the Pacific Community. Available at: www.spc.int/updates/blog/2019/01/genetic-resources-for-food-and-nutrition-security-from-the-pacific-to-the

SPTO. (2018) *Annual Review of Visitor Arrivals in Pacific Island Countries 2017*. South Pacific Tourist Organization, Suva, Fiji. Available at: www.corporate.southpacificislands.travel/wp-content/uploads/2017/02/2017-AnnualTourist-Arrivals-Review-F.pdf

Taylor, M., Hunter, D., Rao, V.R., Jackson, G.V.H., Sivan, P., Guarino, L. (2010) 'Taro collecting and conservation in the Pacific region', in Ramanatha Rao, V., Matthews, P. J., Eyzaguirre, P.B., Hunter, D. (eds) *The Global Diversity of Taro Ethnobotany and Conservation*, pp 150–167, Bioversity International, Rome, Italy.

Thaman, R. (1993) *Land, Plants, Animals and People: Community-based Biodiversity Conservation (CBBC) as a Basis for Ecological, Cultural and Economic Survival in the Pacific Islands*, University of the South Pacific, Suva.

Thaman, R. (2000) 'Traditional environmental knowledge and community-based biodiversity conservation in Fiji: Current status and priorities for its protection and utilisation', USP Geography Working Paper No 7. University of the South Pacific, Suva, Fiji.

Thaman, R. (2002) 'Threats to Pacific Island biodiversity and biodiversity conservation in the Pacific Islands', *Development Bulletin*, vol 58, pp 23–27.

Thaman, R., Faulalo, K. (2016) 'Supportive policies and enabling actions for ecosystem-based agriculture in the Pacific Islands', in *Mainstreaming Ecosystem Services and Biodiversity into Agricultural Production and Management in the Pacific Technical guidance Document*, pp 126–129, Food and Agricultural Organisation of the United Nations, Secretariat of the Convention on Biological Diversity, Secretariat of the Pacific Regional Environment Programme, Secretariat of the Pacific Community, Rome, Italy.

Timothy, D.J., Ron, A.S. (2013) 'Heritage and cuisines, regional identity and sustainable tourism', in C.M. Hall, S. Gössling (eds) *Sustainable Culinary Systems: Local Foods, Innovation, Tourism and Hospitality*, pp 275–290, Routledge, Abingdon.

Tran, X. (2017) 'Days of Eating Earth: My Twelve Months of Devotion: Traditional Knowledge and Culture as the missing link in fixing Japan's food issues'. Unpublished document, Auckland University of Technology, Auckland.

UNWTO. (2019) 'International Tourist Arrivals Reach 1.4 billion Two Years Ahead of Forecasts', United Nations World Tourism Organization. Available at: www2.unwto.org/press-release/2019-01-21/international-tourist-arrivals-reach-14-billion-two-years-ahead-forecasts

Viet, R. (2007) *Tourism, food imports, and the potential of import-substitution policies in Fiji*, Unpublished report prepared for the Ministry of Agriculture, Fisheries & Forestry, Suva, Fiji.

10

START ME UP! FOOD BIODIVERSITY AND YOUTH-LED INNOVATIONS

Eliot Gee and Hyunju Lee

Introduction

This chapter explores the incorporation of local food biodiversity into food and agriculture innovations, with attention given to emerging youth-led and focused practices. The phrase 'innovation', though often associated with cutting-edge technology, is equally applicable to new linkages between pre-existing methods, practices and actors. The chapter highlights opportunities and challenges of mobile apps, urban agriculture, youth movements and more. These varied examples are assessed to suggest how innovative approaches – social and technological – can be best leveraged to ensure fair and effective promotion and protection of local food biodiversity across the producer–consumer spectrum.

Agreeing on innovation

Innovation is frequently invoked as the key to disrupting 'business as usual' in unsustainable food systems, regardless of scale or sector. The term has been used by governing bodies such as the European Commission, which has launched partnership initiatives premised on 'food innovation as a driver of smart regional growth' (Stančová and Cavicchi, 2019), while international networks such as Young Professionals for Agricultural Development (YPARD) lead talks on 'youth as the drivers of innovation' (at FAO, 2018). From the gastronomic side, the New Nordic Movement is often referred to as a 'culinary innovation' (Byrkjeflot et al., 2013), while an 'innovative 4th Industrial Revolution' has been cited as the way to save Amazonian ecosystems (Nobre and Nobre, 2018).

This chapter focuses on how different forms of innovation – particularly those that involve or are led by youth – might be productively applied to conservation and sustainable use of food biodiversity. It should be noted that the common positioning of 'innovation' against 'business as usual' obscures the misuse of many new practices and technologies that have perpetuated environmental degradation and biodiversity loss (Diaz et al., 2019). Thus, this chapter assesses 'true innovation', according to the Latin root from which it is derived: 'restoration' or 'renewal'. We are therefore interested in practices that restore natural biodiversity and renew local ecosystems and communities. While novel technologies and inventions undeniably play an important role in this, they are dependent on a strong implementing social foundation, a view recently echoed by the Food and Agriculture Organization of the UN (Box 10.1).

BOX 10.1 FAO'S WORKING DEFINITION OF INNOVATION

Representatives from 92 countries convened at the *International Symposium on Agricultural Innovation for Family Farmers* at FAO's Rome headquarters in November 2018. Discussion topics ranged from 'blockchain technology in the coffee value chain' to 'enhancing traditional practices as innovation'. For the purposes of the seminar, a working definition of agricultural innovation was used that emphasized the importance of context and process:

> The process whereby individuals or organisations bring new or existing products, processes or ways of organisation into use for the first time in a specific context in order to increase effectiveness, competitiveness, resilience to shocks or environmental sustainability and thereby contribute to food security and nutrition, economic development or sustainable natural resource management.
>
> *FAO, 2018*

This chapter is organized around four themes that represent entry points for promoting biodiversity for food and nutrition. Possibilities within each broad category – novel technology, holistic systems, international networks and education initiatives – are outlined with supporting examples (see Figure 10.1).

Connecting new technologies and local food biodiversity

- Technological innovations with biodiversity-sensitive potential exist for both producers and consumers.
- Mobile applications and blockchain technology contain potential and risk.
- Start-ups and entrepreneurship involve young people and new perspectives.

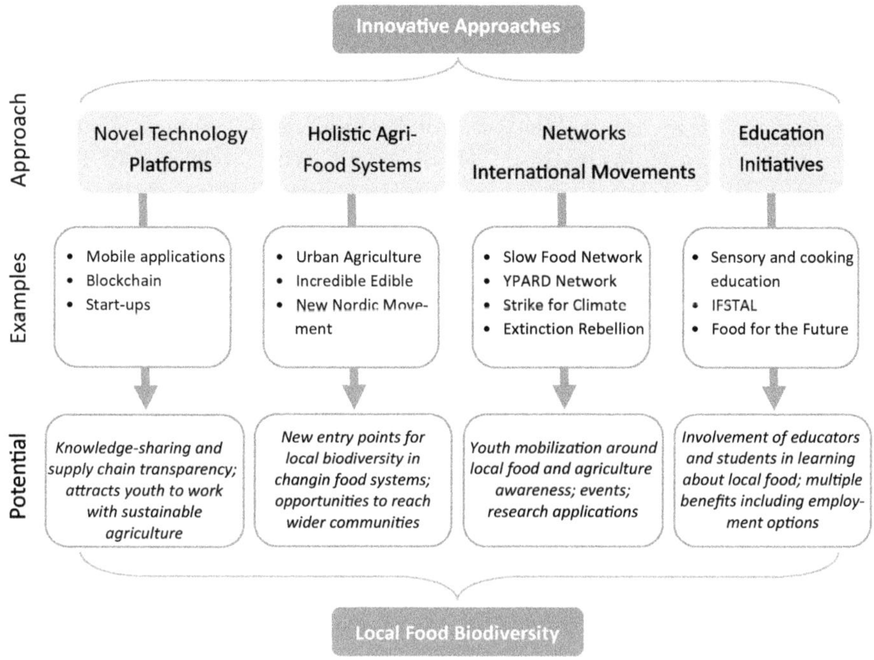

FIGURE 10.1 Innovative approaches and examples for local food biodiversity

Mobile apps: for consumers and farmers

Amongst the proliferation of mobile applications, a significant proportion supply food-based information. A range of informative resources allow consumers to find a local crop in season (*Seasonal Food Guide*), identify plants through a global photo database (*Pl@ntNet*), learn about wild edible crops (*Wild Edibles, Vild Mad*), plan and track grocery purchases (*Green Egg Shopper*) and even trace fresh crops back to farms using blockchain technology (*Harvest-Mark*) (Nierenberg, 2015). All of these can in some way incorporate the use of food biodiversity. For example, *Pl@ntNet* was developed by French research organizations (CIRAD, INRA, Inria and IRD) to crowdsource botanical knowledge, bolstered by visual recognition software and other online databases; this could be upgraded to include edible species' nutrition data and associated traditional knowledge. *Vild Mad* is currently attempting to match consumer interests in discovering different landscapes, wild edibles and recipes in Denmark. This explorative approach could be further developed with randomized software that introduces users to potential food sources according to their season and region.

On the producer side, a growing number of applications offer 'smart phone farmers' unprecedented access to crop, field and market information. This can be best seen in the rapid spread of farmer apps across Africa. For example,

Farmerline was designed by Ghanaian entrepreneurs to link farmers and businesses with location-customized information (including certification requirements, sustainability training and food branding) provided in local languages. The app relies on a platform called *MergData* that has projects on moringa, cacao and other local crops aiming to link 'natural wealth' with 'wellbeing of farmers' (Mergdata, 2018). Another app, *uLima*, was launched in Kenya in 2018, granting users immediate access to crop libraries including input information of the specific crop, real-time local weather information and alerts, market prices, and forum networks amongst other farmers, with an in-app tool called *My Farm* providing day-to-day tracking, guidance, and notifications from pre-planting to post-harvest. These apps have strong potential to boost farmer livelihoods and could be adapted to more directly include information focused on local varieties and their markets. Likewise, farmers could be involved in compiling local crop knowledge, and linked to consumers and local institutions/buyers. But first, appropriate training is needed to empower farmers with the skills to use these tools to their best advantage.

The potential of these apps is just beginning to be understood, but they are already attracting young app developers and entrepreneurs at the forefront of what has been referred to as a '4th Industrial Revolution' (Lombard, 2017). This emerging start-up culture, particularly in West Africa, has attracted the interest of food industry giants who offer funding for hackathons and business pitches. Although this increases the resources available to would-be innovators, it also opens the door to co-option of profit-driven work that might exploit smallholders (Maughan, 2018). Many sustainable innovations, despite being initiated by entrepreneurial 'emerging Davids', are often absorbed by 'Greening Goliaths' with less ambitious goals (Hockerts and Wüstenhagen, 2010).

Breaking down blockchain

At its best, blockchain's decentralized digital record-keeping is intended to democratize supply chains/databases; essentially, every transaction/action taken accumulates as a line of traceable, verifiable data stored on all computers connected to the online system (Mooney, 2018). A recent survey of nine agri-food apps making inroads in East Africa (Bolt, 2019) shows that while high interest in blockchain is accompanied by numerous reservations, it is already being implemented at the financial level in Kenya, with particular attention to farmer resilience. Just as cryptocurrencies like Bitcoin have promised to cut out banks and notaries for investors, blockchain applied to food systems could theoretically eliminate intermediaries and offer smallholders unprecedented access to smart contracts, investment, insurance, credit and emerging markets (Bolt, 2019).

The implications of technology that can immediately track the flow of money, products and knowledge remain largely unexplored for food biodiversity. It is possible that blockchain could increase local product traceability or create a global platform for local biodiversity knowledge-sharing. For example, The Amazonia Third Way Initiative, with support of the World Economic Forum, has proposed an 'innovation ecosystems' development approach that registers the intellectual property of native biodiversity through blockchain technology (Nobre and Nobre, 2018). The proposed 'Amazon biobank' would compile biological assets, traditional knowledge, indigenous skills and other resources. While similar in content to the BFN Project's list of species on the SiBBr database (see Part II, C.1), this information would be dispersed through distributed digital ledgers with the goal of streamlining the compilation and sharing of relevant knowledge and allow ongoing, open-source updating of trustworthy information by anybody with a smartphone (Nobre and Nobre, 2018).

Unfortunately, the hype and hypothetical appeal of blockchain technology is often shattered by the realities of its application. The common assertion that blockchain improves transparency and inclusivity can be questioned if one looks closer at issues of access, privatization and consolidation of power and resources. Although there are a few instances of blockchain use by governments or NGOs advocating for agro-ecological practice (such as an Indian state government partnering with Swedish start-up *Chromaway*), the majority of blockchain use remains firmly in the grasp of multinational commodity traders, contributing to vertical integration across the food supply chain (Mooney, 2018). As with many agri-food innovations, rhetoric such as 'exploitation' and 'extraction' is favoured over sustainable community-minded practice (Maughan, 2018). Critics warn that blockchain's potential to promote open-source genomic information has already been undermined by privatized bio-patents that threaten to 'commodify all of the world's biological diversity (known and unknown)' (Mooney, 2018).

Sustainable start-ups and entrepreneurial culture

While these novel technologies may not offer fix-all solutions, they can connect different actors who are motivated to improve global and community-based food systems. Successful start-ups, such as some of the previously described app developers, demonstrate to young people that it is possible to make a profit in sustainable agriculture. Entrepreneurial bootcamps such as *ImpactHub*, or immersive trainings and summer schools hosted by *Future Food Institute*, bring together young developers around the world to workshop new approaches to themes including healthy diets and local agriculture. Many of these groups are willing to embrace any and all methods, for example the self-described 'agro-nostic' *Thought for Food* conducts inclusive career-building summits intended to build a 'community of changemakers' with backgrounds in science, agronomy, engineering, agriculture, architecture, business and public policy. Partnerships between the private sector, UN agencies, NGOs, and government bring youth entrepreneurs into international development through events such as the first 'SDG Hackathon'[1], hosted in 2019 by FAO and the *Future Food Institute.* The open-minded enthusiasm embodied by most start-ups represents an opportunity to incorporate a diversity of alternative approaches to food system change.

TAKEAWAYS (1):

- Unprecedented quantities of information are available to consumers and farmers through mobile applications, including resources on local crops, wild edibles, markets and agricultural conditions.
- Blockchain technology represents both a possibility and a risk in bringing more biodiversity and smallholder farmers into the food supply chain and gathering genetic data.
- The enthusiasm surrounding technological innovation can be used to mobilize effective research, policy and awareness-building activities for food biodiversity mainstreaming.

Holistic agri-food systems

- Trends towards more sustainable and localized food systems – connecting production more closely to consumption – offer opportunities for integrating a variety of local foods.
- Young producers and consumers take an active role in activities such as urban agriculture and the New Nordic movement.

Recent reports stress the need for systematic change in order to tackle the burden of malnutrition and environmental degradation (Diaz et al., 2019; Willett et al., 2019), necessitating a widespread shift in how we view our food and its sources. This shift is reflected in current food trends that recognize a need for systematic change:

- Reducing environmental impact: at the global and local level.
- Health concerns: malnutrition, diet.
- Demographic changes: feeding an urban population.
- Vegetarianism/veganism: dietary change reflecting concerns over health, ethics and environmental impact.
- Growing/Eating local: re-valuing local farmers and food communities.
- Looking to traditional foodways: valorizing 'roots' and indigenous foods.

Intersecting trends of sustainable, ethical and healthy foods represent a massive opportunity to expand the market for locally-grown biodiversity through sustainable, holistic food systems. For example, the rise of farm-to-fork movements offer an entry point for local produce in supply chains, shortening the distance between producers and consumers, boosting local economies, preserving small-scale farmlands and local cultivar diversity. Likewise, centralized food hubs – organizational arrangements for aggregating, distributing and marketing local products within a region – build capacity for local food biodiversity. Food system reforms include moving towards planetary health diets, reducing the cost of healthy food choices and restructuring the current distribution system to reduce waste and shorten supply chains.

Many system-minded approaches, such as the examples in the following boxes (10.2–3), have the potential to heavily involve youth and bring a fresh perspective to biodiversity mainstreaming for improving diets and related activities. However, there is the challenge, especially with food industry trends, of clearly representing the many dimensions of local foods that cannot be easily boiled down into a single sellable concept. For example, to different people local food might evoke a connection to indigenous culture, while to others it might imply the race to find the next superfood (see Chapter 8 on sustainable gastronomy).

Urban agriculture for local foods

Over half of the world's population (55%) lived in cities in 2018, with a projected 68% by 2050 (UN DESA, 2019). Urbanization is a major factor in food and nutrition insecurity, with urban trends towards hyper-processed, homogenous diets. Fortunately, cities represent fertile ground for innovative local food production with examples including community-based gardening, rooftop and roadside gardening, as well as vertical and hydro/aquaponics growing systems.

In recent years, start-ups have transformed urban spaces to grow salads and herbs, providing fresh produce all year long while reducing land, water and transport costs in metropolises such as London (*Infarm*, *Growing Underground*), New York (*Square Roots*, *Aerofarms*) and Kenya (*Ukulima Tech*). With a focus on leafy greens (which are best suited for cost-efficient indoor growing), some of these models have enabled the incorporation of heirloom varieties and atypical greens that may have been too delicate in the past to survive outside of niche and faraway markets. Now it has become feasible to sell a diversified selection to a wider urban audience at competitive prices. This has particular potential given the increasing funding attained by African agricultural start-ups: USD$19 million of investments from 2016–18 alone (Shapshak, 2018), which can be combined with government-sponsored projects such as Kibera's sack farms (local food production initiated in order to tackle food insecurity and unemployment in Africa's largest urban slum (Mayoyo, 2015)). Additional urban agriculture youth programmes, such as Camp Green in Uganda, 4-H Club in Liberia or Jaden Tap Tap in Haiti, provide training for youth to become farmers and agricultural entrepreneurs while creating urban farm spaces in previously struggling areas (RUAF, 2014, 2018).

Despite the aforementioned importance of urban agriculture, food planning is often omitted in urban planning (Morgan, 2009). A balance of local seasonal foods with fairly traded global foods is critical in addressing this gap. Communities built around local food biodiversity can play an important role in shifting to a 'City Region Food Systems' approach that maximizes the sustainable production of food within a region, while also sourcing foods outside the region that are produced with a similar set of embedded characteristics (Blay-Palmer et al., 2018; FAO, 2019; Hamm, 2019; RUAF, 2014). This theme is further explored in Chapter 6 of this book, which explores urban-rural linkages; for a bottom-up example see Box 10.2.

BOX 10.2 TRANSFORMING NEIGHBOURHOODS WITH THE INCREDIBLE EDIBLE NETWORK

The Incredible Edible Network started in 2012 following the example of Todmorden, a small town in the United Kingdom, which has been previously documented by an earlier entry of this series on *Issues in Agricultural Biodiversity* (Paull, 2013). Todmorden pioneered an open-ended model bringing neighbours together through actions around local food and creating social spaces where people work together on urban gardens, cook-offs and seed swaps. Regional pride, the notion of the commons and a desire for urban restoration is central in powering this grassroots movement, which involves the collaboration of local businesses and volunteers following the tagline: 'If you eat, you're in' (Paull, 2013). Projects connect local production to nutrition, repurpose urban and peri-urban space and foster cross-generational community, spreading to more towns through word of mouth and positive buzz online. Since the previous case study (Paull, 2013), the Incredible Edible Network has coagulated around three 'plates' of community, business and learning, while spreading to more than 100 groups in the UK and over 700 groups worldwide, with new projects including allotments on shopping centre rooftops, collaborations with refugees, revival of parks as food forests and many other partnerships with local schools and businesses.

Incredible Edible Network, 2018

BOX 10.3 THE NEW NORDIC MOVEMENT

Since a dozen Scandinavian chefs established the New Nordic Manifesto in 2004, the paradigm of regional food cuisine has profoundly changed. Rather than overtly naming biodiversity, the culinary movement has focused on promoting the right to good quality food, which emphasizes the human agency in fostering an agricultural landscape rich in local, seasonal produce that protects the land and the health of consumers (Bügel et al., 2016). The union of sustainable gastronomy (see Chapter 12) with practical policies and programmes (reaching an estimated 27 million people) has received significant international attention for promoting healthy school meal procurement as part of education as well as agriculture and nutrition awareness (Halloran et al., 2018). In the realm of education, the more recent Children's Nordic Kitchen Manifesto emphasized the taste and positive food experiences, moving discussion of food beyond nutrition to encompass experimentation, including the discovery of varied food sources depending on the place and time (Wistoft and Qvortrup, 2018, see also Box 10.6 on education). This can include wild edible plants, as can be seen with two projects, Vild mad (Wild food) and Byhøst (Urban Harvest) which supply students with an interactive map and species-based information on harvesting. All of this helps democratize food and involve youth as active cooks and mindful consumers.

While there is debate regarding whether this model can be adapted to other parts of the world, others have already tried. The Indian Food Manifesto aims to, among other goals, 'support traditional produce' and 'revive India's heritage foods' (Tasting India, 2017). However, while examples such as this pick up private sector support, it is not clear whether other regions could establish the strong level of government–private sector partnerships, as the Nordic countries have, that can support substantial implementation.

Nina Olsen Lauridsen, Bioversity

BOX 10.4 BRINGING INDIGENOUS KNOWLEDGE INTO THE FOOD SYSTEM

Indigenous knowledge is intertwined with local foods, and the focus point for many projects trying to link traditional growing and cooking practices with food security:

- *Totomoxtle* (Southwest Mexico) is an art project using veneer material made with husks of heirloom corn. By casting threatened varieties in a new way, the project (which is supported by the International Maize and Wheat Improvement Center, CIMMYT) raises awareness about corn biodiversity and helps generate income for local farmers and craftswomen who have been trying to regenerate traditional agricultural practices (Laposse, 2019).
- *Jatun Sach'a* (Bolivia) is a development project that aims to benefit 12,000 families by promoting Andean crops, such as yucca and majo, in addition to sustainable management of coffee and cocoa (UNODC 2012).
- *Wanted: Food for the Future* (Belgium, Indonesia, Peru, Tanzania) is an international project that focuses on climate-smart traditional food sources (covered in detail later in this chapter in Box 10.7).

TAKEAWAYS (2):

- Environmental and food trends emphasize the need for more unified local food production and consumption.
- Particularly in urban areas, these trends engage youth and represent varied opportunities to promote local foods, gastronomy and marketing.

Networks for biodiversity: gaining youth support through international movements

- Online networks bring together youth from around the world.
- These virtual communities can materialize into events and even research results.

Biodiversity mainstreaming for food and nutrition can be supported through building momentum around social organizations and movements that favour linkages around indigenous and traditional foodways. Longstanding examples such as *Via Campesina* already have extensive experience working with NGOs to promote peasant seed systems while involving youth and indigenous farmers to champion the concept of food biodiversity (Via Campesina, 2000). Work relevant to biodiversity for food and nutrition has significant potential to be picked up and disseminated by similar organizations that are concerned with the rights of local people and foods (one example of this is the Indigenous Partnership for Agrobiodiversity and Food Sovereignty, which trains indigenous youth through an annual fellowship programme divided between Northern India and Rome). Global groups such as The Global Youth Biodiversity Network (GYBN, established by the Convention on Biological Diversity), Young Professionals for Agricultural Development (YPARD) or the Slow Food Youth Network (SFYN) actively gather young people who are passionate about sustainable themes, including biodiversity. Dedicated WhatsApp and Facebook groups bring local knowledge and experience into a wider international social and professional online network, mobilizing millions of youth. In 2018–19 alone, sweeping activism campaigns by Extinction Rebellion and the Strike for Climate Movement demonstrated that these grassroots communities should not be underestimated; many are rapidly taking tangible form in events, rallies and conferences in places from India to England (Boeunda, 2019). The mobilization of youth around climate change may also enable new levels of awareness regarding biodiversity loss (Strike for Climate leader Greta Thunberg has herself invoked the danger of biodiversity loss on multiple occasions); already, widely circulating articles connect the two issues; for example, 'The Rapid Decline Of The Natural World Is A Crisis Even Bigger Than Climate Change', a reference to 2019's IPBES report, counted 38,000 readers in several weeks (Vidal, 2019). Once linked to substantial information on healthy and sustainable foods, this momentum could broaden the scope of public awareness, and lead to greater policy advocacy.

BOX 10.5 INVOLVING YOUTH IN BIODIVERSITY RESEARCH WITH THE ARK OF TASTE

Slow Food's massive biennial *Terra Madre-Salone del Gusto* convention explicitly promotes agro-biodiversity through interactive exhibitions and talks with themes on seed saving, culinary revival movements, regenerative agriculture, and a strong emphasis on regional food knowledge. This recurring platform facilitates awareness-raising activity, attracts prominent funders and increases support for research and policies. With high participation by 'gastronome' students and graduates from the University of Gastronomic Sciences (UNISG), there is also a research component, most notably with the *Ark of Taste* project (see Box 10.5).

With the goal of documenting indigenous and traditional foods, Slow Food members have been mobilized to initiate community canvassing and surveys from Europe to Africa. At Terra Madre 2018, a team including young Italian and Kenyan UNISG researchers launched the first edition of Kenya's *Ark of Taste* book (see Figures 10.2 and 10.3), which compiles traditional plants, including numerous African Leafy Vegetables. The focus of the book's biodiversity conservation efforts is for cultural and environmental reasons. Therefore, an opportunity exists for nutritional value and food composition data (as researched by BFN Kenya, see Part II, Section C) to be incorporated into these entries, drawing a connection to the pre-existing traditional and medicinal knowledge surrounding cultivation and consumption.

FIGURES 10.2 & 10.3 Panel hosted by researchers for the Ark of Taste in Kenya book, at Terra Madre

Source: Eliot Gee, 2018

FIGURES 10.2 & 10.3 (Cont.)

These movements, though global in scope, are characterized by strong local knowledge and community-adapted activity that is shared amongst members. By traversing multiple scales, they contribute a fresh perspective that paves the way for food biodiversity conservation and sustainable use.

TAKEAWAYS (3):

- Traditional foodways and associated knowledge can be re-valued as vital future resources through youth-focused movements.
- Food biodiversity can be presented as a central connecting element between human diets and environmental sustainability.
- Interested youth are connected to each other as well as to NGOs and international development with tangible outputs such as the Youth-UN Global Alliance's Youth Guide to Biodiversity.
- Events offer engaging ways to raise awareness and conduct hands-on activities, with research applications.
- Information is disseminated globally through wide-reaching social media networks.

Education at the root of biodiversity mainstreaming

- Diverse approaches to education instil critical awareness and values that translate into productive innovations.

Education may not immediately sound novel, yet proponents of agro-innovation in Africa (Juma, 2015) suggest that methods such as school gardens, community outreach and experiential learning are crucial to build capacity for local agriculture, conservation and youth interest in sustainable entrepreneurship. Schools offer future decision-makers and 'innovators' the conceptual foundation that will inform what they see as problems and viable solutions to a healthier, sustainable future.

The school environment can function as an educational platform and a market for food biodiversity (for details on school procurement see Chapter 7 of this book; UNSCN, 2017; Tartanac et al., 2018). School-based interventions can involve government, NGOs and the private sector; for example, US salad company *Sweetgreen* who recently pledged $1 million to partner with NGO *FoodCorps* in 'Reimagining School Cafeterias' to promote fresh local produce. In Mexico, the *Let's Food Cities* project has collaborated with municipalities inspired by the Milan Urban Food Policy pact to bring in local food biodiversity procurement and nutrition awareness classes (Let's Food Cities, 2018).

With increased exposure to new foods, students can be involved in every step of food production from seed to table. This process in which students learn about food and agri-food systems represents an important entry point for biodiversity-aware activities, for example early sensory education such as the Australian hands-on *Stephanie Alexander Kitchen Garden Foundation* (similar examples are covered in Box 10.4). As they age, students can take the lead in this process, as with *People and Planet*, which ranks UK universities on sustainability criteria including procurement (People and Planet, 2018). Ultimately, this perspective can instil healthy food habits and knowledge that students will carry in to their adult lives, as can be seen with the paradigm shift of the New Nordic Movement (Box 10.3).

At the higher education level, The University of Gastronomic Sciences (associated with Slow Food and mentioned earlier) has led the push for serious academic and professional engagement with food biodiversity. Other platforms that connect diverse students and experts, such as the UK-based Interdisciplinary Food System Teaching and Learning (IFSTAL) Programme, can feed into collaborative projects that create new linkages. Other examples include *Eating City*, a private–public partnership that gathers young researchers and professionals to participate in publications with themes such as public food procurement and nutrition security.

Many sustainable enterprises grow out of this momentum that balance social and environmental awareness, business and education. For example, Timor-Leste Food Lab (a Slow Food member) runs youth mentoring programmes that advocate for 'native and heritage ingredients' and changing food habits to promote 'biodiversity and indigenous knowledge'. The Timor-Leste Food Innovators Exchange (TLFIX) connects diverse people to raise awareness about the use and nutrition of local foods.

BOX 10.6 DIVERSE APPROACHES TO FOOD EDUCATION

Nutrition and food education can take many forms, from theoretical sharing of nutrition principles to physical kitchen skills. 'Les Classes du Goût', set up by Jacques Puisais for school-aged children to discover the five senses, has expanded to found an international non-profit organization, SAPERE ALSBL, which includes classes in eight counties worldwide in order to help children develop their ability to make balanced eating choices through simple and fun activities. Sensory-based food education like Flavour School in the UK lets students experience the five tastes (sour, sweet, salty, bitter and umami) and express their experiences in verbal and written language through simple and low-cost activities within their school curriculum. Students are also encouraged to experience sharing food together and respecting others' experiences. Similarly, the *Stephanie Alexander Kitchen Garden Foundation* aims to help Australian children to form positive lifelong food habits by supporting teachers to deliver hands-on food education. Students actively participate in the food system from growing and harvesting fruits and vegetables in the garden to preparing and sharing foods in the kitchen and dining table.

In teaching children or adults how to prepare and cook their own food, well-designed cookery education can make students excited and interested in the act of sourcing and making their food, which is important to encourage students to make healthy choices in their diets. Depending on families' socio-economic status, cooking at home might not be an option. There are a number of alternatives: cookery education for both parents/guardians and children; providing children with specific types of vegetables to bring back home; and growing food together with local farmers, parents/guardians and children.

TAKEAWAYS (4):

- Multidisciplinary approaches to food learning can engage youth and offer entry points for consumption, production and entrepreneurship.

Conclusion: innovative action

Beyond the fields of research and international development, 'nutritious food biodiversity' or 'neglected and underutilized species' may not be immediately recognized terms, but the concepts are well-situated to take a more prominent place in the public and policy awareness. Consumers are routinely exposed to 'superfoods' and 'foods for the future' as well as concepts of 'eat local' and 'farm to fork/table'. Now more than ever, these themes are poised to enter the public and policy sphere, backed up with an interdisciplinary research evidence base. The EAT-Lancet Report's 'Planetary

Health Diet' (Willett et al., 2019) has generated widespread public debate around the role of dietary variation in improving both health and the environment; Food Forever's *2020 for 2020* calls for chefs to support heritage food sources; Future 50 Foods, a joint publication by WWF and Knorr, details climate-friendly crops from Bambara groundnut to seaweeds, with simple explanations and preparation instructions that could be easily supplemented with nutritional evidence.

The Strike for Climate Movement initiated by then 15-year-old student Greta Thunberg in 2018 has shown the public that youth are ready to rally around environmental issues and the Sustainable Development Goals (UN Climate Change, 2019). Similarly, Extinction Rebellion[2], an international apolitical network, has been growing its network dramatically worldwide and brought international attention to climate and ecologic issues including biodiversity loss since it started in October 2018. Slow Food International has already raised the public profile of plant, animal and food biodiversity; there is potential to place this in the centre of many food trends, educational programs, start-ups and other networks. It is a challenge to comprehensively share the nutrition aspect and viably upscale activities, but given proper policy and private sector support, an opportunistic case-by-case approach could successfully connect these themes (see Box 10.7 for one example).

BOX 10.7 INNOVATION THROUGH PRACTICE WITH FOODS FOR THE FUTURE

Elsewhere in this book, the Biodiversity for Food and Nutrition Project methodology is described as a partnership-based approach to local food promotion. This can be locally, nationally or internationally-minded, or a combination of all three.

Aspects of this approach can be seen elsewhere, for example with 'Wanted: Food for the Future', an initiative spearheaded by Belgian NGO Rikolto that prioritizes market development for healthy, sustainable local crop production of East African pulses, Andean grains and Indonesian seaweed. Project partners include MESA, a 'Locally Rooted, Globally Connected' NGO that empowers young Peruvian 'stewards' through community-based development training in US exchange programmes. After learning about the application of biodynamic or Community Shared Agriculture (CSA) models, youth return to their communities with ideas for new projects, such as a business that creates new processing and marketing for muña, a traditional Peruvian medicinal plant researched by Slow Food's Ark of Taste. Another partner, SOLID International, focuses on social enterprise, promoting a local Peruvian variety of quinoa while training youth and farmers on business practices to commercialize cañihua, kiwicha, tarwi and other Andes crops. The range of 'Wanted: Food for the Future' partners reflect how community development through local foods can be extended into international value chains and capacity building.

It is possible that trends promoting ancient grains and traditional varieties can help make a wider diversity profitable for producers whether they are in the global north or south. 'Plants of the past' can be relatively easily rebranded as 'plants of the future' as consumers – particularly younger, urban and affluent eaters, but progressively more people – become interested in sustainably diversifying their diets. Likewise, there are encouraging opportunities for young people and farmers in developing regions to use local species for social enterprises and other business opportunities. Novel technologies could play a role in spreading awareness and building farmer and market capacity around neglected and underutilized species, if there is a sufficient supportive structure to ensure equitable access and use (Bolt, 2019; Maughan, 2018; Mooney, 2018). As this chapter has suggested, innovation is a contested term that includes more than simply technology, but also the systematic changes, networks and approaches that can reintroduce a greater and healthier diversity into our diets (see Table 10.1).

TABLE 10.1 Innovative entry points to food biodiversity mainstreaming

Challenges	*Opportunities*
• Promoting context-dependent rather than single-fix dietary alternatives • Priority should be placed on approaches that could decrease, rather than increase, farmer dependence on external inputs • Access: increasing sustainable production and consumption without feeding into further socio-economic stratification, establishing rural-urban relationships that flow both directions • Preventing novel technologies from being monopolized by the private sector • Need to strengthen the links between youth, government, and international development agencies.	• Many consumers are receptive to ancient grains, traditional foods and associated knowledge and culture • Communities are mobilizing around the growing awareness of environmental damage including biodiversity loss • Social media and online communities are capable of rapidly mobilizing people, particularly youth • Knowledge-sharing can be facilitated by novel technologies such as mobile apps and blockchain • Local products can also be traced more easily, enabling low-cost certification • Context-based food hubs can offer viable alternatives to multinational innovation strategies • Multiple approaches to education can introduce food biodiversity to children early on • Social entrepreneurship and start-ups disrupt the status quo and make local agriculture more appealing to youth.

USEFUL LINKS

Eating City concept papers: www.eatingcity.org/concept-papers/
Future Food Institute: https://futurefood.network/
Flavour School: www.flavourschool.org.uk/
Global Youth Biodiversity Network (CBD): www.gybn.org/
Kitchen Garden Foundation: www.kitchengardenfoundation.org.au/
The Indigenous Partnership for Agrobiodiversity and Food Sovereignty website (specifically, see TIP Fellowship Program): http://theindigenouspartnership.org/
Interdisciplinary Food Systems Teaching and Learning (IFSTAL): www.ifstal.ac.uk/
Rediscovered: The inspiring stories of 25 rediscovered super foods from across the globe. www.thelexicon.org/rediscovered/
Thought for Food: https://thoughtforfood.org/
University of Gastronomic Sciences (UNISG) website: www.unisg.it/en/
WWF/Knorr (2019). Future 50 Foods Report: www.knorr.com/content/dam/unilever/knorr_world/global/online_comms_/knorr_future_50_report-1603451.pdf

Acknowledgements

Thank you to Michelle Lopez and Nina Olsen Lauridsen for their contributions while at Bioversity International, and to Diego Valencia (Thought for Food), Inge Overmeer (SOLID) and Maria de Los Angeles Franco (MESA) for sharing material and inspiration from their work.

Notes

1 SDG Hackathon: http://foodinnovationprogram.org/hack-sdgs-the-first-hackathon-on-united-nations-2030-agenda-sustainable-development-goals/
2 Extinction Rebellion: https://rebellion.earth/the-truth/demands/

References

Blay-Palmer, A., Santini, G., Dubbeling, M., Renting, H., Taguchi, M., Giordano, T. (2018) 'Validating the city region food system approach: Enacting inclusive, transformational city region food systems', *Sustainability*, vol 10, no 5, article 1680. doi: 10.3390/su10051680

Boeunda, A. (2019) 'Kids striking against climate change: "We're fighting for our lives"', *National Geographic*, 15 March 2019. Available at: www.nationalgeographic.com/environment/2019/03/kids-climate-march-strikes-around-the-world-fridaysforfuture/ (Accessed 25 March 2019).

Bolt, J. (2019) 'Financial resilience of Kenyan smallholders affected by climate change, and the potential for blockchain technology', *CCAFS Flagship Climate-Smart Technologies and*

Practices, Incentives and Innovative Finance for scaling Climate Smart Agriculture Up and Out, Wageningen Environmental Research Report. Available at: https://library.wur.nl/WebQuery/wurpubs/fulltext/472583

Bügel, S., Hertwig, J., Kahl, J., Lairon, D., Paoletti, F., Strassner, C. (2016) 'The new Nordic diet as a prototype for regional sustainable diets', in Meybeck, A., Redfern, S. (eds) *Sustainable value chains for sustainable food systems: A workshop of the FAO/UNEP Programme on Sustainable Food Systems*, Rome, 8–9 June 2016, pp 109–116, The Food and Agriculture Organization of the United Nations, Rome, Italy. Available at: www.fao.org/3/a-i6511e.pdf

Byrkjeflot, H., Strandgaard, J., Svejenova, S. (2013) 'From label to practice: The process of creating new Nordic cuisine', *Journal of Culinary Science & Technology*, vol 11, no 1, pp 36–55. doi: 10.1080/15428052.2013.754296

Diaz, S., Settele, J., Brondízio, E., Ngo, H., Guèze, M., Agard, J., Arneth, A., Balvanera, P., Brauman, K., Butchart, S., Chan, K. (2019) 'Summary for policymakers of the global assessment report on biodiversity and ecosystem services of the Intergovernmental Science-Policy Platform on Biodiversity and Ecosystem Services.' Available at: www.ipbes.net/news/ipbes-global-assessment-summary-policymakers-pd

FAO. (2018) 'The international symposium on agricultural innovation for family farmers:Voices from the symposium', Food and Agriculture Organization of the United Nations, *YouTube* www.youtube.com/playlist?list=PLzp5NgJ2-dK5vzFZnDZxzNBA63jiLozMk (Accessed 21 Jan 2019).

FAO. (2019) 'City region food systems programme', *Food and Agriculture Organization of the United Nations*. Available at: www.fao.org/in-action/food-for-cities-programme/approach-old/crfs/en/

Halloran, A., Fischer-Møller, M.F. Skylare, E. and Persson, M. (2018) *Solutions menu: A Nordic guide to sustainable food policy*, Ministers. Available at: https://norden.diva-portal.org/smash/get/diva2:1214792/FULLTEXT01.pdf

Hamm, M. (2019) 'City region food systems: What? Why? How?', Centre for Food Policy, *YouTube*. Available at: www.youtube.com/watch?v=F9UKSA7TjzE

Hockerts, K., Wüstenhagen, R. (2010) 'Greening Goliaths versus emerging Davids—Theorizing about the role of incumbents and new entrants in sustainable entrepreneurship', *Journal of Business Venturing*, vol 25, no 5, pp 481–492. doi: 10.1016/j.jbusvent.2009.07.005

Incredible Edible Network. (2018) 'Incredible Edible Network Impact Report 2018', *Incredible Edible Network*. Available at: www.incredibleedible.org.uk/wp-content/uploads/2018/10/IE-Impact-Report-Final-Web.pdf

Juma, C. (2015) *The New Harvest: Agricultural Innovation in Africa*, Oxford University Press, New York.

Laposse, F. (2019) 'Totomoxtle'. Available at: www.fernandolaposse.com/projects/totomoxtle/

Let's Food Cities. (2018) 'El sistema alimentario de Guanajuato'. Available at: http://letsfoodcities.com/es/lets-food-cities-3/las-ciudades/guanajuato-mexico-2

Lombard, W.A. (2017) 'The 4th industrial revolution: Is it here?', *FarmBiz*, vol 3, no 12, pp 6–7. doi: 10520/EJC-c05137ec9

Maughan, C. (2018) 'The dark side of innovation for family farmers: Reflections on an international symposium on innovation', *Agroecology Now!* Available at: www.agroecologynow.com/darksideinnovation/ (Accessed 14 January 2019).

Mayoyo, P. (2015) 'How to grow food in a slum: Lessons from the sack farmers of Kibera', *The Guardian*. Available at: www.theguardian.com/global-development-professionals-network/2015/may/18/how-to-grow-food-in-a-slum-sack-farmers-kibera-urban-farming

Mergdata. (2018) 'Mergdata case studies: Case studies on Mergdata use cases'. Available at: https://mergdata.com/case-studies/

Mooney, P. (2018) "Blocking the chain: Industrial food chain concentration, big data platforms and food sovereignty solutions', *ETC Group*. Available at: www.etcgroup.org/sites/www.etcgroup.org/files/files/blockingthechain_english_web.pdf

Morgan, K. (2009) 'Feeding the city: The challenge of urban food planning', *International Planning Studies*, vol 14, no 4, pp 341–348. doi: 10.1080/13563471003642852

Nierenberg, D (2015) 'Twenty-three mobile apps changing the food system', *Food Tank*. Available at: https://foodtank.com/news/2015/01/twenty-three-mobile-apps-changing-the-food-system/

Nobre, I., Nobre, C.A. (2018) 'The Amazonia third way initiative: the role of technology to unveil the potential of a novel tropical biodiversity-based economy', in Loures, L.C. (ed) *land use – Assessing the past, envisioning the future*, IntechOpen. doi: 10.5772/intechopen.80413

Paull, J. (2013) 'Case study 10 – "Please pick me": How Incredible Edible Todmorden is repurposing the commons for open source food and agricultural biodiversity', in Fanzo, J., Hunter, D., Borelli, T., Mattei, F. (eds) *Diversifying food and diets: Using agricultural biodiversity to improve nutrition and health*, pp 336–345, Routledge, Abingdon.

People and Planet. (2018) 'People & Planet University League – How sustainable is your university?' Available at: https://peopleandplanet.org/university-league

RUAF. (2014) *GROW the city: Innovations in urban agriculture* Urban Agriculture Magazine no.28 December 2014. Available at: www.ruaf.org/sites/default/files/UAM28%281%29.pdf

RUAF. (2018) *Youth in food: Opportunities for education and employment*, Urban Agriculture Magazine no.35 November 2018. Available at: www.ruaf.org/sites/default/files/RUAF_ UAM%2035_web.pdf

Shapshak, T. (2018) 'African agri-tech startups boom with 110% growth since 2016' *Forbes*. Available at: www.forbes.com/sites/tobyshapshak/2018/05/09/african-agri-tech-startups-boom-with-110-growth-since-2016/#e619184433cd

Stančová, K.C., Cavicchi, A. (2019) 'Food: innovation and diversification paths', in *Smart specialisation and the agri-food system: A European perspective*, pp 1–24, Palgrave Pivot, Cham.

Tartanac, F., Swensson, L.F.J., Polo Galante, A., Hunter, D. (2018) 'Institutional food procurement for promoting sustainable diets', in Burlingame, B., Dernini, S. (eds) *Sustainable diets: The transdisciplinary imperative*, pp 240–247, CABI, Wallingford UK.

Tasting India. (2017) 'The Indian Food Manifesto'. Available at: www.tastingindiasymposium.com/the-indian-food-manifesto/

UN Climate Change. (2019) 'UN climate press release: Experts, leaders agree on key ways to mobilize accelerated action for climate, sustainable development', *UN Climate Change*. Available at: https://unfccc.int/news/experts-leaders-agree-on-key-ways-to-mobilize-accelerated-action-for-climate-sustainable-development

UN DESA - United Nations, Department of Economic and Social Affairs, Population Division (2019) 'World urbanization prospects: The 2018 revision' (ST/ESA/SER.A/420) United Nations, New York. Available at: https://population.un.org/wup/Publications/Files/WUP2018-Report.pdf

UNODOC. (2012) 'Jatun Sach'a – Benefiting thousands of farmers' families in Bolivia', *United Nations Office on Drugs and Crime*. Available at: www.unodc.org/unodc/en/

frontpage/2012/January/jatun-sacha-benefiting-thousands-of-farmers-families-in-bolivia.html
UNSCN. (2017) 'Schools as a system to improve nutrition: A new statement for school-based food and nutrition interventions, United Nations System Standing Committee on Nutrition', *United Nations System Standing Committee on Nutrition*. Available at: www.unscn.org/uploads/web/news/document/School-Paper-EN-WEB-nov2017.pdf
Via Campesina. (2000) 'Biodiversity and genetic resources', Via Campesina III International Conference, Bangalore, India. Available at: https://viacampesina.org/en/biodiversity-and-genetic-resources12/ (Accessed 8 January 2019).
Vidal, J (2019) 'The rapid decline of the natural world is a crisis even bigger than climate change', *Huffington Post*, 15 March 2019. Available at: www.huffingtonpost.ca/entry/nature-destruction-climate-change-world-biodiversity_n_5c49e78ce4b06ba6d3bb2d44
Willett, W., Rockström, J., Loken, B., Springmann, M., Lang, T., Vermeulen, S., Garnett, T., Tilman, D., DeClerck, F., Wood, A., Jonell, M., Clark, M., Gordon, L.J., Fanzo, J., Hawkes, C., Zurayk, R., Rivera, J.A., De Vries, W., Sibanda, L.M., Afshin, A., Chaudhary, A., Herrero, M., Agustina, R., Branca, F., Lartey, A., Fan, S., Crona, B., Fox, E., Bignet, V., Troell, M., Lindahl, T., Singh, S., Cornell, S.E., Reddy, K.S., Narain, S., Nishtar, S., Murray, C.J.L. (2019) 'Food in the anthropocene: the EAT-lancet commission on healthy diets from sustainable food systems', *The Lancet Commissions*, vol 393, no 10170, pp 447–492. doi: 10.1016/S0140-6736(18)31788-4
Wistoft, K., Qvortrup, L. (2018) 'When the kids conquered the kitchen: Danish taste education and the new Nordic kitchen', *Gastronomica: The Journal of Critical Food Studies*, vol 8, no 4, pp 82–93. doi: 10.1525/gfc.2018.18.4.82

11

NEGLECTED NO MORE

Reframing the food systems narrative using agricultural biodiversity

Teresa Borelli, Eliot Gee and Danny Hunter

Part I of this book reminds us that the way we currently grow our food is contributing to humanity overstepping several "planetary boundaries" that regulate the stability and resilience of the Earth system (Rockström et al., 2009; Steffen et al., 2015). Our food systems are currently caught in a vicious cycle. Production statistics highlight that our food systems are overproducing energy-dense foods such as sugar, grains, fats and oils (KC et al., 2018), while we are failing to tap into the planetary wealth of diverse fruits, vegetables, pulses and minor grains, particularly the nutrient-dense varieties among these food groups, which could generate the desired win-win scenario needed for healthy people and a healthy planet.

It is time to change the narrative around food systems. As Part II and III of this book have illustrated, there are many bright spots which can help facilitate transitions to diversified and resilient food systems. Resilient and sustainable initiatives that put biodiversity for food and nutrition at their core are emerging through local leadership and creative thinking to address the critical challenges of biodiversity loss, environmental degradation and malnutrition. The GEF-supported BFN Project, described in detail in this book, demonstrates how four countries were able to address sectoral barriers, nurture partnerships and produce tangible biodiversity for food and nutrition solutions to safeguard biodiversity, ensure healthier diets and protect the environment, while creating new business opportunities and promoting rural development. In doing so, they have provided a roadmap and much needed guidance for other countries to do likewise.

As the BFN Project has demonstrated, many of these forgotten genetic resources could play a key role in addressing local challenges linked to malnutrition, climate change, poverty and shrinking food biodiversity. Many of these locally important food species provide valuable macro- and micronutrients, as well as beneficial bioactive non-nutrients that contribute to dietary health (Hunter et al., 2019) and towards achieving Sustainable Development Goal 2 (End hunger, achieve food security and improved nutrition and promote sustainable agriculture) by 2030. Often, they grow on resource-poor soils, are pest and disease resistant and require limited inputs, putting less pressure on the environment (Padulosi et al., 2011). Yet many countries fail to adequately value and conserve this diversity disregarding their use for a variety of socio-economic or political reasons.

Recent global reports and scientific literature on food system transitions call for holistic, and alternative, models that focus on sustainable production and rural development, while empowering consumers to make better decisions around diets that are healthier for people and for the planet (FOLU, 2019; HLPE, 2017, 2019; Vermeulen et al., 2019). Although it is understood that solutions to engender this "great food transformation" will need to be tailored to different contexts, reports converge towards a single set of shared goals and key recommendations. Many of these have been addressed by the comprehensive approach used by BFN to better mainstream biodiversity for food and nutrition into practices and programmes.

The BFN Project has demonstrated the often-superior nutritional value and additional benefits of much of our underutilized food biodiversity by using innovative research partnerships and approaches to increase the knowledge, appreciation, awareness and utilization of this diversity, encompassing both cultivated and wild edible species. BFN partners in Brazil, Kenya, Sri Lanka and Turkey have prioritized a rich diversity of neglected and underutilized species and varieties for healthier diets and improved nutrition, and used this knowledge to mainstream these species and their diversity into production and consumption systems, including linking biodiversity to school meals and public food procurement, into food-based dietary guidelines, and into markets including sustainable gastronomy and tourism – innovative pathways for shaping consumer preferences.

These innovations and approaches are visionary in meeting the challenges of environmental sustainability and resilience, improved diet-related health and nutrition and improved livelihoods in the 2030 sustainable development context, as well as putting the conservation of biodiversity on a much stronger footing. In Brazil and Kenya, for example, the project's mainstreaming efforts resulted in the endorsement of national policy instruments that promote the conservation and sustainable use of food biodiversity (Hunter et al., 2019). A food system transformed in this way contributes significantly to reducing biodiversity loss, greenhouse gas emissions, contamination and shortages of water, ecosystems pollution, and land degradation. In addition, it provides income generation benefits, especially for the youth, as a result of diversification options; it also contributes to empowerment of vulnerable groups including women and indigenous people and local communities, reviving traditional knowledge and cultural heritage, as well as supporting rural development and strengthening local economies.

Biodiversity for food and nutrition – the way forward

Achieving this transformative change requires certain key actions and pathways, including the diversification of food production and consumption, as proposed by FAO and others (Bioversity International, 2017; FAO, 2014, 2018a, b, 2019; FOLU, 2019; HLPE, 2017, 2019; IPES-Food, 2016; Vermeulen et al., 2019; Willett et al., 2019). A first critical pathway is to urgently safeguard and conserve the remaining globally important genetic diversity (including varieties, landraces and breeds, and wild relatives of crops and livestock) *in situ* and *ex situ* in order to halt and reverse biodiversity loss. The second pathway is to provide support for more comprehensive biodiversity mainstreaming innovations as well as investments in sustainable and diverse production landscapes and practices that value, conserve, and creatively and sustainably use genetic resources for food and agriculture. For instance, support for biodiversity-friendly management practices in crop, livestock, forestry and aquaculture production systems that make use of nutrient-rich species and varieties, using the power of policy incentives and public food purchasing to drive diversification. The final pathway in this transformative change is the promotion of sustainable and healthy diets, empowering consumers to improve the healthiness and biodiversity of their diets, and thus creating greater demand and desirability for food biodiversity.

In 2020, the countries that are parties to the Convention on Biological Diversity will meet to adopt a Post-2020 Global Biodiversity Framework as a stepping stone towards the 2050 Vision of Living in harmony with nature, whereby "By 2050, biodiversity is valued, conserved, restored and wisely used, maintaining ecosystem services, sustaining a healthy planet and delivering benefits essential for all people." These global consultations offer a unique opportunity for parties and partners to commit and mobilize, provide real political will and leadership and coalesce action around the transformation of agriculture, food systems and diets based on conservation and use of biodiversity.

One area of focus in the negotiations of the post-2020 biodiversity strategy might include efforts to strengthen and broaden the scope of existing Aichi Biodiversity Target 7 to include considerations of mainstreaming biodiversity across the full spectrum of food systems. Any new target would embrace a food systems perspective, which puts biodiversity at its core and makes more explicit the links to other relevant Aichi Biodiversity Targets and Sustainable Development Goals. Efforts to strengthen the role of the CBD's *Cross-cutting initiative on biodiversity for food and nutrition* in implementing and monitoring progress in this context would greatly facilitate biodiversity mainstreaming for healthy diets and improved nutrition.

Following COP13 in Mexico, FAO established the *Biodiversity Mainstreaming Platform*, which it jointly spearheads with the CBD, other UN organizations and partners. The Platform facilitates an integrated approach across agricultural sectors and countries, to promote good practices, collect data, develop metrics and indicators, promote sharing of knowledge and cross-fertilization of ideas, and capitalizes on FAO as a forum for policy dialogue. Both the Platform, and the FAO *Strategy on Biodiversity Mainstreaming across Agricultural Sectors*, recognize the importance of biodiversity mainstreaming for safeguarding nutrition and health. Working through the platform could facilitate country implementation of the *Voluntary Guidelines on Mainstreaming Biodiversity into Policies, Programmes and National and Regional Plans of Action on Nutrition* and the forthcoming *Voluntary Guidelines on Food Systems and Nutrition* (to be endorsed by CFS in 2020), as well as other relevant instruments, tools and voluntary guidelines for transforming food systems. The platform also provides a mechanism to guide countries on mainstreaming biodiversity for food and nutrition into relevant national development strategies and plans including National Biodiversity Strategy and Action Plans (NBSAPs), Multi-sectoral Nutrition Plans, National Agricultural Plans, National Climate Change Adaptation Programmes of Action (NAPAs) and National Climate Change Adaptation Plans (NAPs).

Another opportunity to ensure the momentum of the BFN Project is the potential for an Action Network on Biodiversity, Diets and Nutrition to support the implementation of the UN Decade on Action for Nutrition. Action Networks are an established mechanism for strengthening collaboration among countries to implement their commitments to the UN Decade of Action on Nutrition 2016–2025. They represent informal coalitions of countries focusing on a specific nutrition-related action area to foster policy dialogue, exchange good practices, illustrate successes and challenges, and facilitate and enhance local action for improving food systems, diets and nutrition for all. An Action Network on Biodiversity, Diets and Nutrition would support implementation of key recommendations on diversification of diets and food systems from the ICN2 Framework of Action and Rome Declaration on Nutrition, contributing significantly to the UN Decade of Action for Nutrition. Such an action network would also help keep the momentum of BFN countries going, and use this as a platform to form a coalition of action with other like-minded countries and initiatives with an interest in the role of food biodiversity for diversifying diets and improving nutrition.

The nature-based solutions highlighted by the BFN Project provide some of the "best available evidence" for the sustainable management and use of food biodiversity to address the societal challenges we face today. Examples include the mainstreaming of biodiversity for food and nutrition to diversify farming and food systems, supply chains and public food procurement, as well as innovative strategies such as gastronomy and food innovations that incentivize demand and desirability for food biodiversity. Only when we are actively using biodiversity for sustainable farming and food systems that deliver diverse, nutritious foods can we hope to effectively safeguard this green gold for future generations and reverse current alarming rates of biodiversity loss.

References

Bioversity International. (2017) *Mainstreaming Agrobiodiversity in Sustainable Food Systems: Scientific Foundations for an Agrobiodiversity Index*, Bioversity International, Rome, Italy.

FAO. (2014) *Building a Common Vision for Sustainable Food and Agriculture: Principles and Approaches*, FAO, Rome.

FAO. (2018a) *Transforming Food and Agriculture to Achieve the SDGs: 20 Interconnected Actions to Guide Decision-Makers*, FAO, Rome, Italy.

FAO. (2018b) *Biodiversity for Sustainable Agriculture: FAO's work on biodiversity for food and agriculture*, FAO, Rome, Italy.

FAO. (2019) Scaling Up Agroecology to Achieve the Sustainable Development Goals. *Proceedings of the 2nd FAO International Symposium on Agroecology*, 3–5 April 2018, Rome. FAO, 2019, 416p. Licence: CC BY-NC-SA 3.0 IGO.

FOLU. The Food and Land Coalition. (2019). Growing Better: Ten Critical Transitions to Transform Food and Land Use. The Food and Land Coalition.

HLPE. (2017) *Nutrition and food systems.* A report by the High Level Panel of Experts of Food Security and Nutrition of the Committee on World Food Security. Report 12. Rome. www.fao.org/3/a-i7846e.pdf

HLPE. (2019) Agroecological and other innovative approaches for sustainable agriculture and food systems that enhance food security and nutrition. A report by the High Level Panel of Experts on Food Security and Nutrition of the Committee on World Food Security, Rome.

Hunter, D., Borelli, T., Beltrame, D.M.O., Oliveira, C.N.S., Coradin, L., Wasike, V.W., Wasilwa, L., Mwai, J., Manjella, A., Samarasinghe, G.W.L., Madhujith, T., Nadeeshani, H.V.H., Tan, A., Tuğrul Ay, S., Güzelsoy, N., Lauridsen, N., Gee, E., Tartanac, F. (2019) 'The potential of neglected and underutilized species for improving diets and nutrition', *Planta*, vol 250, no 3, pp709–729. doi: 10.1007/s00425-019-03169-4.

IPES-Food. (2016) From uniformity to diversity: A paradigm shift from industrial agriculture to diversified agroecological systems. IPES-Food – International Panel of Experts on Sustainable Food Systems, Brussels, Belgium. www.ipes- ood.org/_img/upload/files/UniformityToDiversity_FULL.pdf

KC, K.B., Dias, G.M., Veeramani, A., Swanton, C.J., Fraser, D., Steinke, D., Lee, E., Wittman, H., Farber, J.M., Dunfield, K., McCann, K. (2018) 'When too much isn't enough: Does current food production meet global nutritional needs?', *PLoS One*, vol 13, no 10, pp e02025683.

Padulosi, S., Hunter, D., Jarvis, A., Heywood, V. (2011) 'Underutilized crops and climate change—current status and outlook', In Yadav S. et al. (ed) *Crop adaptation to climate change*. Wiley-Blackwell, Ames, pp 507–521.

Rockström, J., Steffen, W, Noone, K., Persson, Å., Chapin, F. S., Lambin, E., Lenton, T.M., Scheffer, M., Folke, C., Schellnhuber, H., Nykvist, B., De Wit, C.A., Hughes, T., van der Leeuw, S., Rodhe, H., Sörlin, S., Snyder, P.K., Costanza, R., Svedin, U., Falkenmark, M., Karlberg, L., Corell, R.W., Fabry, V.J., Hansen, J., Walker, B., Liverman, D., Richardson, K., Crutzen, P., Foley, J. (2009) 'Planetary boundaries: Exploring the safe operating space for humanity', *Ecology And Society*, vol 14, no 2, p 32.

Steffen, W., Richardson, K., Rockström, J., Cornell, S.E., Fetzer, I., Bennett, E.M., Biggs, R., Carpenter, S.R., de Vries, W., de Wit, C.A., Folke, C., Gerten, D., Heinke, J. M., Mace, G.M., Persson, L.M., Ramanathan, V., Reyers, B., Sörlin, S. (2015) 'Planetary boundaries: Guiding human development on a changing planet', *Science*, vol 347, no 6223, pp 1259855. doi: 10.1126/science.1259855.

Vermeulen, S., Park, T., Khoury, C.K., Mockshell, J., Béné, C., Thi, H.T., Heard, B., Wilson, B. (2019) Changing diets and transforming food systems. CCAFS Working Paper no. 282. Wageningen, the Netherlands: CGIAR Research Program on Climate Change, Agriculture and Food Security (CCAFS).

Willett, W., Rockström, J., Loken, B., Springmann, M., Lang, T., Vermeulen, S., Garnett, T., Tilman, D., DeClerck, F., Wood, A., Jonell, M., Clark, M., Gordon, L.J., Fanzo, J., Hawkes, C., Zurayk, R., Rivera, J.A., De Vries, W., Sibanda, L.M., Afshin, A., Chaudhary, A., Herrero, M., Agustina, R., Branca, F., Lartey, A., Fan, S., Crona, B., Fox, E., Bignet, V., Troell, M., Lindahl, T., Singh, S., Cornell, S.E., Reddy, K.S., Narain, S., Nishtar, S., Murray, C.J.L. (2019) 'Food in the Anthropocene: The EAT-Lancet Commission on healthy diets from sustainable food systems', *The Lancet Commissions*, vol 393, no 10170, pp 447–492. doi: 10.1016/S0140-6736(18)31788-4.

LIST OF PROJECT PARTNERS

Country	*Name*	*Affiliation*	*Contribution to the BFN Project*
Brazil	• Prof. Dr. Raquel A. C. Santiago (Coordinator) • 46 Professors, researchers and students • Quilombola communities	• Federal University of Goiás (UFG)	• Compilation of food composition data • Lab analysis • Recipe development • Assessment and capacity building with Quilombola communities (fruit pulp processing, "Meeting on Health and Quality of Life of Quilombolas in Goiás", "Guidelines for Quilombola Schools" • Tasting events (malls, schools, food fairs – Agro Centro Oeste fair, Disco Xepa with Slow Food)
	• Prof. Dr. Adriana Siqueira (Coordinator) • 36 Professors, researchers and students	• Federal University of Ceará (UFC) • State University of Ceará (UECE) • EMBRAPA Tropical Fruits	• Compilation of food composition data • Lab analysis • Recipe development • Assessment and capacity building with Quilombola communities (gastronomy course – International Congress of Gastronomy and Food Science, workshop about inclusion of native biodiversity foods)
	• Prof. Dr Vanuska Lima da Silva (Coordinator) • 17 Professors, researchers and students	• Federal University of Rio Grande do Sul (UFRGS) • EMBRAPA Temperate Climate	• Compilation of food composition data • Lab analysis • Recipe development • Assessment and capacity building with Quilombola communities

(*Continued*)

(Cont.)

Country	*Name*	*Affiliation*	*Contribution to the BFN Project*
	• Prof. Dr Veridiana de Rosso (Coordinator, compilation and lab analysis) • Prof. Dr Semiramis Martins and Prof. Dr Andrea Guerra (recipes) • 7 Professors, researchers and students • 8 Chefs who are contributing with the recipe books	• Federal University of São Paulo (UNIFESP) • Mackenzie University	• Compilation of food composition data • Lab analysis • Recipe development • Assessment of knowledge and utilization of BFN priority species in school feeding in municipalities of the Southeast region (MSc dissertation by Camila Delgado)
	• Dr. Jaime Aguiar (Coordinator) • 4 Researchers	• National Institute of Amazonian Research	• Lab analysis
	• Prof. Dr. Deborah Bastos (Coordinator) • 1 Professor and 2 graduate students	• São Paulo State University (USP)	• Compilation of food composition data and biodiversity indicators
	• Prof. Francisco Silva (Coordinator) • 1 Professor and 1 undergraduate student	• Federal University of Pará (UFPA)	• Compilation of food composition data
	• 20 Professors and professionals working at the "Collaborating Center for the School Feeding Programme" – CECANE	• Federal University of Santa Catarina (UFSC)	• Integration of capacity building activities on BFN previously conducted by the CECANE with PNAE actors (farmers, nutritionists and municipal managers • School Feeding Meeting of Santa Catarina state in 2015 – BFN was the central theme, 300 people participated (PNAE actors)

• 10 Professors and professionals working at the "Collaborating Center for the School Feeding Programme" (CECANE) • 30 Professors and students working with the extension project "The Food is Ours"	• Federal University of Paraná (UFPR)	• Integration of BFN capacity building activities previously conducted by the CECANE with PNAE actors (farmers, nutritionists and municipal managers) • Integration of BFN concerns in the extension project "The Food is Ours": www.acomidaenossa.ufpr.br/
• National Steering Committee (NSC)	• Ministry of the Environment (MMA) • Ministry for Agrarian Development (MDA) • Ministry of Social Development and Fight Against Hunger (MDS) • Ministry of Agriculture, Livestock and Food Supply (MAPA) • Ministry of Education (MEC) • Ministry of Health (MS) • National Supply Company (CONAB) • National Fund for Education Development (FNDE) • National Food and Nutrition Security Council (CONSEA) • Brazilian Agricultural Research Corporation (EMBRAPA) • National Federation of Nutritionists (FNN) • Partner universities	• Facilitate partnerships

(*Continued*)

(Cont.)

Country	*Name*	*Affiliation*	*Contribution to the BFN Project*
Sri Lanka	• Dr. W.L.G. Samarasinghe (National Project Coordinator) • Ms. A.B. Sartaj (Scientific Assistant) • Ms. Nethmini Samaradiwakara (Scientific Assistant) • Ms. Thilini Abeywickrama (Project Assistant)	• BFN Project Sri Lanka	• Technical expertise and guidance for sample collection • Monitoring the sub project implementation and relevant work
	• Prof. Terrence Madhujith (Principal Investigator in project execution) • Prof. Swarna Wimalasiri (Supporting project execution) • Ms. Harshani Nadeeshani (Research Assistant) • Ms. Amali Kulasinghe (Research Assistant) • Ms.Tharika Dilhani (Research Assistant)	• Department of Food Science and Technology, Faculty of Agriculture, University of Peradeniya	• Carrying out the sub project activities (sample collection, analysis, data validation and reporting)
	• Dr. Renuka Silva	• Faculty of Livestock Fisheries and Nutrition, Wayamba University of Sri Lanka	• Co-Investigator in project execution

Kenya	• Aurillia Manjella	• Bioversity International	• Collection of foods and implementation of the ACIAR supported activities in Busia
	• David Khachia (Director) • Dr. Osia Mwanje (CEC) • Ruth Mukhonhgo (Chief Officer) • Florence Kigunzu (Agriculture Officer) • David Ombalo • Jane Wambugu	• Ministry of Agriculture, Livestock, and Fisheries – Busia County • Project Site Implementation Committee (PROSICO)	• Technical oversight of Busia project activities – Extension • Coordination and implementation support of BFN in Busia • Policy development • Training and coordination of the food analysis and food composition table
	• Dr. Maurice Siminyu (CEC) • Scholastic Nabade (Nutrition Officer) • Gladys Mugambi (NSC) • Nancy Njine • John Mwai • Veronicah Kirogo	• Ministry of Health – Busia County • Project Site Implementation Committee (PROSICO)	• Coordination and implementation support of BFN in Busia • Sample collection, training and coordination of the food analysis and food composition table
	• Francis Odeyo • Joseph Etyang	• Ministry of Education – Busia County • Project Site Implementation Committee (PROSICO)	• Coordination and Implementation support of BFN in Busia
	• Prof. Judith Kimiywe (NSC) • Prof. Anselimo Makokha (NSC) • Prof. Mary Abukutsa (NSC) • Dr. Peter Chege • Juliana Kioo	• Jomo Kenyatta University of Agriculture and Technology (JKUAT)	• Sample collection, training and coordination of the food analysis
	• Susan Momanyi-Nyasimi	• Kisii University	• Sample collection, training and coordination of the food analysis
	• Dr. Obonyo Ouche (Principal)	• Mundika High School	• Oversight of the ACIAR schools project at Mundika high school

(*Continued*)

(Cont.)

Country	*Name*	*Affiliation*	*Contribution to the BFN Project*
	• Collins Ogutu • Angela Theuri • Angela Kimani • Emiliana Mbelenga (Consultant)	• FAO • FAO-Kenya	• Sample collection, training and coordination of the food analysis
	• Dr. Lusike Wasilwa (Country Representative for BFN Project) • Dr. Victor Wasike (NSC) • John Ndungu (NSC) • Violet Kirigua (NSC) • Miriam Otipa • David Karanja	• Kenya Agricultural & Livestock Research Organization (KALRO)	• Training and coordination of the food analysis and food identification • Marketing and socio-economic studies • Implementation of policy-related activities, awareness-raising activities, and compilation of indigenous recipes
	• Dr. Joseph Mutanga (NSC) • Ruth Adeka	• National Museums of Kenya (NMK)	• Input on indigenous knowledge • Compilation of indigenous recipes in Busia
	• William Buluma	• Sustainable Income Generating Investment Group (SINGI) • Project Site Implementation Committee (PROSICO)	• Coordination and implementation support of BFN in Busia
	• Phillip Ndemwa	• Kenya Medical Research Institute (KEMRI)	• Sample collection, training and coordination of the food analysis
	• Elizabeth Wafula	• International Potato Center (CIP)	• Sample collection, training and coordination of the food analysis

	• Erick Ngethe	• World Agroforestry Centre (ICRAF)	• Sample collection, training and coordination of the food analysis
	• Christine Chege	• International Center for Tropical Agriculture (CIAT-Kenya)	• Sample collection
	• Cyprian Kabbis	• SGS (K) Ltd	• Training and coordination of the food analysis
	• Maryanne Wanza • Dorothy Murugu	• Bounty consultancy	• Training and coordination of the food analysis • Compilation of the recipe book
Turkey	• Dr. Hasan GEZGİNÇ (National Coordinator, NSC, NSAC, ISC) • Birgül GÜNER (National Assistant Coordinator, NSC, NSAC, ISC) • Hilal YÜCE ARSLAN (National Project Management Unit)	• General Directorate of Agricultural Research and Policies of the Ministry of Agriculture and Forestry • Department of Field Crops	• Project coordination and management
	• Dr. Ayfer TAN (Aegean Regional Coordinator, NSC, NSAC, ISC) • Dr. Neşe ADANACIOĞLU, lerzan AYKAS	• Department of Biodiversity and Genetic Resources, Aegean Agricultural Research Institute	• Technical coordination and implementation of project activities related to project components (knowledge base, policy and regulatory framework, increased awareness and outscaling) • Organization of the Alaçatı Herb Festival
	• Dr. Saadet TUGRUL AY (Mediterranean Regional Coordinator, NSC, NSAC)	• West Mediterranean Agricultural Research Institute	• Technical coordination and implementation of project activities related to project components (knowledge base, policy and

(Continued)

(Cont.)

Country	Name	Affiliation	Contribution to the BFN Project
	• Dr. Ahu ÇINAR • Orçun ÇINAR		regulatory framework, increased awareness and outscaling)
	• Dr. Kürşat ÖZBEK (Black Sea Regional Coordinator, NSC, NSAC) • Sevinç KARABAK (Socio-Economy Coordinator, NSC, NSAC) • Durmuş DENİZ	• Field Crops Central Research Institute, Ankara	• Technical coordination and implementation of project activities related to project components (knowledge base, policy and regulatory framework, increased awareness and outscaling) • Coordination and implementation of socio-economic analysis • Collaboration with regional coordinators and teams
	• Nurcan AYŞAR GÜZELSOY (Food and Nutritional Coordinator, NSC, NSAC) • Food and Nutritional Team	• Central Research Institute of Food and Feed Control	• Coordination and implementation of food composition analysis • Collaboration with regional coordinators and teams
	• Aysun YENICE • Sebla İLERİ	• Alaçatı Art and Culture Association	• Awareness activities in collaboration with the Aegean Regional Coordinator • Organization of Alaçatı Herb Festival
	• Aslı ALBAYRAK	• Çeşme Municipality	• Awareness activities in collaboration with Aegean Regional Coordinator • Organization of Alaçatı Herb Festival
	• Ahmet KECECI	• Directorate of Çeş1me Agriculture and Forestry County	• Awareness activities in collaboration with Aegean Regional Coordinator

• Regional partners from other ministries	• Regional authorities of Ministry of Education and schools	• Collaboration for the case studies, policy and regulatory framework • Increased awareness
• Private sector	• Various	• Marketing of target species • Cultivation • Collaboration with regional coordinators and teams
• Farmers	• Various	• Cultivation and production of target species and some priority species • Collaboration with regional coordinators and teams

INDEX

Note: References in *italics* are to figures, those in **bold** to tables.

For Product Safety Concerns and Information please contact our EU
representative GPSR@taylorandfrancis.com
Taylor & Francis Verlag GmbH, Kaufingerstraße 24, 80331 München, Germany

www.ingramcontent.com/pod-product-compliance
Ingram Content Group UK Ltd.
Pitfield, Milton Keynes, MK11 3LW, UK
UKHW021505240926
475702UK00035B/387

9780367141509